Developing Students' Statistical Reasoning

Developing Students' Statistical Reasoning

Connecting Research and Teaching Practice

by

Joan B. Garfield
University of Minnesota, USA

and

Dani Ben-Zvi
University of Haifa, Israel

With Contributions from

Beth Chance and Elsa Medina
California Polytechnic State University, San Luis Obispo, USA

Cary Roseth
Michigan State University, USA

Andrew Zieffler
University of Minnesota, USA

Springer

Dr. Joan B. Garfield
University of Minnesota
Dept. of Educational Psychology
178 Educational Sciences
56 East River Road
Minneapolis, MN, 55455
USA
jbg@umn.edu

Dr. Dani Ben-Zvi
University of Haifa
Faculty of Education
31905 Haifa
Mount Carmel
Israel
dbenzvi@univ.haifa.ac.il

ISBN: 978-1-4020-8382-2 e-ISBN: 978-1-4020-8383-9

Library of Congress Control Number: 2008929495

Printed on acid-free paper

9 8 7 6 5 4 3 2 1

springer.com

Foreword

In theory, there is no difference between theory and practice.
But in practice, there is.

Yogi Berra

In the last decade, statistics education has received growing recognition as a discipline and has begun to accumulate a body of research. However, this research has not always had a meaningful and visible impact on student learning and classroom practice. Even among those who consider statistics education their primary professional focus, there has not been widespread effort to carefully consider existing research in terms of its implications for content and pedagogy in practice. This is a situation very much deserving of our attention if we wish to be taken seriously as teaching professionals.

The Teaching Professional Challenge

Consider the following scenario:

> You visit your doctor for advice about how to best treat a relatively common illness. The doctor says "Well, I always just tell my patients to take an aspirin and get some sleep. It's not very effective, but some people get better. I know that there have been some recent research studies looking at the effectiveness of this treatment and also looking at some alternative approaches, but frankly I just haven't had the time to read them carefully or to think about how they might apply to my patients. So, why don't you go home, take an aspirin and get some sleep."

How would you react? I am guessing that you wouldn't just say thank you and head home for a nap. I know I would be looking for a new doctor and would have been extremely disappointed by a doctor that I felt did not live up to professional expectations. We expect professionals to stay abreast of research in their respective fields and for research to inform their practice.

Granted, the scenario above is contrived to make a point, but hopefully you can already see where this is headed. While we expect professionals in other fields to stay current and to be able to understand the implications of research on practice, for some reason this has not been an expectation of teaching professionals. It seems to me that this is an embarrassing situation and something we should be working hard to change. This presents an enormous challenge, but also a great and exciting opportunity.

Where Do We Begin?

My hope is that statistics educators will take up the "teaching professional" challenge, and that is why I am delighted to see the publication of this book. It provides an excellent starting point, providing a forum for sharing existing research results relevant to statistics education and offering guidance in terms of implications for classroom practice. Statistics education related research from across many disciplines has now been summarized in one coherent work, creating a solid foundation for future statistics education research. This affords statistics educators easy access to this body of work and insight into implications for classroom practice, and it also provides great motivation for statistics education researchers to develop research questions with potential classroom impact. But even more important, this book provides a model for linking research to practice that I hope will motivate statistics educators to continue the process of seeking out (or conducting) relevant research and using what is learned to make informed classroom decisions.

Moving Statistics Education Forward – Translating Research into Practice

There are great opportunities on the horizon, should we choose to take advantage of them – opportunities to facilitate changes that will lead to improved student understanding as well as opportunities to increase the perceived value and legitimacy of statistics education as an important discipline. This book makes an important and timely contribution by demonstrating how we, as statistics educators, can begin to take advantage of these opportunities. What is needed now is the motivation, the energy, and the dedication to acknowledge the value of statistics education research by putting it into practice in ways that improve student learning.

Read this book and then use it as a model to help you put research into practice!

Roxy Peck

California Polytechnic State University
San Luis Obispo, USA

Preface

Statistics education has emerged as an important discipline that supports the teaching and learning of statistics. The research base for this new field has been increasing in size and scope, but has not always been connected to teaching practice nor accessible to the many educators teaching statistics at different educational levels. Our goal in writing this book was to provide a useful resource for members of the statistics education community that facilitates the connections between research and teaching. Although the book is aimed primarily at teachers of an introductory statistics course at the high school or college level, we encourage others interested in statistics education to consider how the research summaries, ideas, activities, and implications may be useful in their line of work as well.

This book builds on our commitment over the past decade to exploring ways to understand and develop students' statistical literacy, reasoning, and thinking. Despite living and teaching in two different countries, we have worked together to understand and promote the scholarship in this area. After co-chairing five international research forums (SRTL), co-editing a book (*The Challenge of Developing Literacy, Reasoning, and Thinking,* Ben-Zvi & Garfield, 2004b), and serving as guest editors for two special issues of SERJ (*Statistics Education Research Journal*), it seemed the right time to finally write a book together that is built on our knowledge, experience, and passion for statistics education. It has been a great experience working on this book, which required ongoing reading, writing, discussing, and learning. It took 3 years to write and revise this book, which required visits to each other's homes, phone calls, and innumerable email exchanges. We now offer our book to the statistics education community and hope that readers will find it a useful and valuable resource.

The book is divided into three parts. Part I consists of five chapters on important foundational topics: the emergence of statistics education as a discipline, the research literature on teaching and learning statistics, practical strategies for teaching students in a way that promotes the development of their statistical reasoning, assessment of student outcomes, and the role of technological tools in developing statistical reasoning.

Part II of the book includes nine chapters, each devoted to one important statistical idea: data, statistical models, distribution, center, variability, comparing groups, sampling and sampling distributions, statistical inference, and covariation. These chapters present a review and synthesis of the research literature related to the statistical topic, and then suggest implications for developing this idea through carefully structured sequences of activities.

Part III of the book focuses on what we see as the crucial method of bringing about instructional change and implementing ideas and methods in this book: through collaboration. One chapter focuses on collaborative student learning, and the other chapter discusses collaborative teaching and classroom-based collaborative research.

Although we have presented the chapters in Part II in the order in which we think they may be introduced to students, we point out that readers may want to read these chapters in an alternative order. For example, one of our reviewers suggested reading Chapters 6, 9, and 10 before the rest of the chapters in Part II. Another suggestion was to read Chapters 1 and 2 and then skip to Part II, then return to Chapters 3, 4, 5, 15 and 16.

The suggested activities in Part II have in many cases been adapted from activities developed by others in the statistics education community, and we present a table of all activities described in Part II that credits a source, where possible. However, we note that sometimes it was impossible to track down the person who developed an activity used in the book, because no one was willing to take credit for it. If we have failed to credit a creator of an activity used in our book, we apologize for this oversight. We also note that the activities and lessons are based on particular technological tools we like, such as *TinkerPlots* and *Fathom*, and certain Web applets. However, we acknowledge that instructors may choose to use alternative technological tools that they have access to or prefer.

This project would have been impossible without the help of several dedicated and hard working individuals. First and most importantly, we would like to thank four individuals who made important contributions to writing specific chapters. Beth Chance took the lead in developing the chapters on assessment and technology (assisted by Elsa Medina) and provided insightful feedback on the research chapter. Cary Roseth wrote the majority of the chapter on collaborative learning and provided feedback on the chapter on collaborative teaching and research as well. Andy Zieffler provided valuable assistance in writing the chapter on covariational reasoning and also gave helpful advice on additional chapters of the book.

The lessons and activities described in this book were developed and modified over the past 4 years as part of a collaborative effort of the undergraduate statistics teaching team in the Department of Educational Psychology at the University of Minnesota. This team has included Beng Chang, Jared Dixon, Danielle Dupuis, Sharon Lane-Getaz, and Andy Zieffler. We greatly appreciate the contributions of these dedicated graduate students and instructors in developing and revising the lessons, and particularly the leadership of Andy Zieffler in coordinating this group and providing feedback on needed changes. We also appreciate Andy's work in constructing the Website that posts the lesson plans and activities described in Part II, and the funding for this (as part of the AIMS project) by the National Science Foundation (DUE 0535912).

We gratefully acknowledge the tremendously valuable feedback offered by six reviewers of our earlier manuscript: Gail Burrill, Michelle Everson, Randall Groth, Larry Lesser, Roxy Peck, and Mike Shaughnessy. We hope they will see how we have utilized and incorporated their suggestions. We also appreciate feedback and encouragement offered by the original reviewers of first chapters and prospectus. We also gratefully acknowledge the important contributions of advisers to the NSF AIMS project who offered feedback on lessons and activities as well as on the Website. They are Beth Chance, Bill Finzer, Cliff Konold, Dennis Pearl, Allan Rossman, and Richard Schaeffer. Bob delMas, co-PI of the AIMS grant has also offered helpful advice. In addition, Rob Gould, the external evaluator for the AIMS project offered valuable insights and advice.

We are grateful to Springer Publishers for providing a publishing venue for this book, and to Harmen van Paradijs, the editor, who skillfully managed the publication on their behalf. We owe deep thanks to Shira Yuval and Idit Halwany, graduate students from the University of Haifa who assisted in editing and checking details in the manuscript.

Lastly, many thanks go to our spouses Michael Luxenberg and Hava Ben-Zvi and to our children – Harlan and Rebecca Luxenberg, and Noa, Nir, Dagan, and Michal Ben-Zvi. They have been our primary sources of energy and support.

Joan B. Garfield — University of Minnesota, USA
Dani Ben-Zvi — University of Haifa, Israel

Contents

Part I
The Foundations of Statistics Education

Chapter 1
The Discipline of Statistics Education[1]

The case for substantial change in statistics instruction is built on strong synergies between content, pedagogy, and technology.

Moore (1997, p. 123)

Overview

This chapter introduces the emerging discipline of statistics education and considers its role in the development of students who are statistically literate and who can think and reason about statistical information. We begin with information on the growing importance of statistics in today's society, schools and colleges, summarize unique challenges students face as they learn statistics, and make a case for the importance of collaboration between mathematicians and statisticians in preparing teachers to teach students how to understand and reason about data. We examine the differences and interrelations between statistics and mathematics, recognizing that mathematics is the discipline that has traditionally included instruction in statistics, describe the history of the introductory college course, introduce current instructional guidelines, and provide an overview of the organization and content of this book.

The Growing Importance of Statistics

No one will debate the fact that quantitative information is everywhere and numerical data are increasingly presented with the intention of adding credibility to advertisements, arguments, or advice. Most would also agree that being able to provide good evidence-based arguments and to be able to critically evaluate data-based claims are important skills that all citizens should have, and therefore, that all students should learn as part of their education (see Watson, 2006). It is not surprising therefore that statistics instruction at all educational levels is gaining more students and drawing more attention.

The study of statistics provides students with tools and ideas to use in order to react intelligently to quantitative information in the world around them. Reflecting this need to improve students' ability to think statistically, statistics and statistical reasoning are becoming part of the mainstream school curriculum in

[1] This chapter is partly based on the following paper: Ben-Zvi, D., & Garfield J. (in press). Introducing the emerging discipline of statistics education. *School Science and Mathematics*.

many countries (e.g., Mathematics Curriculum Framework for Western Australia, http://www.curriculum.wa.edu.au/; Mathematics National Curriculum for England, http://www.nc.uk.net/; National Council of Teachers of Mathematics in the U.S., 2000). At the U.S. college level enrollments in statistics courses continue to grow (Scheaffer & Stasney, 2004). Statistics is becoming such a necessary and important area of study, Moore (1998) suggested that it should be viewed as one of the liberal arts, and that it involves distinctive and powerful ways of thinking. He wrote: "Statistics is a general intellectual method that applies wherever data, variation, and chance appear. It is a fundamental method because data, variation, and chance are omnipresent in modern life" (p. 134).

The Challenge of Learning and Teaching Statistics

Despite the increase in statistics instruction at all educational levels, historically the discipline and methods of statistics have been viewed by many students as a difficult topic that is unpleasant to learn. Statisticians often joke about the negative comments they hear when others learn of their profession. It is not uncommon for people to recount tales of statistics as the worst course they took in college. Many research studies over the past several decades indicate that most students and adults do not think statistically about important issues that affect their lives. Researchers in psychology and education have documented the many consistent errors that students and adults make when trying to reason about data and chance in real world problems and contexts. In their attempts to make the subject meaningful and motivating for students, many teachers have included more authentic activities and the use of new technological tools in their instruction. However, despite the attempts of many devoted teachers who love their discipline and want to make the statistics course an enjoyable learning experience for students, the image of statistics as a hard and dreaded subject is hard to dislodge. Currently, researchers and statistics educators are trying to understand the challenges and overcome the difficulties in learning and teaching this subject so that improved instructional methods and materials, enhanced technology, and alternative assessment methods may be used with students learning statistics at the school and college level.

In our previous book (Ben-Zvi & Garfield, 2004b) we list some of the reasons that have been identified to explain why statistics is a challenging subject to learn and teach. Firstly, many statistical ideas and rules are complex, difficult, and/or counterintuitive. It is therefore difficult to motivate students to engage in the hard work of learning statistics. Secondly, many students have difficulty with the underlying mathematics (such as fractions, decimals, proportional reasoning, and algebraic formulas) and that interferes with learning the related statistical concepts. A third reason is that the context in many statistical problems may mislead the students, causing them to rely on their experiences and often faulty intuitions to produce an answer, rather than select an appropriate statistical procedure and rely on data-based evidence. Finally, students equate statistics with mathematics and expect the focus to be on numbers, computations, formulas, and only one right answer. They are uncomfortable with the messiness of data, the ideas of randomness and chance, the different possible interpretations based on different assumptions, and the extensive use of writing, collaboration and communication skills. This is also true of many mathematics teachers who find themselves teaching statistics.

The Goals of This Book

Despite the growing body of research related to teaching and learning statistics, there have been few direct connections made between the research results and practical suggestions for teachers. Teachers of statistics may be looking for suggestions from the research literature but find it hard to locate them since studies are published in journals from other disciplines that are not readily accessible. In addition, many research studies have been conducted in settings that do not seem to easily transfer to the high school or college classroom (e.g., studies in a psychology lab, or studies in a teaching experiment at an elementary school), or have been carried out using methods with which most statisticians are not familiar (e.g., studies involving collection and analysis of extensive qualitative data). Statisticians, in contrast, are more familiar with randomized controlled experiments and often look for studies using these methods, set in high school or college classrooms, to provide results to inform their teaching.

We find it fascinating that statistics education has been the focus for researchers in many disciplines, perhaps because statistical reasoning is used in many disciplines and provides so many interesting issues and challenges. Today, researchers in mathematics education study children's understanding of statistical concepts as well as how they learn to use data analysis meaningfully. They also study how K-12 teachers understand statistical ideas and methods and how this affects the way they teach children. Researchers in psychology explore judgments and decisions made in light of uncertainty, and the use of intuitions and heuristics in dealing with uncertainty. Researchers in educational measurement study the assessment of statistical anxiety and attitudes towards statistics, as well as factors that predict student achievement in statistics courses such as mathematics background and motivation. Only recently has there been a core set of researchers looking at understanding of and reasoning about particular statistical concepts, how they might be developed through carefully planned sequences of activities, and how this might take place in the classrooms.

Our main goal in writing this book was to build on relevant research that informs the teaching and learning of statistics to enhance two aspects of teachers' knowledge: their knowledge of what it means for students to understand and reason about statistical concepts, and the pedagogical methods for developing understanding and reasoning about these concepts. We try to summarize the research and highlight the important statistical concepts for teachers to emphasize, as well as reveal the interrelationships among concepts. We also make specific suggestions regarding how to plan and use classroom activities, integrate technological tools, and assess students' learning in meaningful ways.

The goals listed above are aimed to help instructors of statistics deal with the challenges they face, to suggest ways that may help to make statistics instruction more effective, and to engage students in reasoning and thinking statistically. By compiling and building on the various research studies that shed light on the difficulties students have learning statistics, we are able to offer suggestions for how students may be guided to construct meaning for complex statistical ideas, concepts and procedures. The research literature is often difficult for statistics teachers to find and access, synthesize, and apply to their teaching practice. Therefore, we try in this book to provide links between the research literature and teaching practice. We include examples, activities, and references to useful resources. We also incorporate many uses of instructional software and Web tools and resources. Finally, we offer an accompanying Website with materials to supplement each chapter (see http://www.tc.umn.edu/~aims).

We begin this first chapter with a brief introduction and historical perspective of the emerging field of statistics education and the development of the introductory college course, we present arguments for how statistics differs from mathematics, focus on the importance of statistical reasoning, and provide an overview of the subsequent chapters in this book.

The Development of the Field of Statistics Education

Statistics education is an emerging field that grew out of different disciplines and is currently establishing itself as a unique field of study. The two main disciplines from which statistics education grew are statistics and mathematics education. As early as 1944 the American Statistical Association (ASA) developed the Section on Training of Statisticians (Mason, McKenzie, & Ruberg, 1990) that later (1973) became the Section on Statistical Education. The International Statistical Institute (ISI) similarly formed an education committee in 1948. The early focus in the statistics world was on training statisticians, but this later broadened to include training, or education, at all levels. In the 1960s an interest emerged in the mathematics education field about teaching students at the pre-college level how to use and analyze data. In 1967 a joint committee was formed between the American Statistical Association (ASA) and the National Council of Teachers of Mathematics (NCTM) on Curriculum in Statistics and Probability for grades K-12. In the early 1970s many instructional materials began to be developed in the USA and in other countries to present statistical ideas in interesting and engaging ways, e.g., the series of books *Statistics by Example*, by Mosteller, Rourke, and Thomas (1973e) and Mosteller, Kruskal, Pieters, and Rising (1973a, 1973b, 1973c, 1973d), and *Statistics: A Guide to the Unknown* by Tanur et al. (1972), which was recently updated (Peck et al., 2006).

In the late 1970s the ISI created a task force on teaching statistics at school level (Gani, 1979), which published a report, *Teaching Statistics in Schools throughout the World* (Barnett, 1982). This report surveyed how and where statistics was being taught, with the aim of suggesting how to improve and expand the teaching of this important subject. Although there seemed to be an interest in many countries in including statistics in the K-12 curriculum, this report illustrated a lack of coordinated efforts, appropriate instructional materials, and adequate teacher training.

By the 1980s the message was strong and clear: Statistics needed to be incorporated in pre-college education and needed to be improved at the postsecondary level. Conferences on teaching statistics began to be offered, and a growing group of educators began to focus their efforts and scholarship on improving statistics education. The first *International Conference on Teaching Statistics* (ICOTS) was convened in 1982 and this conference has been held in a different part of the world every 4 years since that date (e.g., ICOTS-7, 2006, see http://www.maths.otago.ac.nz/icots7/icots7.php; ICOTS-8, 2010, see http://icots8.org/).

In the 1990s there was an increasingly strong call for statistics education to focus more on statistical literacy, reasoning, and thinking. One of the main arguments presented is that traditional approaches to teaching statistics focus on skills, procedures, and computations, which do not lead students to reason or think statistically. In their landmark paper published in the *International Statistical Review*, which included numerous commentaries by leading statisticians and statistics educators, Wild and Pfannkuch (1999) provided an empirically-based comprehensive description of the processes involved in the statisticians' practice of data-based enquiry from problem formulation to conclusions. Building on the interest in this topic, *The International Research Forums on Statistical Reasoning, Thinking, and Literacy* (SRTL) began in 1999 to foster current and innovative research studies that examine the nature and development of statistical literacy, reasoning, and thinking, and to explore the challenge posed to educators at all levels – and to develop these desired learning goals for students. The SRTL Forums offer scientific gatherings every 2 years and related publications (for more information see http://srtl.stat.auckland.ac.nz). Additional explanations and reference to publications explicating the nature and development of statistical literacy, reasoning and thinking are summarized in Chapters 2 and 3 and other relevant chapters in this book.

One of the important indicators of a new discipline is scientific publications devoted to that topic. At the current time, there are three journals. *Teaching Statistics*, which was first published in 1979, and the *Journal of Statistics Education*, first published in 1993, were developed to focus on the teaching of statistics as well as statistics education research. While recently the *Statistical Education Research Journal* (first published in 2002) was established to exclusively publish research in statistics education. More information on books, conferences, and publications in statistics education is provided at the resources section in the end of this book.

Collaborations Among Statisticians and Mathematics Educators

Some of the major advances in the field of statistics education have resulted from collaborations between statisticians and mathematics educators. For example, the *Quantitative Literacy Project* (QLP) was a decade-long joint project of the

American Statistical Association (ASA) and the National Council of Teachers of Mathematics (NCTM) headed by Richard Scheaffer that developed exemplary materials for secondary students to learn data analysis and probability (Scheaffer, 1990). The QLP first produced materials in 1986 (Landwehr & Watkins, 1986). Although these materials were designed for students in middle and high school, the activities were equally appropriate for college statistics classes, and many instructors began to incorporate these activities into their classes. Because most college textbooks did not have activities featuring real data sets with guidelines for helping students explore and write about their understanding, as the QLP materials provided, additional resources continued to be developed. Indeed, the QLP affected the nature of activities in many college classes in the late 1980s and 1990s and led to the *Activity Based Statistics* project, also headed by Richard Scheaffer, designed to promote the use of high quality, well structured activities in class to promote student learning (Scheaffer, Watkins, Witmer, & Gnanadesikan, 2004a, 2004b).

Members of the statistics and mathematics education disciplines have also worked together on developing the *Advanced Placement* (AP) statistics course. This college level introductory statistics course was first taught to high school students in the U.S. in 1996–97 and the first exam was given in 1997, as part of the College Board Advanced Placement Program. Currently hundreds of high school mathematics teachers and college statistics teachers meet each summer to grade together the open-ended items on the AP Statistics exam, and have opportunities to discuss teaching and share ideas and resources.

More recent efforts to connect mathematics educators and statisticians to improve the statistical preparation of mathematics teachers include the ASA *TEAMS* project (see Franklin, 2006). A current joint study of the International Association for Statistical Education (IASE) and the International Congress on Mathematics Instruction (ICMI) is also focused on this topic (see http://www.ugr.es/~icmi/iase_study/).

Statistics and Mathematics

Although statistics education grew out of statistics and mathematics education, statisticians have worked hard to convince others that statistics is actually a separate discipline from mathematics. Rossman, Chance, and Medina (2006) describe statistics as a mathematical science that uses mathematics but is a separate discipline, "the science of gaining insight from data." Although data may seem like numbers, Moore (1992) argues that data are "numbers with a context." And unlike mathematics, where the context obscures the underlying structure, in statistics, context provides meaning for the numbers and data cannot be meaningfully analyzed without paying careful consideration to their context: how they were collected and what they represent (Cobb & Moore, 1997).

Rossman et al. (2006) point out many other key differences between mathematics and statistics, concluding that the two disciplines involve different types of reasoning and intellectual skills. It is reported that students often react differently to learning mathematics than learning statistics, and that the preparation of teachers of statistics requires different experiences than those that prepare a person to teach mathematics, such as analyzing real data, dealing with the messiness and variability of data, understanding the role of checking conditions to determine if assumptions are reasonable when solving a statistical problem, and becoming familiar with statistical software.

How different is statistical thinking from mathematical thinking? The following example illustrates the difference. A statistical problem in the area of bivariate data might ask students to determine the equation for a regression line, specifying the slope and intercept for the line of best fit. This looks similar to an algebra problem: numbers and formulas are used to generate the equation of a line. In many statistics classes taught by mathematicians, the problem might end at this stage. However, if statistical reasoning and thinking are to be developed, students would be asked questions about the context of the data and they would be asked to describe and interpret the relationship between the variables, determining whether simple linear regression is an appropriate procedure and model for these data. This type of reasoning and thinking is quite different from the mathematical reasoning and thinking required to calculate the slope and intercept using algebraic formulas. In fact, in many classes, students may not be asked to calculate the quantities from formulas, but rather rely on technology to produce the numbers. The focus shifts to asking students to interpret the values in context (e.g., from interpreting the slope as rise over run to predicting change in response for unit-change in explanatory variable).

In his comparison of mathematical and statistical reasoning, delMas (2004) explains that while these two forms of reasoning appear similar, there are some differences that lead to different types of errors. He posits that statistical reasoning must become an explicit goal of instruction if it is to be nourished and developed. He also suggests that experiences in the statistics classroom focus less on the learning of computations and procedures and more on activities that help students develop a deeper understanding of stochastic processes and ideas. One way to do this is to ground learning in physical and visual activities to help students develop an understanding of abstract concepts and reasoning.

In order to promote statistical reasoning, Moore (1998) recommends that students must experience firsthand the process of data collection and data exploration. These experiences should include discussions of how data are produced, how and why appropriate statistical summaries are selected, and how conclusions can be drawn and supported (delMas, 2002). Students also need extensive experience with recognizing implications and drawing conclusions in order to develop statistical thinking. (We believe future teachers of statistics should have these experiences as well). We have tried to imbed these principles in specific chapters of this book (Chapters 6–14) suggesting how different statistical concepts may be taught as well as in the overall chapters on pedagogical issues in teaching statistics (Chapters 3–5).

Recommendations such as those by Moore (1998) have led to a more modern or "reformed" college-level statistics course that is less like a mathematics course and more like an applied science. The next sections provide some background on this course and the trajectory that led to this change.

Changes in the Introductory College Statistics Course

In his forward to *Innovations in Teaching Statistics* (Garfield, 2005), George Cobb described how the modern introductory statistics course has roots that go back to early books on statistical methods (i.e., Fisher's, 1925 *Statistical Methods for Research Workers* and Snedecor's, 1937 *Statistical Methods*). Cobb wrote:

> By 1961, with the publication of *Probability with Statistical Applications* by Fred Mosteller, Robert Rourke, and George Thomas, statistics had begun to make its way into the broader academic curriculum, but here again, there was a catch: in these early years, statistics had to lean heavily on probability for its legitimacy. During the late 1960s and early 1970s, John Tukey's ideas of exploratory data analysis (EDA) brought a near-revolutionary pair of changes to the curriculum, first, by freeing certain kinds of data analysis from ties to probability-based models, so that the analysis of data could begin to acquire status as an independent intellectual activity, and second, by introducing a collection of "quick-and-dirty" data tools, so that, for the first time in history, students could analyze real data without having to spend hours chained to a bulky mechanical calculator. Computers would later complete the "data revolution" in the beginning statistics curriculum, but Tukey's EDA provided both the first technical breakthrough and the new ethos that avoided invented examples. 1978 was another watershed year, with the publication of two other influential books, *Statistics*, by David Freedman, Robert Pisani, and Roger Purves, and *Statistics: Concepts and Controversies*, by David S. Moore. I see the publication of these two books 25 years ago as marking the birth of what we regard, for now at least, as the modern introductory statistics curriculum.
>
> The evolution of content has been paralleled by other trends. One of these is a striking and sustained growth in enrollments. Two sets of statistics suffice here: (1) At two-year colleges, according to the Conference Board of the Mathematical Sciences, statistics enrollments have grown from 27% of the size of calculus enrollments in 1970, to 74% of the size of calculus enrollments in 2000. (2) The Advanced Placement exam in statistics was first offered in 1997. There were 7,500 students who took it that first year, more than in the first offering of an AP exam in any subject. The next year more than 15,000 students took the exam, the next year more than 25,000, and the next, 35,000.
>
> Both the changes in course content and the dramatic growth in enrollments are implicated in a third set of changes, a process of democratization that has broadened and diversified the backgrounds,

interests, and motivations of those who take the courses. Statistics has gone from being a course taught from a book like Snedecor's, for a narrow group of future scientists in agriculture and biology, to being a family of courses, taught to students at many levels, from high school to post-baccalaureate, with very diverse interests and goals.

Guidelines for Teaching Introductory Statistics

In the early 1990s a working group headed by George Cobb as part of the Curriculum Action Project of the Mathematics Association of America (MAA) produced guidelines for teaching statistics at the college level (Cobb, 1992) to be referred to as the new guidelines for teaching introductory statistics. They included the following recommendations:

1. ***Emphasize Statistical Thinking***

   Any introductory course should take as its main goal helping students to learn the basic elements of statistical thinking. Many advanced courses would be improved by a more explicit emphasis on those same basic elements, namely:

   The need for data

   Recognizing the need to base personal decisions on evidence (data), and the dangers inherent in acting on assumptions not supported by evidence.

   The importance of data production

   Recognizing that it is difficult and time-consuming to formulate problems and to get data of good quality that really deal with the right questions. Most people don't seem to realize this until they go through this experience themselves.

   The omnipresence of variability

   Recognizing that variability is ubiquitous. It is the essence of statistics as a discipline and it is not best understood by lecture. It must be experienced.

   The quantification and explanation of variability

   Recognizing that variability can be measured and explained, taking into consideration the following: (a) randomness and distributions; (b) patterns and deviations (fit and residual); (c) mathematical models for patterns; (d) model-data dialogue (diagnostics).

2. ***More Data and Concepts, Less Theory and Fewer Recipes***

   Almost any course in statistics can be improved by more emphasis on data and concepts, at the expense of less theory and fewer recipes. To the maximum extent feasible, calculations and graphics should be automated.

3. ***Foster Active Learning***

   As a rule, teachers of statistics should rely much less on lecturing, much more on the alternatives such as projects, lab exercises, and group problem solving and discussion activities. Even within the traditional lecture setting, it is possible to get students more actively involved. (Cobb, 1992).

The three recommendations were intended to apply quite broadly, whether or not a course had a calculus prerequisite, and regardless of the extent to which students are expected to learn specific statistical methods. Moore (1997) described these recommendations in terms of changes in *content* (more data analysis, less probability), *pedagogy* (fewer lectures, more active learning), and *technology* (for data analysis and simulations).

Influence of the *Quantitative Literacy Project*

In his reflections on the past, present and future of statistics education, Scheaffer (2001) described the philosophy and style of the "new" statistics content that was embedded in the revolutionary Quantitative Literacy Project (QLP) materials, described earlier in this chapter. The QLP attempted to capture the spirit of modern statistics as well as modern ideas of pedagogy by following a philosophy that emphasized understanding and communication. Scheaffer, the leader of this project, described the guiding principles of QLP as:

1. Data analysis is central.
2. Statistics is not probability.
3. Resistant statistics (such as median and interquartile range) should play a large role.
4. There is more than one way to approach a problem in statistics.
5. Real data of interest and importance to the students should be used.
6. The emphasis should be on good examples and building intuition.
7. Students should write more and calculate less.
8. The statistics taught in the schools should be important and useful in its own right, for all students.

Scheaffer (2001) noted that these principles are best put into classroom practice with a teaching style emphasizing a hands-on approach that engages students to *do* an activity, *see* what happens, *think* about what they just saw, and then *consolidate* the new information with what they have learned in the past. He stressed that this style required a laboratory in which to experiment and collect data, but the "laboratory" could be the classroom itself, although the use of appropriate technology was highly encouraged. Scheaffer recognized that the introductory college course and the Advanced Placement High School course should also model these principles. Many of the activities developed from these principles have been adapted into lessons that are described in Part II of this book.

Changes in the Introductory Statistics Course

Over the decade that followed the publication of the Cobb report (1992), many changes were implemented in the teaching of statistics. Many statisticians became involved in the reform movement by developing new and improved versions of the introductory statistics course and supplementary material. The National Science Foundation funded many of these projects (Cobb, 1993). But what effect did the report and new projects have on the overall teaching of statistics?

In 1998 and 1999, Garfield (reported in Garfield, Hogg, Schau, and Whittinghill, 2002) surveyed a large number of statistics instructors from mathematics and statistics departments and a smaller number of statistics instructors from departments of psychology, sociology, business, and economics. Her goal was to determine how the introductory course was being taught and to explore the impact of the educational reform movement. The results of this survey suggested that major changes were being made in the introductory course, that the primary area of change was in the use of technology, and that the results of course revisions generally were positive, although they required more time from the course instructor than traditional methods of teaching. Results were surprisingly similar across departments, with the main differences found in the increased use of graphing calculators, active learning and alternative assessment methods in courses taught in mathematics departments in two year colleges, the increased use of Web resources by instructors in statistics departments, and the reasons cited for why changes were made (more mathematics instructors were influenced by recommendations from statistics education). The results were also consistent in reporting that more changes were anticipated, particularly as more technological resources became available.

Scheaffer (2001) wrote that there seems to be a large measure of agreement on what content to emphasize in introductory statistics and how to teach the course. This is reflected in the course guide for the Advanced Placement statistics course which organized the content into four broad areas:[2]

- Exploring Data: Describing patterns and departures from patterns.
- Sampling and Experimentation: Planning and conducting a study.
- Anticipating Patterns: Exploring random phenomena using probability and simulation.
- Statistical Inference: Estimating population parameters and testing hypotheses.

However, Scheaffer noted further important changes needed, and encouraged teachers of statistics to:

- Deepen the discussion of exploratory data analysis, using more of the power of revelation, residuals, re-expression, and resistance as recommended by the originators of this approach to data.

[2] For more details on the AP statistics course, see http://www.collegeboard.com/student/testing/ap/sub_stats.html?stats.

- Deepen the exposure to study design, separating sample surveys (random sampling, stratification, and estimation of population parameters) from experiments (random assignment, blocking, and tests of significant treatment differences).
- Deepen the understanding of inferential procedures for both quantitative and categorical variables, making use of randomization and resampling techniques.

The introductory statistics course today is moving closer to these goals, and we hope that the lessons in this book will help accomplish Scheaffer's recommendations.

The goals for students at the elementary and secondary level tend to focus more on conceptual understanding and attainment of statistical literacy and thinking and less on learning a separate set of tools and procedures. New K-12 curricular programs set ambitious goals for statistics education, including developing students' statistical literacy, reasoning and understanding (e.g., NCTM, 2000; and Project 2061's Benchmarks for Science Literacy, American Association for the Advancement of Science, 1993).

As demands for dealing with data in an information age continue to grow, advances in technology and software make tools and procedures easier to use and more accessible to more people, thus decreasing the need to teach the mechanics of procedures but increasing the importance of giving more people a sound grasp of the fundamental concepts needed to use and interpret those tools intelligently. These new goals, described in the following section, reinforce the need to reexamine and revise many introductory statistics courses, in order to help achieve the important learning goals for students.

Current Guidelines for Teaching the Introductory Statistics Course

In 2005 the Board of Directors for the American Statistical Association endorsed a set of six guidelines for teaching the introductory college statistics course (the Guidelines for Assessment and Instruction in Statistics Education (GAISE) Project, Franklin & Garfield, 2006). These guidelines begin with a description of student learning goals which are reprinted here:

Goals for Students in an Introductory Course: What It Means to Be Statistically Educated

The desired result of all introductory statistics courses is to produce statistically educated students, which means that students should develop statistical literacy and the ability to think statistically. The following goals represent what such a student should know and understand. Achieving this knowledge will require learning some statistical techniques, but the specific techniques are not as important as the knowledge that comes from going through the process of learning them. Therefore, we are not recommending specific topical coverage.

1. ***Students should believe and understand why***

- Data beat anecdotes.
- Variability is natural and is also predictable and quantifiable.
- Random *sampling* allows results of surveys and experiments to be extended to the population from which the sample was taken.
- Random *assignment* in comparative experiments allows cause and effect conclusions to be drawn.
- Association is not causation.
- Statistical significance does not necessarily imply practical importance, especially for studies with large sample sizes.
- Finding no statistically significant difference or relationship does not necessarily mean there is no difference or no relationship in the population, especially for studies with small sample sizes.

2. ***Students should recognize:***

- Common sources of bias in surveys and experiments.
- How to determine the population to which the results of statistical inference can be extended, if any, based on how the data were collected.
- How to determine when a cause and effect inference can be drawn from an association, based on how the data were collected (e.g., the design of the study)
- That words such as "normal," "random" and "correlation" have specific meanings in statistics that may differ from common usage.

3. ***Students should understand the parts of the process through which statistics works to answer questions, namely:***

- How to obtain or generate data.
- How to graph the data as a first step in analyzing data, and how to know when that's enough to answer the question of interest.
- How to interpret numerical summaries and graphical displays of data – both to answer questions and to check conditions (in order to use statistical procedures correctly).
- How to make appropriate use of statistical inference.
- How to communicate the results of a statistical analysis.

4. ***Students should understand the basic ideas of statistical inference:***

- The concept of a sampling distribution and how it applies to making statistical inferences based on samples of data (including the idea of standard error)
- The concept of statistical significance including significance levels and P-values.
- The concept of confidence interval, including the interpretation of confidence level and margin of error.

5. ***Finally, students should know:***

- How to interpret statistical results in context.
- How to critique news stories and journal articles that include statistical information, including identifying what's missing in the presentation and the flaws in the studies or methods used to generate the information.
- When to call for help from a statistician.

To achieve these desired learning goals, the following recommendations are offered:

1. Emphasize statistical literacy and develop statistical thinking.
2. Use real data.
3. Stress conceptual understanding rather than mere knowledge of procedures.
4. Foster active learning in the classroom.
5. Use technology for developing conceptual understanding and analyzing data.
6. Use assessments to improve and evaluate student learning.

Although these guidelines are stated fairly simply here, there is more elaboration in the GAISE report along with suggestions and examples (Franklin & Garfield, 2006). It is also important to note that today's introductory statistics course is actually a family of courses taught across many disciplines and departments. There are many different versions out there of the introductory course. For example, some courses require a calculus prerequisite, and others do not require any mathematics beyond high school elementary algebra. Some courses cover what we might consider more advanced topics (like ANOVA and multiple regression), and others do not go beyond a simple two sample t-test. Some course aim at general literacy and developing informed consumers of data, while other courses are focused on preparing users and producers of statistics. There is continuing debate among educators as to just what topics belong in the introductory course and what topics could be eliminated in light of the guidelines, advances in the practice of statistics, and new technological tools.

Despite the differences in the various versions of the introductory statistics course, there are some common learning goals for students in any of these courses that are outlined in the new guidelines. For example, helping students to become statistically literate and to think and reason statistically. However, working to achieve these goals requires more than guidelines. It requires a careful study of research on the development and understanding of important statistical concepts, as well as literature on pedagogical methods, student assessment, and technology tools used to help students learn statistics. We have written this book to provide this research foundation in an accessible way to teachers of statistics.

Connecting Research to Teaching Practice

Now that statistics education has emerged as a distinct discipline, with its own professional journals and conferences and with new books being published on teaching statistics (e.g., Garfield, 2005; Gelman & Nolan, 2002; F. Gordon, & S. Gordon,

1992; Moore, 2000), it is time to connect the research, to reform recommendations (such as the Guidelines for Assessment and Instruction in Statistics Education – GAISE, 2005a, 2005b) in a practical handbook for teachers of statistics. That is what we aim to do in this book.

Our book is structured around the big ideas that are most important for students to learn, such as data, statistical models, distribution, center, variability, comparing groups, sampling, statistical inference, and covariation. In doing so, we offer our reflections and advice based on the relevant research base in statistics education. We summarize and synthesize studies related to different aspects of teaching and learning statistic and the big statistical ideas we want students to understand.

A Caveat

A question raised by reviewers of an earlier version of this book was "how do we know that the materials and approaches described in this book are actually effective?" While statisticians and statistics educators would like "hard data" on the effectiveness of the suggestions and materials we provide, we cannot provide data in the form of results from controlled experiments. It is hard to even imagine what such an experiment might look like, since the materials we provide may be implemented in various ways. While we have seen the activities and methods used effectively over several semesters of teaching introductory statistics, we have also seen even the most detailed lesson plans implemented in different ways, where not all activities are used, where there is more discussion on one topic than another, or when the teacher does more talking and explaining than was indicated in the lesson plan. These differences in implementation of the materials is partly due to the fact that our materials are flexible and encourage discussion and exploration, but also to the power of a teacher's beliefs about teaching and learning statistics, the constraints under which they teach, and the nature of different classroom communities and the students that make up these communities.

Despite the lack of evidence from controlled experiments, we do have a strong foundation in research as well as current learning theories for our pedagogical method, which is described in detail in Chapter 3. Again, from our experience observing teachers using these materials, they appear to encourage the development of students' statistical reasoning. We have seen students develop confidence in using and communicating their statistical reasoning as they are guided in the activities described in our book. The materials that we describe, and which can be accessed in the accompanying Website, provide detailed examples of how the pedagogical methods may be used in a statistics course, and are based on our understanding of the research and its implications for structuring sequences of activities for developing key statistical ideas.

Our most important and overarching goal is to provide a bridge between educational research and teaching practice. We encourage readers to reflect on the key aspects of the sample activities we describe as well as on the overall pedagogical principles they reflect.

Overview of This Book

The following chapter (Chapter 2) provides a brief introduction to the diverse research literature that addresses the teaching and learning of statistics. We begin with the earliest studies by psychologist Jean Piaget and then summarize research studies from various disciplines organized around the major research questions they address. We conclude this chapter with pedagogical implications from the research literature. Chapter 3 addresses instructional issues related to teaching and learning statistics and the role of learning theories in designing instructions. We propose a Statistical Reasoning Learning Environment, (SRLE) and contrast this approach to more traditional methods of teaching and leaning statistics. Chapter 4 discusses current research and practice in the areas of assessing student learning, and Chapter 5 focuses on the use of technology to improve student learning of statistics.

Part II of the book consists of nine chapters (Chapters 6 through 14), each focusing on a specific statistical topic (data, statistical models, distribution, center, variability, comparing groups, samples and sampling, statistical inference, and covariation). These chapters all follow a similar structure. They begin with a snapshot of a research-based activity designed to help students develop their reasoning about that chapter's topic. Next is the rationale for this activity, discussion of the importance of understanding the topic, followed by an analysis of how we view the place of this topic in the curriculum of an introductory statistics course. Next, a concise summary of the relevant research related to teaching and learning this topic is provided, followed by our view of implications of this research to teaching students to reason about this topic.

To connect research to practice, we offer in each chapter a table that provides a bridge between the research and a possible sequence of practical teaching activities. This list of ideas and activities can be used to guide the development of students' reasoning about the topic. Following this table are descriptions of a set of sample lesson plans and their associated activities that are posted on the accompanying Website in full detail. The purpose of this brief description of the lessons is to explain the main ideas and goals of the activities and emphasize the flow of ideas and their relation to the scholarly literature. The use of appropriate technological tools is embedded in these sample activities.

The two chapters in Part III (Chapters 15 and 16) focus on one of the most important ways to make positive changes happen in education, via collaboration. Chapter 15 discusses the value and use of collaboration in the statistics classroom to facilitate and promote student learning. Chapter 16 focuses on collaboration among teachers. The first part of this chapter makes the case for collaboration among teachers of statistics as a way to implement and sustain instructional changes and as a way to implement a *Statistical Reasoning Learning Environment* described in Chapter 3. The second part of the chapter describes collaboration among teachers as researchers in order to generate new methods to improve teaching and learning and to contribute to the knowledge base in statistics education. The goal of these final chapters is to convince readers that collaboration is an essential way to bring about instructional change, to create new knowledge, and most important of all, to improve student learning of statistics.

Supplementary Website for This Book

There is a Website with supplementary materials, produced by the NSF-funded *Adapting and Implementing Innovative Material in Statistics* (AIMS) Project (see http://www.tc.umn.edu/~aims). These materials (which are described in detail in the Introduction to Part II of this book) include a set of annotated lesson plans, classroom activities, and assessment items.

Summary

Statistics education has emerged as an important area of today's curriculum at the high school and college level, given the growth of introductory courses, desired learning outcomes for students, and the endorsement of new guidelines for teaching. We hope that this book will contribute to the growth and visibility of the field of statistics education by providing valuable resources and suggestions to all the many dedicated teachers of statistics. By making the research more accessible and by connecting the research to teaching practice, our aim is to help advance the field, improve the educational experience of students who study statistics, overturn the much maligned image of this important subject, and set goals for future research and curricular development.

Chapter 2
Research on Teaching and Learning Statistics[1]

People have strong intuitions about random sampling; ... these intuitions are wrong in fundamental respects;... these intuitions are shared by naive subjects and by trained scientists; and ... they are applied with unfortunate consequences in the course of scientific inquiry.
(Tversky & Kahneman, 1971, p. 105)

Overview

This chapter provides an overview of current research on teaching and learning statistics, summarizing studies that have been conducted by researchers from different disciplines and focused on students at all levels. The review is organized by general research questions addressed, and suggests what can be learned from the results about each of these questions. The implications of the research are described in terms of eight principles for learning statistics from Garfield (1995), which are revisited in light of results from current studies.

Introduction: The Expanding Area of Statistics Education Research

Today, statistics education can be viewed as a new and emerging discipline, when compared to other areas of study and inquiry. This new discipline has a research base that is often difficult to locate and build upon. For many people interested in reading this area of scholarship, statistics education research can seem to be an invisible, fragmented discipline. This is because studies related to this topic of interest have appeared in publications from diverse disciplines, and are more often thought of as studies in those disciplines (e.g., psychology, science education, mathematics education, or in educational technology) than in the area of statistics education. In 2002, the *Statistics Education Research Journal* (http://www.stat.auckland.ac.nz/~iase/serj) was established, and the discipline now has its first designated scientific

[1] This chapter is partly based on the following paper: Garfield, J., & Ben-Zvi, D. (2007). How students learn statistics revisited: A current review of research on teaching and learning statistics. *International Statistical Review, 75*(3), 372–396. (http://isi.cbs.nl/ISReview/abst753-8-Garfield_Ben-Zvi.pdf).

journal, which focuses exclusively on high-quality research. This should make it easier for future researchers to become acquainted with the discipline and locate studies for literature reviews and for teachers of statistics to look for research relevant to the teaching and learning of statistics.

In addition to SERJ, research studies related to statistics education have been published in the electronic *Journal of Statistics Education* (JSE), conference proceedings such as The International Conference on the Teaching Statistics (ICOTS, http://www.stat.auckland.ac.nz/~iase/conferences), International Group for the Psychology of Mathematics Education (PME, http://igpme.org), The Mathematics Education Research Group of Australasia (MERGA, http://www.merga.net.au), The International Congress on Mathematics Education meetings (ICME, http://www.mathunion.org/ ICMI), and The International Statistical Institute (ISI, http://isi.cbs.nl). The numerous presentations and publications from these conferences reflect the fact that there now exists an active group of educators, psychologists, and statisticians who are involved in scholarship related to the teaching and learning of statistics. In addition, more graduate students are completing dissertations in various departments that relate to teaching and learning statistics. Over 47 doctoral dissertations have been reported since 2000 (see http://www.stat.auckland.ac.nz/iasedissert).

There is much to learn from the abundant current literature that offers important contributions to understanding the nature of statistical reasoning and what it means to understand and learn statistical concepts. In this chapter, we provide first an overview of the foundational research conducted primarily by psychologists on how people make judgments and decisions when faced with uncertainty. Much of the literature summarized in previous reviews (e.g., Garfield & Ahlgren, 1988; Garfield, 1995; Shaughnessy, 1992; Shaughnessy, Garfield, & Greer, 1996) summarized this line of research.

We then provide an overview of the more current research, summarizing a sampling of studies that have been conducted by researchers from different disciplines (psychology, mathematics education, educational psychology, and statistics education). We organize these summaries according to the general research questions addressed, and suggest our view of what can be learned from the results. We describe some of the research methods used in these studies, along with their strengths and limitations. We then provide a summary of a newer focus of research that examines the development of statistical literacy, reasoning, and thinking. We provide some general implications from the research in terms of teaching and assessing students and highlight eight principles for learning statistics. These implications provide a basis for a pedagogical model, which we name as the *Statistical Reasoning Learning Environment* (SRLE, described in detail in Chapter 3) as well as provide a foundation for the specific research summaries and implications in Part II of this book.

Foundational Studies on Statistical Reasoning and Understanding

The initial research in the field was undertaken during the 1950s and 1960s by Piaget and Inhelder (1951, 1975). This early work focused on the developmental growth and structure of people's probabilistic thinking and intuitions. Although the researchers of that period were not motivated by any interest in probability and statistics as part of the school curriculum, this work inspired much of the later research that did focus on learning and teaching issues (e.g., Fischbein, 1975).

During the 1970s, a new area of research emerged, conducted primarily by psychologists studying how people make judgments and decisions when faced with uncertainty. A seminal collection of these studies was published in 1982 (Kahneman, Slovic, and Tversky, 1982). The researchers in this area focused on identifying many incorrect ways people reason, labeling "heuristics" (explained below) and biases, and then studying what factors affected these errors. Although a few psychologists also designed training activities to overcome some misconceptions, these methods were not necessarily embedded in a course or curriculum and focused on a particular type of reasoning.

Most of the published research in this area consists of studies of how adults understand or misunderstand particular statistical ideas. An influential series of studies by Kahneman et al. (1982) revealed some prevalent ways of thinking about statistics, called "heuristics," that are inconsistent with a correct understanding. In psychology, "heuristics" are simple, efficient rules of thumb hard-coded by evolutionary processes, which have been proposed to explain how people make decisions, come to judgments and solve problems, typically when facing complex problems or incomplete information. These rules work well under most circumstances, but in certain cases lead to systematic cognitive biases. Some salient examples of these faulty "heuristics" are summarized below.

Representativeness: People estimate the likelihood of a sample based on how closely it resembles the population. Use of this heuristic also leads people to judge small samples to be as likely as large ones to represent the same population. For example: Seventy percent Heads is believed to be just as likely an outcome for 1000 tosses as for 10 tosses of a fair coin.

Gambler's fallacy: Use of the representativeness heuristic leads to the mistaken view that chance is a self-correcting process. People mistakenly believe past events will affect future events when dealing with random activities. For example, after observing a long run of heads, most people believe that now a tail is "due" because the occurrence of a tail will result in a more representative sequence than the occurrence of another head.

Base-rate fallacy: People ignore the relative sizes of population subgroups when judging the likelihood of contingent events involving the subgroups, especially when empirical statistics about the probability are available (called the "base rate"). For example, when asked the probability of a hypothetical student taking history (or economics), when the overall proportion of students taking these courses is 70, people ignore these "base rate" probabilities, and instead rely on information provided about the hypothetical student's personality to determine which course is more likely to be chosen by that student.

Availability: Strength of association is used as a basis for judging how likely an event will occur. For example, estimating the divorce rate in your community by recalling the divorces of people you know, or estimating the risk of a heart attack among middle-aged people by counting the number of middle-aged acquaintances who have had heart attacks. As a result, people's probability estimates for an event are based on how easily examples of that event are recalled.

Conjunction fallacy: The conjunction of two correlated events is judged to be more likely than either of the events themselves. For example, a description is given of a 31-year old woman named Linda who is single, outspoken, and very bright. She is described as a former philosophy major who is deeply concerned with issues of discrimination and social justice. When asked which of two statements are more likely, fewer pick A: Linda is a bank teller, than B: Linda is a bank teller active in the feminist movement, even though A is more likely than B.

Additional research has identified misconceptions regarding correlation and causality (Kahneman et al., 1982), conditional probability (e.g., Falk, 1988; Pollatsek, Well, Konold, & Hardiman, 1987), independence (e.g., Konold, 1989b), randomness (e.g., Falk, 1981; Konold, 1991), the Law of Large Numbers (e.g., Well, Pollatsek, & Boyce, 1990), and weighted averages (e.g., Mevarech, 1983; Pollatsek, Lima, & Well, 1981).

A related area of work in psychology has identified a way of thinking referred to as the "outcome orientation." Konold (1989a) described this way of reasoning as the way people use a model of probability that leads them to make yes or no decisions about single events rather than looking at the series of events. For example: A weather forecaster predicts the chance of rain to be 70% for 10 days. On 7 of those 10 days it actually rained. How good were his forecasts? Many students will say that the forecaster did not do such a good job, because it should have rained on all days on which he gave a 70% chance of rain. They appear to focus on outcomes of single events rather than being able to look at series of events – 70% chance of rain means that it should rain. Similarly, a forecast of 30% rain would mean it would not rain. Fifty percent chance of rain is interpreted as meaning that you cannot tell either way. The power of this notion is evident in the college student who, on the verge of giving up, made this otherwise perplexing statement: "I don't believe in probability; because even if there is a 20% chance of rain, it could still happen" (Falk & Konold, 1992, p. 155). Later work by Konold and colleagues documented the inconsistency of student reasoning as they responded to similar assessment items, suggesting that the context of a problem may affect students' use (or lack of use) of intuitions or reasoning strategies (see Konold, Pollatsek, Well, & Gagnon, 1997).

Subsequent research to the foundational studies on faulty heuristics, biases, and misconceptions focused on methods of training individuals to reason more correctly. Some critics (e.g., Gigerenzer, 1996; Sedlmeier, 1999) argued that the cause of many identified misconceptions was actually people's inability to use proportional reasoning, required by many of these problems that involved probabilities. They suggested to use a frequency approach (using counts and ratios rather than percents and decimals), and observed that subjects performed better on similar tasks when using frequencies rather than fractions or decimals.

Recognizing these persistent errors, researchers have explored ways to help college students and adults to correctly use statistical reasoning, sometimes using specific training sessions (e.g., Fong, Krantz, & Nisbett, 1986; Nisbett, 1993; Sedlmeier, 1999; Pollatsek et al., 1987). Some of these studies involve a training component that takes place in a lab setting and involves paper and pencil assessments of learning the concept of interest. Lovett (2001) collected participants' talk-aloud protocols to find out what ideas and strategies students were using to solve data analysis problems. She found that feedback could be given to help students improve their ability to select appropriate data analyses.

Researchers continue to examine errors and misconceptions related to statistical reasoning. Many of these studies focus on topics related to probability (e.g., Batanero & Sánchez, 2005; O'Connell, 1999; Fast, 1997; Hirsch & O'Donnell, 2001; Tarr & Lannin, 2005). However, other studies have examined misconceptions and errors related to additional topics such as contingency tables (Batanero, Estepa, Godino, & Green, 1996), sampling distributions (Yu & Behrens, 1995), significance tests (e.g., Falk & Greenbaum, 1995), and a variety of errors in statistical reasoning (e.g., Garfield, 2003; Tempelaar, Gijselaers, & van der Loeff, 2006).

What Can We Learn from These Studies?

The main message from this body of research seems to be that inappropriate reasoning about statistical ideas is widespread and persistent, similar at all age levels (even among some experienced researchers), and quite difficult to change. There are many misconceptions and faulty intuitions used by students and adults that are stubborn and difficult to overcome, despite even the best statistics instruction. In addition, students' statistical reasoning is often inconsistent from item to item or topic to topic, depending on the context of the problem and students' experience with the context. Although some types of training seem to lead to positive results, there is no strong evidence that the results were sustained beyond the training sessions or could be generalized beyond the specific types of problems used.

Recent Research on Teaching and Learning Statistics

In this section, we provide an overview of the more current research that has been conducted by researchers from different disciplines, organized according to the general research questions addressed. We also suggest our view of what can be learned from the results.

How Do School Students Come to Understand Statistics and Probability?

In contrast to studies on misconceptions and faulty heuristics that looked at particular types of training to overcome or correct these types of problems, another line of inquiry has focused on how to develop good statistical reasoning and understanding, as part of instruction in elementary and secondary mathematics classes. Researchers began to take an interest in studying how children understand basic concepts related to data analysis when these topics began to be added to the K-12 mathematics curricula (in the 1980s and 1990s, e.g., NCTM, 2000). These studies revealed many difficulties students have with concepts that were believed to be fairly elementary such as the mean (Bright & Friel, 1998; Konold et al., 1997; Mokros & Russell, 1995; Rubin, Bruce, & Tenney, 1991; Russell & Mokros, 1996; Shaughnessy, 1992, 2007). Not surprisingly, most of the research examining school children's understanding of data analysis has been conducted by mathematics education researchers who have focused their studies on foundational concepts and their interconnections, such as data, distribution, center, and variability (e.g., Bakker & Gravemeijer, 2004; Cobb, McClain, & Gravemeijer, 2003b). The focus of these studies was to investigate how students begin to understand these ideas and how their reasoning develops when using carefully designed activities assisted by technological tools.

Studies focused on students in K-12 classes, investigating how they come to understand statistical ideas such as data (e.g., Ben-Zvi & Arcavi, 2001), distribution (Pfannkuch, 2006b; Prodromou & Pratt, 2006; Reading & Reid, 2006; Watson, 2005), variability (Bakker, 2004b; Ben-Zvi, 2004b; delMas & Liu, 2005; Hammerman & Rubin, 2004; Reading, 2004), and probability (e.g., Abrahamson, Janusz, & Wilensky, 2006; Pratt, 2007). Some involved teaching experiments (Steffe & Thompson, 2000; Cobb, Confrey, diSessa, Lehrer, & Schauble, 2003a) conducted over several weeks, where a team of researchers and teachers teach and/or closely observe the class to see how particular activities and tools help develop understanding of a statistical concept or set of concepts (e.g., Cobb, 1999; Saldanha & Thompson, 2003; Shaughnessy, Ciancetta, Best, & Canada, 2004).

Interest in probability continues among mathematics educators, primarily documenting the difficulties students have understanding these concepts at different grade levels, the common misconceptions about probability, and the role of computer tools to help students develop reasoning about chance and uncertainty (see Jones, Langrall, & Mooney, 2007). In a recent compilation of research studies in this area, Jones (2005) remarks that there is a need to study the evolution and development of students' reasoning and to find ways to link ideas of chance and data, rather than studying probability as a formal mathematics topic. This type of work is currently underway by Konold who is developing the *Model Chance* software, and conducting studies using the software with young children (Konold, Kazak, Dahlstrom-Hakki, Lehrer, & Kim, 2007).

What Can We Learn from These Studies?

The studies focused on developing students' reasoning about data and chance suggest that these ideas are often more complex and difficult for students to learn than was assumed. Studies involving elementary and secondary school students that focus on understanding of particular concepts (e.g., Ben-Zvi, Gil, & Apel, 2007; Cobb et al., 2003b; Pfannkuch, 2006b) show that carefully designed sequences of activities using appropriate technological tools can help students improve reasoning and understanding over substantial periods of time (Ben-Zvi, 2000). These studies suggest some possible sequences of activities that can help students develop ideas of important concepts such as distribution, variability, and covariation, and offer implications for the types of instructional activities and technological tools that may facilitate students' learning and reasoning. More implications are included later in this chapter.

How Do Preservice and Practicing Teachers Develop Understanding of Statistics?

A newer line of research that has also been the focus of studies by mathematics educators is the study of preservice or practicing teachers' knowledge of statistics and probability, and how that understanding develops in different contexts (e.g., Leavy, 2006; Makar & Confrey, 2005; Pfannkuch, 2006b). Some studies of preservice K-12 teachers focus on undergraduate college students majoring in elementary or mathematics education and how they understand and reason about statistics (e.g., Groth & Bergner, 2005).

In one study, Groth and Bergner (2005) examined the use of metaphors as a way to reveal student understanding, and were disappointed to note that students (preservice teachers) who have completed a course in statistics have limited and often incorrect notions of the idea of sample. Leavy (2006) examined preservice teachers' reasoning about the concept of distribution. In her one group pretest–posttest design that took place during a semester-long mathematics methods course, participants worked in small groups on two statistical inquiry projects requiring the collection, representation, analysis, and reporting of data involving comparing distributions. She found that many teachers appeared to be gaining in their ability to reason about distributions in comparing groups while others failed to use the relevant content they had learned when comparing groups of data (see also Ciancetta, 2007).

Stohl (2005) summarizes studies that examine teachers' understanding and teaching of probability. She addresses problems resulting from mathematics teachers' more computational approach to thinking about probability and suggests ways to better prepare teachers to understand and teach this challenging topic.

Some studies on practicing teachers examine how these "students" learn and reason about statistics as a result of workshops or in-service courses (e.g., Mickelson

& Heaton, 2004). Makar and Confrey (2005) and Hammerman and Rubin (2004) suggest that teachers' understanding of basic statistical analysis, such as comparing two groups, can be very confused (e.g., wanting to compare individual data points rather than group trends). However, they have found that with carefully designed instruction using innovative visualization software (such as *Fathom*, Key Curriculum Press, 2006; http://www.keypress.com/fathom, or *TinkerPlots*, Konold & Miller, 2005; http://www.keypress.com/tinkerplots), they can be guided to reason more statistically. Studies also focus on of how teachers teach and how their knowledge affects their teaching of statistics (e.g., Canada, 2004, 2006; Makar & Confrey, 2005; Rubin, Hammerman, Campbell, & Puttick, 2005; Pfannkuch, 2006b).

What Can We Learn from These Studies?

The studies focused on preservice and in-service K-12 teachers suggest that both have many difficulties understanding and teaching core ideas of probability and statistics. The studies suggest further explorations are needed in the issues of developing teacher knowledge of statistics as well as methods of helping teachers to understand the big ideas of statistics. A current joint IASE-ICMI study is focused on this issue (see http://www.ugr.es/~icmi/iase_study). Efforts such as the TEAM project (Franklin & Mewborn, 2006) have attempted to bring mathematics educators and statisticians together to create new ways to prepare future K-12 teachers of statistics, by making sure that these students have a course in statistics as part of their requirements, taught in methods that emphasize conceptual understanding, data exploration, and use of appropriate technology. How to help practicing teachers develop a better knowledge of statistics is still an area that needs to be further explored.

How Do College Students Learn Statistics?

Researchers across many disciplines have long been interested in the teaching and learning of statistics in college classes perhaps because of the tremendous numbers of students who enroll in introductory statistics course as a requirement for their degree programs. Some of the studies on college level students examined a particular activity or intervention; others have looked at use of a technological tool or teaching method (e.g., Noll, 2007). Several statisticians who teach statistics have focused their attention on studying students' learning in their classes (e.g., Chance, 2002; Lee, Zeleke, & Wachtel, 2002; Wild, Triggs, & Pfannkuch, 1997). Most of these studies involve the researchers' own classes, sometimes examining one class, or involving multiple classes at the same institution.

Because of the large number and variety of studies in college settings, this section is subdivided into several subsections that correspond to important questions regarding the teaching and learning of statistics after secondary school.

How Can Technology be Used to Promote Statistical Reasoning?

One of the major areas of current interest is the role technological tools (such as computers, graphing calculators, software, and Internet) can play in helping students develop statistical literacy and reasoning. Research on simulation training indicates that even a well-designed simulation is unlikely to be an effective teaching tool unless students' interaction with it is carefully structured (Lane & Peres, 2006). Simulations, however, can play a significant role in enhancing students' ability to study random processes and statistical concepts (Lane & Peres, 2006; Lane & Tang, 2000; Mills, 2004).

Using a collaborative classroom research model that implemented activities and gathered data in three different institutions, delMas, Garfield, and Chance (1999) studied the development of reasoning about sampling distributions, using a simulation program and research-based activities. They found that student performance on a specially designed posttest, to assess students' reasoning about sampling distributions, improved as the activity was changed to imbed assessments within the activity. They also found that having students make and test conjectures about different empirical sampling distributions from various populations. Lunsford, Rowell, and Goodson-Espy (2006) replicated this study in a different type of undergraduate course and found similar results.

Lane and Tang (2000) compared the effectiveness of simulations for teaching statistical concepts to the effectiveness of a textbook; while Aberson, Berger, Healy, Kyle, and Romero (2000) studied the impact of a Web-based, interactive tutorial used to present the sampling distribution of the mean on student learning.

In a study of students' reasoning about the standard deviation, delMas (2005) had students manipulate a specially designed software tool to create histograms with the highest or lowest possible standard deviation, given a set of fixed bars. He identified some common ways students understand and misunderstand the standard deviation, such as thinking of "spread" as spreading butter, being evenly distributed in a graph. He also found that students had difficulty reasoning about bars in a histogram having density, in that they represent several points on a particular interval on a graph.

How Effective is Online Instruction?

Another topic of interest to statistics educators has been the use of online instruction either in a Web-based course or "hybrid/blended" course, in which a significant amount of the course learning activity has been moved online, making it possible to reduce the amount of time spent in the classroom. For example, Utts (2003) and Ward (2004) found no differences in course performance for students in a hybrid versus a traditional course, and concluded that hybrid courses were not resulting in decreased student performance, although Utts noted lower evaluations by students in the hybrid courses. However, no significant differences in course performance do not imply that there were no real differences in student outcomes for the compared instructional methods.

What Do Students Remember After Taking Statistics?

Mathews and Clark (2003) and Clark, Karuat, Mathews, and Wimbish (2003) investigated high achieving students (an A grade in their first statistics course) from four tertiary institutions on their understanding of the mean, standard deviation, and the Central Limit Theorem. Their interviews of students within the first 6 weeks of the term after the completion of the statistics course revealed that students tended to have relatively unsophisticated understandings of the concepts of mean and standard deviation and fragmentary recall of the Central Limit Theorem.

How Effective is Active Learning in Teaching Statistics?

Keeler and Steinhorst (1995), Giraud (1997), and Magel (1998) investigated different methods of cooperative learning in teaching statistics at their institutions, and found generally positive results. Keeler and Steinhorst (1995) found that when students worked in pairs, the final grades were higher and more students stayed in the course than in previous semester. Giraud (1997) found that using cooperative groups in class to work on assignments led to higher test grades than students in a lecture class. Magel (1998) found that implementing cooperative groups in a large lecture class also led to improved test scores compared to grades from a previous semester that did not use group work.

Meletiou and Lee (2002) organized their curricula along a Project-Activities-Cooperative Learning-Exercises model emphasizing statistical thinking and reasoning and an orientation toward investigating conjectures and discovery of results using data. Students were assessed on their understanding at the beginning and end of the course. Increased understanding was observed on tasks requiring statistical reasoning such as deducing whether a set of data could have been drawn at random from a particular population.

How Can Formal Statistical Ideas be Developed from Informal Ideas?

Building on collaborative classroom research methods, Garfield, delMas,and Chance (2007) used Japanese Lesson Study to design, test, and revise a lesson to help students develop reasoning about variability, building formal ideas from informal ideas. Japanese Lesson Study builds on the idea that teachers can conduct their own classroom research by carefully examining a particular problem in their class, trying an activity or set of activities to develop student learning, and then to evaluate, reflect, and revise the activity.

A group of novice and experienced teachers designed a lesson to help reveal and build on students' informal intuitions about variability, which was taught, observed, analyzed, and revised. Their study suggested a sequence of activities to help students develop a deep understanding of the concept of variability and measures such as range, interquartile range, and standard deviation. Schwartz, Sears, and Chang (2007) used a similar approach to develop what they referred to as students' prior knowledge, using specific activities to motivate and engage students to develop more formal reasoning about particular statistical concepts.

Can Training Improve Students' Statistical Problem Solving?

In one type of study, students are trained in a particular type of procedure to see if this affects their performance on different outcome measures. For example, Quilici and Mayer (2002) taught college students to sort statistics word problems on the basis of structural features (i.e., whether the problem could be solved by *t*-test, correlation, or chi-square statistics) rather than surface features (i.e., the problem's cover story). In this study, college students displayed a higher level of structural awareness (i.e., sorting word problems on the basis of structural features) at the end rather than the beginning of their first statistics course. Recognizing that the problem one is working on can be solved using the same method as a problem one already knows is an important skill in statistical problem solving. Lovett (2001) collected participants' talk-aloud protocols to find out what ideas and strategies students were using to solve data analysis problems. She found that feedback could be given to help students improve their ability to select appropriate data analyses. Meyer and Lovett (2002) developed a computer-based training program to provide scaffolding to guide students in analyzing data to solve statistical problems.

What is the Role of Affect in Learning Statistics?

Several researchers have explored factors related to students' success in statistics classes. Most of these studies have examined noncognitive variables, such as students' attitudes and anxiety about statistics, (e.g., Schau & Mattern, 1997a). This work has sometimes included development of instruments to assess student outcomes (e.g., attitudes, anxiety, and reasoning). Studies have also examined relationships between different student characteristics (e.g., mathematics background, statistics attitudes, or anxiety) and course outcomes for students taking statistics in education or psychology courses (e.g., Elmore & Vasu, 1986; Wisenbaker & Scott, 1997). In addition, some of these studies examined what graduate students in education, psychology, or the social sciences know and understand while or after learning statistics (e.g., O'Connell & Corter, 1993; Earley, 2001; Finney, 2000; Huberty, Dresden, & Bak, 1993).

How Does Students' Reasoning Develop During a Statistics Course?

In a recent study, Zieffler (2006) studied the growth in the students' reasoning about bivariate data over an introductory statistics course. He found that most of the growth in this reasoning, as measured by four administrations of a bivariate reasoning scale, happened before students formally studied a unit on bivariate data. His results suggested that perhaps the course that was designed to help students develop their general statistical reasoning, was helping them reason well about distributions of bivariate data before they formally studied that topic. He recommended the use of similar longitudinal studies, administering a set of items at three or more points of time in order to model the growth of student during instruction, a suggestion also included in the recent report on statistics in mathematics education research (Scheaffer, 2007). In a related study, Zieffler, Garfield, delMas, and Gould (2007) explore the growth in students' statistical reasoning throughout a 14-week class that embedded a sequence of simulation activities designed to develop student's inferential reasoning. They found that students' reasoning did not develop in a consistent linear way throughout the course. They also found that in some cases students' reasoning about concepts (such as sampling distribution) developed before the formal study of that topic, supporting the previous results by Zieffler (2006) about bivariate reasoning.

What Can We Learn from These Studies?

The many studies that focus on teaching and learning statistics at the college level continue to point out the many difficulties college students have learning, remembering, and using statistics, and point to some modest successes. These studies also serve to illustrate the many practical problems faced by college statistics instructors such as how to incorporate active or collaborative learning in a large class, whether or not to use an online or "hybrid" course, or how to select one type of software tool as more effective than another.

Many of these studies set out to answer a question such as "which is better?" However, these studies reveal that it is difficult to determine the impact of a particular teaching method or instruction tool on students' learning in a course due to limitations in study design or assessments used. While teachers would like research studies to convince them that a particular teaching method or instructional tool leads to significantly improved student outcomes, that kind of evidence is not actually available in the research literature. The results of many of the comparative studies are usually limited to that particular course setting and cannot be generalized to other courses. For example, if one study compared a particular type of active learning to a "traditional" course, results cannot be generalized to active learning vs. a "traditional" course, because of the variety of methods of active learning and the variety of "traditional" courses.

Though not based on comparative experiments, some recent classroom research studies, while not trying to be comparative, suggest some practical implications for teachers. For example, developing a deep understanding of statistics concepts is quite challenging and should not be underestimated. Research suggests that it takes time, a well thought out learning trajectory, and appropriate technological tools, activities, and discussion questions to develop deep understanding. Good reasoning about important concepts can be developed very carefully using activities and tools given enough time, and revisiting of these ideas.

The research studies on attitudes and anxiety suggest that there are few strong (or large) predictors of how well students do in a statistics course, that there is little change in attitudes from beginning to end of a first course in statistics (and sometimes negative changes) and that difficulties in students' reasoning and poor attitudes are fairly widespread. The evidence does not show that if students are good in mathematics or have good attitudes, they will be likely to succeed in statistics, which is contrary to many teachers' beliefs. Instead, students who may not be strong in mathematics may work hard, enjoy the subject matter, and do very well in an introductory statistics course. Variables such as motivation, conscientiousness, and desire to learn may be better predictors.

These studies suggest the types of negative value judgments students place on the study of statistics, how difficult they perceive the subject to be, and how useful, as pertaining to either one's course or the field in general. Nevertheless, the studies suggest that teachers need to cultivate more positive beliefs about the value of statistics and statistical literacy by being aware that students come to statistics courses with a great variety of expectations and perspectives on what statistics is about, and their own ability or lack of ability to succeed in the course.

One consistent problem in many of the quantitative studies focused on college students has to do with the lack of high quality and consistent measures used to assess student learning outcomes. It is very common for these studies to use final exam scores or course grades as outcome measure. These measures are often problematic because they are used with establishing evidence of validity and reliability and do not necessarily measure outcome of general value to the wider community (Garfield, 2006). In the past few years, new instruments have been carefully developed and studied (e.g., delMas, Garfield, Ooms, & Chance, 2007), which may lead to less reliance on teacher made measures.

In recent years, there has also been more attention paid to distinguishing and defining learning outcomes in introductory statistics courses, and the frequently used terms for these outcomes refer to statistical literacy, statistical reasoning, and statistical thinking. Clarifying desired learning outcomes can also help researchers better develop and use appropriate measures in their studies, and to align these measures with learning goals valued by the statistics education community.

Distinguishing Between Statistical Literacy, Reasoning, and Thinking

Although statistics is now viewed as a unique discipline, statistical content is most often taught in the mathematics curriculum (K–12) and in departments of mathematics (college level). This has led to exhortations by leading statisticians, such as Moore (1998), about the differences between statistics and mathematics. These arguments challenge statisticians and statistics educators to carefully define the unique characteristics of statistics and in particular, the distinctions between statistical literacy, reasoning, and thinking (Ben-Zvi & Garfield, 2004a). We prefer the following definitions:

Statistical literacy is a key ability expected of citizens in information-laden societies, and is often touted as an expected outcome of schooling and as a necessary component of adults' numeracy and literacy. Statistical literacy involves understanding and using the basic language and tools of statistics: knowing what basic statistical terms mean, understanding the use of simple statistical symbols, and recognizing and being able to interpret different representations of data (Garfield, 1999; Rumsey, 2002; Snell, 1999). There are other views of statistical literacy such as Gal's (2000, 2002), whose focus is on the data consumer: Statistical literacy is portrayed as the ability to interpret, critically evaluate, and communicate about statistical information and messages. Gal (2002) argues that statistically literate behavior is predicated on the joint activation of five interrelated knowledge bases (literacy, statistical, mathematical, context, and critical), together with a cluster of supporting dispositions and enabling beliefs. Watson and Callingham (2003) proposed and validated a model of three levels of statistical literacy (knowledge of terms, understanding of terms in context, and critiquing claims in the media).

Statistical reasoning is the way people reason with statistical ideas and make sense of statistical information. Statistical reasoning may involve connecting one concept to another (e.g., center and spread) or may combine ideas about data and chance. Statistical reasoning also means understanding and being able to explain statistical processes, and being able to interpret statistical results (Garfield, 2002b). We see statistical reasoning as the mental representations and connections that students have regarding statistical concepts.

Statistical thinking involves a higher order of thinking than statistical reasoning. Statistical thinking is the way professional statisticians think (Wild & Pfannkuch, 1999). It includes knowing how and why to use a particular method, measure, design or statistical model; deep understanding of the theories underlying statistical processes and methods; as well as understanding the constraints and limitations of statistics and statistical inference. Statistical thinking is also about understanding how statistical models are used to simulate random phenomena, understanding how data are produced to estimate probabilities, recognizing how, when, and why existing inferential tools can be used, and being able to understand and utilize the context of a problem to plan and evaluate investigations and to draw conclusions (Chance, 2002). Finally, we view statistical thinking as the normative use of statistical models, methods, and applications in considering or solving statistical problems.

Statistical literacy, reasoning, and thinking are unique areas, but there is some overlap with a type of hierarchy, where statistical literacy provides the foundation for

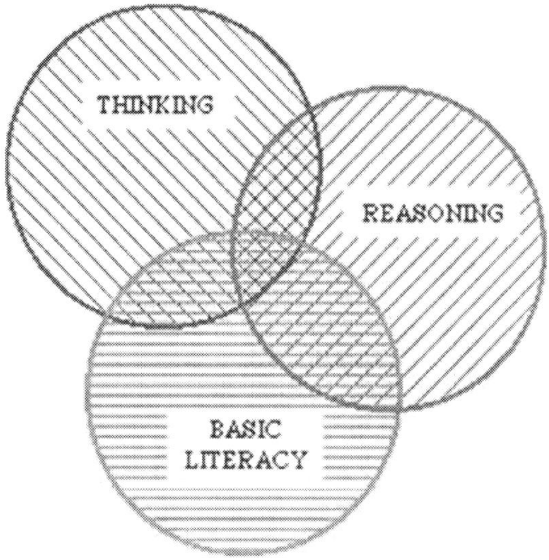

Fig. 2.1 The overlap and hierarchy of statistical literacy, reasoning, and thinking (Artist Website, https://app.gen.umn.edu/artist)

reasoning and thinking (see Fig. 2.1). A summary of additional models of statistical reasoning and thinking can be found in Jones, Langrall, Mooney, and Thornton (2004).

There is a growing network of researchers who are interested in studying the development of students' statistical literacy, reasoning, and thinking (e.g., SRTL – *The International Statistical Reasoning, Thinking, and Literacy Research Forums*, http://srtl.stat.auckland.ac.nz/). The topics of these research studies conducted by members of this community reflect the shift in emphasis in statistics instruction, from developing procedural understanding, i.e., statistical techniques, formulas, computations and procedures, to developing conceptual understanding and statistical literacy, reasoning, and thinking.

Current research studies address this shift by concentrating on some core ideas of statistics, often referred to as the "big ideas". This research focus is parallel to the increasing attention that is being paid in the educational research community to the need to clearly define and focus both research and instruction, and therefore, assessment, on the "big ideas" of a discipline (Bransford, Brown, & Cocking, 2000; Wiggins & McTighe, 1998). The following sections offer a brief summary of some of the research on reasoning about three of the key "big ideas": distribution, center, and variability. The amount of research in these areas illustrates the complexity of studying and developing student's reasoning about these ideas.

Limitations of Research Related to Teaching and Learning Statistics

Given that research related to teaching and learning statistics has been conducted across disciplines, in a fragmented, disconnected way, it is not surprising that several limitations are apparent in this research. For example:

- Too many small studies (e.g., conducted in one class or at one school) with too broad a focus (e.g., "Is a Web-based class as effective as a "regular" class?") that cannot generalize answers to larger questions.
- Lack of connection to theory or previous research (which makes it difficult to frame good questions or interpret results).
- Lack of focused questions on the effects of a particular tool or method or activity on learning to understand a particular concept.
- Lack of appropriate research instruments to measure important student outcomes (e.g., using only instructor-designed final exams or course grades as outcome measures).
- Lack of connections among current research and researchers (due to different disciplines, journals, theoretical backgrounds, problems, etc.).
- Failure to look beyond the researcher's own discipline to other related studies, perhaps thinking that the work in other disciplines does not apply to their questions, or because they do not know what is present there or how to find it.
- Failure to recognize and note that lack of a statistically significant effect does not guarantee no effect (treatments may be causing differences that are not detected).
- Too many studies conducted by a single researcher rather than a collaborative team, with different strengths and backgrounds (collaboration in research is described in Chapter 16).

When teachers of statistics read about the results of a research study and consider how it might affect their teaching, there are some important questions to keep in mind. First of all, they should ask what the learning goals of the study were and how they were measured. If a study uses final exam or course grades, it is useful to know what these represent and that they measure learning goals of importance. For example, a study that uses a final exam as a measure of student learning, where the exam consists primarily of items that can be memorized or computed, most likely is missing important information on student conceptual learning outcomes.

A second question to consider is whether the explanatory variables are well defined and consistent with the researcher's theory and beliefs about learning and teaching. For example, a study that purports to be investigating the effectiveness of active learning in a college statistics class, may be defining active learning as having students work in pairs on completing a worksheet. This activity may not be viewed as active learning by other educators and therefore, the results would not generalize to the broad domain of active learning methods.

A third question to consider is whether the results of the study suggest something useful for other teachers, such as a new way to help students understand a concept or develop understanding. Do the results have specific implications for teaching, rather than broad implications about a general tool or teaching method? Finally, did the researchers carefully review the existing literature and was their study based on previous studies and theories? If so, how do their results support, extend, or contradict previous results and implications?

Given the growing numbers of students taking statistics at all levels and an increasing need for these students to develop their statistical understanding and value statistics, more high-quality research is needed to provide guidance to teachers of statistics. Much of this research will most likely involve some use of qualitative methods, which are described in the following section.

Research Methodologies: Qualitative Methods

Many of the studies summarized in the chapters of Part II of this book, which provide a research base for the proposed lessons and instructional activities, have used qualitative methods. Some studies utilize videotaped classroom observations and teacher or student interviews as a way to gather data. Other sources of qualitative data include students' responses to open-ended questions, field notes of teachers and researchers, and samples of students' work (e.g., graphs constructed, statistics projects).

Some studies combine qualitative data with quantitative data, using comparison of pre-and posttest assessments data along with interviews and/or teaching experiments. A few other studies include some quantitative data in the context of student assessment, but it is hard to find a truly quantitative study in the research today that applies an experimental design or multivariate model.

It may seem surprising that few statistical summaries are actually included in these studies, given that the subject being studied is statistics. And it may seem surprising that the research studies in this book are not traditional designed experiments, involving control groups compared to groups that have received experimental treatment, the gold standard of experimental design. However, statistics education currently, tends to follow the tradition of mathematics and science education, in using mostly qualitative methods to develop an understanding of the nature of students' thinking and reasoning, and to explore how these develop (see Kelly & Lesh, 2000).

Some promising new quantitative methods are beginning to be used in educational research (Scheaffer, 2007) as well as use of Rasch modeling to transform qualitative into quantitative data (see Watson & Callingham, 2003; Watson, Kelly, Callingham and Shaughnessy, 2003) to model levels of statistical literacy. We believe that research in statistics education will continue to involve methodologies that provide qualitative data, and describe some of these methods in the following section.

Teaching Experiments

Although referred to as "experiments," teaching experiments are a type of classroom-based research. They do not involve comparisons of a randomly imposed treatment to a control and do not take place in controlled laboratory settings. Instead, teaching experiments (which are sometimes called design experiments) take place in regular classrooms and are part of students' instruction in a subject. They involve designing, teaching, observing, and evaluating a sequence of activities to help students develop a particular learning goal (Steffe & Thompson, 2000). The primary goal of conducting a teaching experiment is not to assess the effectiveness of the preformulated instructional design, but rather to improve the design by checking and revising conjectures about the trajectory of learning for both the classroom community and the individual students who compose that classroom. Thus, the goal is to integrate the teacher's instructional goals and directions for learning with the trajectory of students' thinking and learning. This can often result in "greater understanding of a learning ecology – a complex, interacting system involving multiple elements of different types and levels – by designing its elements and anticipating how those elements function together to support learning" (Cobb et al., 2003a, p. 9).

This type of research is usually of high intensity (e.g., 20 weeks) and somewhat invasive, in that each lesson in a particular design experiment is observed, videotaped, and analyzed. The structure of teaching experiments varies greatly, but they generally have three stages: preparation for the teaching experiment, the classroom instruction and interaction with students, and debriefing and analyzing the teaching episodes (Ben-Zvi, Garfield, & Zieffler, 2006). These are sometimes referred to as the preparation phase, the actual experimentation phase, and the retrospective analysis (Gravemeijer, 2000). In longitudinal studies, these phases are repeated several times (e.g., Hershkowitz et al., 2002).

The first stage in a teaching experiment is preparation for the actual study. It is during this stage that the research team, which usually includes researchers and teachers, envisions how dialogue and activity will occur as a result of planned classroom activity. The researchers propose a sequence of ideas and knowledge that they hope students will construct as they participate in the activities and classroom dialogue and plan instruction to help move students along this path toward the desired learning goal.

During the actual teaching experiment, the researchers test and modify their conjectures about the statistical learning trajectory as a result of their communication, interaction, and observation of students. The learning environment also evolves as a result of the interactions between the teacher and students as they engage in the content. The research team ideally meets after every classroom session to modify the learning trajectory and plan new lessons. These meetings are generally audio-taped for future reference. Because of the constant modification, classroom lessons cannot be planned in detail too far in advance.

The research team performs the retrospective analysis after an entire teaching experiment has been completed. It is during this stage that the team develops domain-specific instructional theory to help guide future instruction. They also develop new hypothetical learning trajectories for future design experiments.

Action Research

While many of the studies reviewed in this book take place in statistics classes, most do not follow the rigor of the teaching experiment described above. Many take the form of classroom or action research (e.g., delMas et al., 1999). Action research in education is an inquiry process in which teachers typically examine their own educational practice systematically and carefully using the techniques of research (Glanz, 2003). Action research has the potential to generate genuine and sustained improvements in schools. It gives educators new opportunities to reflect on and assess their teaching; to explore and test new ideas, methods, and materials; to assess how effective the new approaches were; to share feedback with fellow team members; and to make decisions about which new approaches to include in the team's curriculum, instruction, and assessment plans. The action research uses an iterative cycle to study the impact of an activity on students' reasoning as the researchers develop a model of reasoning about a statistical concept.

General Implications of the Research Literature for Teaching Statistics

Research studies across the disciplines that relate to statistics education provide valuable information for teachers of statistics. For example, some of the studies reveal the types of difficulties students have when learning particular topics, so that teachers may not only be aware of where errors and misconceptions might occur and how students' statistical reasoning might develop, but also what to look for in their informal and formal assessments of their learning.

We think that the research literature is especially important to consider because it contradicts many informal or intuitive beliefs held by teachers. For example, that students earning a grade of A in a statistics class understand the basic ideas of statistics (e.g., Clark et al., 2003; Mathews & Clark, 2003), or that students' reasoning about statistics is consistent from topic to topic (e.g., Konold, 1995). In addition, even the most clever and carefully designed technological tool or good instructional activity will not necessarily lead students to correctly understand and reason about an abstract statistical concept (e.g., Chance, DelMas, & Garfield, 2004).

Principles for Learning Statistics

After reviewing the research related to teaching and learning statistics over a decade ago, Garfield (1995) proposed 10 principles for learning statistics. Despite the increased number of studies since that article was published, we believe that these principles are still valid. They are also consistent with recent cognitive psychology publications on student learning, such as *How People Learn* (Bransford et al., 2000), which focus on promoting learning for understanding, developing student-centered learning environments, and rethinking what is taught, how it is taught, and how it is assessed. These principles have been regrouped into eight research-supported statements about student learning of statistics.

1. *Students learn by constructing knowledge.* Teaching is not telling, learning is not remembering. Regardless of how clearly a teacher or book tells them something,

students will understand the material only after they have constructed their own meaning for what they are learning. Moreover, ignoring, dismissing, or merely "disproving" the students' current ideas will leave them intact – and they will outlast the thin veneer of course content (Bakker & Gravemeijer, 2004; Lehrer & Schauble, 2007).

Students do not come to class as "blank slates" or "empty vessels" waiting to be filled, but instead approach learning activities with significant prior knowledge. In learning something new, they interpret the new information in terms of the knowledge they already have, constructing their own meanings by connecting the new information to what they already believe (Bransford et al., 2000). Students tend to accept new ideas only when their old ideas do not work, or are shown to be inefficient for purposes they think are important.

2. *Students learn by active involvement in learning activities.* Research suggests that students learn better if they are engaged in, and motivated to struggle with their own learning. For this reason, if no other, students appear to learn better if they work cooperatively in small groups to solve problems and learn to argue convincingly for their approach among conflicting ideas and methods (e.g., Giraud, 1997; Keeler & Steinhorst, 1995; Magel, 1998). Small-group activities may involve groups of three or four students working in class to solve a problem, discuss a procedure, or analyze a set of data. Groups may also be used to work on an in-depth project outside of class. Group activities provide opportunities for students to express their ideas both orally and in writing, helping them become more involved in their own learning. However, just being active and having a good time is not enough to ensure learning. Good learning activities are carefully designed and the teacher has an important role to listen, probe, sum up, and assess the main points (e.g., Chick & Watson, 2002; Courtney, Courtney, & Nicholson, 1994; Perkins & Saris, 2001; Potthast, 1999).
3. *Students learn to do well only what they practice doing.* Practice may mean hands-on activities, activities using cooperative small groups, or work on the computer. Students also learn better if they have experience applying ideas in new situations. If they practice only calculating answers to familiar, well-defined problems, then that is all they are likely to learn. Students cannot learn to think critically, analyze information, communicate ideas, make arguments, tackle novel situations, unless they are permitted and encouraged to do those things over and over in many contexts. Merely repeating and reviewing tasks is unlikely to lead to improved skills or deeper understanding (e.g., Pfannkuch, 2005a; Watson, 2004; Watson & Shaughnessy, 2004).
4. *It is easy to underestimate the difficulty students have in understanding basic concepts of probability and statistics.* Many research studies have shown that ideas of probability and statistics are very difficult for students to learn and often conflict with many of their own beliefs and intuitions about data and chance (delMas et al., 2007; Jones et al., 2007; Shaughnessy, 1992, 2007).
5. *It is easy to overestimate how well their students understand basic concepts.* A few studies have shown that although students may be able to answer some test items correctly or perform calculations correctly, they may still misunderstand

basic ideas and concepts. Also, students who receive top grades in a class may not understand and remember the basic ideas of statistics (e.g., Clark et al., 2003; Mathews & Clark, 2003).

6. *Learning is enhanced by having students become aware of and confront their errors in reasoning.* Several research studies in statistics as well as in other disciplines show that students' errors in reasoning (sometime appearing to be misconceptions) are often strong and resilient – they are slow to change, even when students are confronted with evidence that their beliefs are incorrect (Bransford et al., 2000).

   Students seem to learn better when activities are structured to help students evaluate the difference between their own beliefs about chance events and actual empirical results. If students are first asked to make guesses or predictions about data and random events, they are more likely to care about and process the actual results. When experimental evidence explicitly contradicts their predictions, students should be helped to evaluate this difference. In fact, unless students are forced to record and then compare their predictions with actual results, they tend to see in their data confirming evidence for their misconceptions of probability (e.g., Jones, Langrall, Thornton, & Mogill, 1999; Konold, 1989a; Shaughnessy, 1977). Research in physics instruction also points to this method of testing beliefs against empirical evidence (e.g., Clement, 1987).

7. *Technological tools should be used to help students visualize and explore data, not just to follow algorithms to predetermined ends.* Technology-based instruction appears to help students learn basic statistics concepts by providing different ways to represent the same data set (e.g., going from tables of data to histograms to boxplots) or by allowing students to manipulate different aspects of a particular representation in exploring a data set (e.g., changing the shape of a histogram to see what happens to the relative positions of the mean and median). Instructional software may be used to help students understand abstract ideas. For example, students may develop an understanding of the Central Limit Theorem by constructing various populations and observing the distributions of statistics computed from samples drawn from these populations (e.g., Ben-Zvi, 2000). The computer can also be used to improve students' understanding of probability by allowing them to explore and represent statistical models, change assumptions and parameters for these models, and analyze data generated by applying these models (Biehler, 1991; Jones et al., 2007).

   Innovative new visualization software, such as *Fathom* (Key Curriculum Press, 2006) and *TinkerPlots* (Konold & Miller, 2005) are available to students at all levels to explore data and learn to reason statistically.

8. *Students learn better if they receive consistent and helpful feedback on their performance.* Learning is enhanced if students have ample opportunities to express ideas and get feedback on their ideas. Feedback should be analytical, and come at a time when students are interested in it (see Garfield & Chance, 2000). There must be time for students to reflect on the feedback they receive, make adjustments, and try again before being given a grade. For example, evaluation of student projects may be used as a way to give feedback to students while they

work on a problem during a course, not just as a final judgment when they are finished with the course. Since statistical expertise typically involves more than mastering facts and calculations, assessment should capture students' ability to reason, communicate, and apply their statistical knowledge. A variety of assessment methods should be used to capture the full range of students' learning, e.g., written and oral reports on projects, minute papers (e.g., Angelo & Cross, 1993) reflecting students' understanding of material from one class session, or essay questions included on exams. Teachers should become proficient in developing and choosing appropriate methods that are aligned with instruction and key course goals, and should be skilled in communicating assessment results to students (e.g., delMas et al., 1999).

Summary

There has been a tremendous increase in research studies focused on teaching and learning statistics and probability over the past 15 years. These studies continue to span many different disciplines and differ in focus, theory, methodology, and supporting literature. However, when reviewed together, they suggest the difficulties students have learning statistics and the need to revise traditional methods of teaching. The most recent studies on the development of particular types of learning outcomes and reasoning about special topics offer many implications for changes in curriculum and teaching methods. However, there are still many open questions and much work is needed to offer more specific guidance to teachers of statistics. Since research is now elucidating some conceptual foundations (e.g., notion of distribution, variability, sampling, and statistical inference) for statistics education, the consequence is that statistics education is emerging as a discipline in its own right, not an appendage to mathematics education.

We find the most helpful results to come from collaborative research projects, and encourage future researchers to find collaborators, ideally from different disciplines, to combine expertise in the content area (statistics), student learning (education and/or psychology), and assessment. We provide some examples and suggestions for collaborative research in Chapter 16. Eventually, we hope to see larger studies on particular questions of interest conducted across several institutions, using high quality measurement instruments. We are particularly interested in seeing studies that use newer methods of analysis (e.g., hierarchical linear modeling, analysis of longitudinal data) that allow the careful study of the growth of reasoning and learning over different instructional settings and/or over a period of time. The new guidelines for using statistical methods in mathematics education research (see Scheaffer, 2007) offer many useful suggestions for improving the growing field of statistics education research as well. We look forward to the wealth of results from new studies that will inform our knowledge about how students learn statistics.

The research studies summarized in each chapter in Part II provide more details on the complexity of teaching and learning different statistical topics, explaining why they are so difficult for students to learn. These studies suggest that it is important for teachers to move beyond a focus on skills and computations and the role of teacher as the one who delivers the content. Instead, the suggested role of teachers is to provide a carefully designed learning environment, appropriate technological tools, and access to real and interesting data sets, as well as scaffolding guidance to students' emerging statistical reasoning. Chapter 3 outlines a different instruction model – *Statistical Reasoning Learning Environment* (SRLE) – where the teacher orchestrates class work and discussion and provides timely and nondirective interventions as a representative of the statistics discipline in the classroom. This type of teaching requires a teacher who is aware not only of the complexities and difficulty of the concepts but also of the desired learning goals – such as what effective statistical literacy, reasoning, and thinking look like – so that assessments can be examined and compared to these goals. The research provided in this chapter and the chapters in Part II can help provide this knowledge base for teachers of statistics.

Chapter 3
Creating a Statistical Reasoning Learning Environment

Shorn of all subtlety and led naked out of the protective fold of educational research literature, there comes a sheepish little fact: lectures don't work nearly as well as many of us would like to think.

(Cobb, 1992, p. 9)

Overview

The research studies from the previous chapter suggested very different ways of teaching than traditional lectures, which is how most current statistics instructors learned this subject themselves. Leaving that familiar method to try active learning techniques can be quite challenging. This chapter offers advice on how to deal with many practical issues involved in student learning in an interactive statistics class and describes ways to build what we refer to as a "Statistical Reasoning Learning Environment" (SRLE).

Before addressing these practical instructional issues, we begin by considering theories of learning that guide our actions as teachers. We summarize some of the important aspects of current learning theories and provide a model of instruction based on these theories. We offer suggestions for facilitating successful classroom discussions, building cooperative learning experiences for students (see also Chapter 15), and using technology to support students' construction of statistical knowledge (more in Chapter 5). We also discuss challenges and constraints that make it difficult to teach in a way that promotes statistical literacy and reasoning.

Theories of How Students Learn

Whether or not we are aware of it, our own theories of how students learn guide us as we teach. Many times we tend to teach the way we have experienced instruction; we teach the way we have been taught. Most of us experienced "traditional" methods of teaching, where the instructor lectures and provides notes on an overhead or a board (blackboard or white board), and we have taken notes. Sometimes lectures have been stimulating, and we often have acquired knowledge. Sometimes we have found it easier to read the notes or textbook and that has been sufficient to learn the material. Sometimes this lecture/textbook strategy has been possible because

J.B. Garfield, D. Ben-Zvi, *Developing Students' Statistical Reasoning*,

what has been assessed on exams is our ability to remember what the instructor or textbook has told us.

This sequence has been referred to as "teaching is telling and learning is remembering" and is based on the assumption that knowledge consists of facts and skills Thompson & Zeuli 1999. This theory assumes that the teachers "tell" students the facts and skills they need to know, that students are empty vessels or blank slates in which they receive the knowledge, and then they give it back. The degree to which students give back or remember the correct knowledge is considered the degree to which they have learned.

In fact, despite the literally thousands of efforts to sway instruction and student learning away from "teaching is telling and learning is remembering" since World War II, few have had significant or enduring effects (Cohen & Ball, 1999). In recent years, the new science of learning is beginning to provide knowledge to significantly improve students' abilities to become active learners who seek to understand complex subject matter and are better prepared to transfer what they have learned to new problems and settings. In particular, the theory of learning called "constructivism" describes learning as something quite different than the telling/remembering sequence.

Constructivism

Although there are different versions of constructivism, the basic idea is that people learn by constructing knowledge, rather than by receiving knowledge. In the constructivist theory, the meaning of "knowing" has shifted from being able to remember and repeat information to being able to find and use it (Simon, 1995).

There are four characteristics of learning in constructivism: (a) an emphasis on understanding; (b) a focus on the processes of knowing (e.g., Piaget, 1978; Vygotsky, 1978); (c) the principle that people construct new knowledge and understandings based on what they already know and believe; and (d) the importance of helping people take control of their own learning, predict their performances on various tasks, and to monitor their current levels of mastery and understanding (*metacognition,* e.g., Brown, 1975; Flavell, 1976).

In the most general sense, the contemporary view of learning in accordance with the constructivist theory is that new knowledge and understandings are based on the existing knowledge and beliefs we already have and are grounded in our experiences (e.g., Cobb, 1994; Piaget, 1978; Vygotsky, 1978). We learn by doing. And when we learn, our previous knowledge does not go away; it is integrated with the new knowledge. One implication for teaching is that teachers need to pay attention to the incomplete understandings, the false beliefs, and the naïve renditions of concepts that learners bring with them to a given subject. Teachers then need to build on these ideas in ways that help each student achieve a more mature understanding. If students' initial ideas and beliefs are ignored, the understandings that they develop can be very different from what the teacher intends (Bransford et al., 2000).

Sociocultural Theory

A related important perspective is that of *sociocultural theory* that draws heavily on the work of Vygotsky (1986), as well as later theoreticians (for example, Wertsch, 1991, 1998). Vygotsky described learning as being embedded within social events and occurring as a learner interacts with people, objects, and events in the environment. The sociocultural perspective has a profound implication concerning the important role of *enculturation* processes in learning. Briefly stated, the process of enculturation refers to entering a community or a practice and picking up their points of view. The beginning student learns to participate in a certain cognitive and cultural practice, where the teacher has the important role of a mentor and mediator, or the "enculturator" (cf., Schoenfeld, 1992; Resnick, 1988). This is especially the case with regard to statistical thinking, with its own culture, values, and belief systems, and habits of questioning, representing, concluding, and communicating (see Chapter 1). Thus, for statistical enculturation to occur, specific thinking tools are to be developed alongside cooperative and communicative processes taking place in the classroom.

Building Instruction on Theories of Learning

The implication of these current theories of learning is that good instructional practice consists of designing learning environments that stimulate students to construct knowledge. This involves activities that provide students many opportunities to think, reason, and reflect on their learning, as well as discussing and reflecting with their peers. It does not mean that teachers should never "tell" students anything directly and instead should always allow them to construct knowledge for themselves. Rather it means that learning is enhanced when teachers pay attention to the knowledge and beliefs that learners bring to a learning task, use this knowledge as a starting point for new instruction, and monitor students' changing conceptions as instruction proceeds.

It should be apparent that it is easier to prepare a lecture than it is to design a learning environment, where students engage in activities and discussions and/or collaborative projects, supported by technological tools. While the first approach is teacher centered: "what is it I want to tell my students," "what material do I want to cover," etc., the second approach is more student-centered: "what can I do to promote students learning," "how can I engage students in learning, hands-on activities, developing reasoning, discussing ideas, working in teams," etc. According to this latter approach, the teacher is cast as an instructional coach, a co-learner, an enculturator, or a facilitator, rather than as a conduit of knowledge in a teacher-centered classroom.

Why change from a teacher-centered approach to a student-centered approach? We think the answer is that the second approach is more effective in helping students build a deeper understanding of statistics and to be able to leave a class and use what they have learned in subsequent classes or in the real world. One problem with the "teaching is telling" approach is that students rarely have a chance to develop a deep understanding of what they have "learned," and quickly forget it after they complete a course.

A Statistical Reasoning Learning Environment (SRLE)

An effective and positive statistics classroom can be viewed as a learning environment for developing in students a deep and meaningful understanding of statistics and helping students develop their ability to think and reason statistically. We call this type of classroom the "Statistical Reasoning Learning Environment" (SRLE). By calling it a learning environment, we emphasize that it is more than a textbook, activities, or assignments that we provide to our students. It is the combination of text materials, class activities and culture, discussion, technology, teaching approach, and assessment.

Our model is based on six principles of instructional design described by Cobb & McClain (2004) that on first glance may seem very similar to the six Guidelines for Assessment and Instruction in Statistics Education (GAISE, 2005a, 2005b) recommendations for teaching statistics. They include the use of real data, activities, technology, and assessment; however, the Cobb and McClain's principles go beyond the more general GAISE guidelines, as shown by the bold formatting in their design principles listed below:

1. Focuses on developing central statistical ideas **rather than on presenting set of tools and procedures**.
2. Uses real and motivating data sets **to engage students in making and testing conjectures.**
3. Uses classroom activities **to support the development of students' reasoning**.
4. Integrates the use of appropriate technological tools **that allow students to test their conjectures,** explore and analyze data, and develop their statistical reasoning.
5. **Promotes classroom discourse that includes statistical arguments and sustained exchanges that focus on significant statistical ideas.**
6. Uses assessment to learn what students know and to **monitor the development of their statistical learning as well as to evaluate instructional plans and progress.**

We now elaborate on each of these topics in the SRLE.

1. Focus on Developing Central Statistical Ideas (Content)

There are several key statistical ideas that we would like all students to understand at a deep conceptual level. These ideas serve as overarching goals that direct our teaching efforts and motivate and guide students' learning. These include:

- *Data*: Understanding the need for data in making decisions and evaluating information, the different types of data, and how the methods of collecting data (via surveys) and producing data (in experiments) make a difference in the types of conclusions that can be drawn, knowing the characteristics of good data and how to avoid bias and measurement error. Understanding the role, importance of and distinction between random sampling and random assignment in collecting and producing data.
- *Distribution*: Understanding that a set of data may be examined and explored as an entity (a distribution) rather than as a set of separate cases; that a graph of these (quantitative) data can be summarized in terms of shape, center, and spread; that different representations of the same data set may reveal different aspects of the distribution; that visually examining distributions is an important and necessary part of data analysis, and that distributions may be formed from sets of individual data values or from summary statistics such as means (e.g., sampling distributions of means). Distributions also allow us to make inferences by comparing an obtained sample statistic to a distribution of all possible sample statistics for a particular theory or hypothesis.
- *Variability*: Understanding that data vary, sometimes in predictable ways. There may be sources of variability that can be recognized and used to explain the variability. Sometimes the variability is due to random sampling or measurement error. Other times, it is due to the inherent properties of what is measured (e.g., weights of 4 year olds). An important part of examining data is to determine how spread out the data are in a distribution. It is usually helpful to know a measure of center when interpreting measures of variability, and the choice of these measures depends on the shape and other characteristics of the distribution. Different variability measures tell you different things about the distribution (e.g., standard deviation focuses on typical distance from the mean, range tells the difference between the minimum and maximum value, and IQR reveals the width of the middle half of the data).
- *Center*: Understanding the idea of a center of a distribution as a "signal in a noisy process" (Konold et al., 2002), which can be summarized by a statistical measure (such as mean and median). It is most helpful to interpret a measure of center along with a measure of spread, and these choices often are based on the shape of the distribution and whether or not there are other features such as outliers, clusters, gaps, and skewness.
- *Statistical Models*: Understanding that statistical models may be useful in helping us explain or predict data values. We often compare the data to a model (e.g., the normal distribution or a regression model) and then see how well the data fit the model by examining residuals or deviations from the model. We also use models to simulate data in order to explore properties of procedures or concepts.
- *Randomness*: Understanding that each outcome of a random event is unpredictable, yet we may predict long-term patterns. For example, we cannot predict if a roll of a fair die will be a 2, or any other number, but we can predict that over many rolls about 1/6 will be 2's.
- *Covariation*: Understanding that the relationship between two quantitative variables may vary in a predictable way (e.g., high values with one variable tend to

occur with high values of another). Sometimes this relationship can be modeled with a straight line (the regression line). This allows us to predict values of one variable using values of the other variable. An association does not necessarily imply causation, although there may be a causal relationship (a randomized comparative experiment is needed to determine cause and effect).

- *Sampling*: Understanding that much of statistical work involves taking samples and using them to make estimates or decisions about the populations from which they are drawn. Samples drawn from a population vary in some predictable ways. We examine the variability within a sample as well as the variability between samples when making inferences.
- *Statistical Inference*: Understanding that making estimates or decisions is based on samples of data in observational and experimental studies. The accuracy of inferences is based on the variability of the data, the sample size, and the appropriateness of underlying assumptions such as random samples of data and samples being large enough to assume normally distributed sampling distributions. A P-value is an indicator used to evaluate the strength of evidence against a particular conjecture, but it does not suggest the practical importance of a statistical result. The P-value indicates how likely a sample or experimental result as extreme as what was observed would be given a particular theory or claim and helps to answer the question "is this result due to chance or due to an effect of interest (such as a condition in an experiment)."

While most textbooks present material in a structure based on a logical analyses of the content, students often see the content as a sequential set of tools and procedures and do not see how the concepts are interrelated. For example, learning about distribution early in a course is rarely connected to the ideas of sampling distributions later in a course. We advocate a focus on these key ideas and the interrelations among them and suggest ways to present them throughout a course, revisiting them in different contexts, illustrating their multiple representations and interrelationships, and helping students recognize how they form the supporting structure of statistical knowledge.

2. Use Real and Motivating Data

Data are at the heart of statistical work, and data should be the focus for statistical learning as well. Throughout a course, students need to consider methods of data collection and production and how these methods affect the quality of the data and the types of analyses that are appropriate. Interesting data sets motivate students to engage in activities, especially ones that ask them to make conjectures about a data set before analyzing it.

The GAISE report that provides a set of six guidelines for teaching the introductory college statistics course (see Chapter 1; Franklin & Garfield, 2006) states:

> It is important to use real data in teaching statistics, for reasons of authenticity, for considering issues related to how and why the data were produced or collected, and to relate the analysis to the problem context. Using real data sets of interest to students is also a good way to engage them in thinking about the data and relevant statistical concepts. There are many types of real data including archival data, classroom-generated data, and simulated data. Sometimes hypothetical data sets may be used to illustrate a particular point (e.g., the Anscombe data (1973) illustrates how four data sets can have the same correlation, but strikingly different scatterplots) or to assess a specific concept. It is important to only use created or realistic data for this specific purpose and not for general data analysis and exploration. An important aspect of dealing with real data is helping students learn to formulate good questions and use data to answer them appropriately based on how and why the data were produced.

There are a variety of easy ways to collect data sets, some of which are described in Chapter 6 of this book. Many good data sets with a variety of contexts can be readily accessed on the Web (see resources section of Chapter 6 on Data in the accompanying Website, http://www.tc.umn.edu/~aims). The GAISE report (Franklin & Garfield (2006) cautions teachers to make sure that the questions used with data sets are of interest to students and that not all data sets interest all students, so data should be used from a variety of contexts.

3. Use Classroom Activities to Develop Students' Statistical Reasoning

An important part of the SRLE is the use of carefully designed activities that promote student learning through collaboration, interaction, discussion, data, and interesting problems (e.g., McKeachie, Pintrich, Lin, Smith, & Sharma, 1990; Bransford et al., 2000). The positive effects of active learning have been found for short-term mastery, long-term retention, or depth of understanding of course material; acquisition of critical thinking or creative problem-solving skills; formation of positive attitudes toward the subject being taught; and increasing the level of confidence in knowledge or skills.

The GAISE report (Franklin & Garfield, 2006) states:

> Using active learning methods in class is a valuable way to promote cooperative learning, allowing students to learn from each other. Active learning allows students to solve problems, answer questions, formulate questions of their own, discuss, explain, debate, or brainstorm during class. Thus they are involved in discovering, constructing, and understanding important statistical ideas and modeling statistical thinking. Activities have an added benefit in that they often engage students in learning and make the learning process fun. Other benefits of active learning methods are the practice students get communicating in the statistical language and learning to work in teams. Activities offer the teacher an informal method of assessing student learning and provide feedback to the instructor on how well students are learning. *It is important that teachers not underestimate the ability of activities to teach the material or overestimate the value of lectures, which is why suggestions are provided for incorporating activities even in large lecture classes.*

We favor two different models of class activities in the SRLE. The first engages students in making conjectures about a problem or a data set, as introduced in the preceding section on using real data. This method involves discussing students' conjectures, gathering or accessing the relevant data, using technology to test their conjectures, discussing the results, and then reflecting on their own actions and thoughts. An activity like this could be based on "can students in this class correctly identify Coke or Pepsi in a blind taste test?," or "which human body measurements have a normal distribution?"

The second type of activity is based on cooperative learning, where two or more students are given questions to discuss or a problem to solve as a group. For example, students could be given an activity involving a Web applet for bivariate data where they are asked to figure out a rule describing how individual points that seem to be outliers may affect the correlation and fitting of a regression line for set of bivariate data (e.g., the *Least Squares Regression* and *Guess the correlation* applets in http://www.rossmanchance.com/applets/). They try different locations of a point, seeing the resulting effect on the correlation coefficient and regression line. When using cooperative learning activities, it is important that students work together as a group (and often in pairs using technology), not just compare their answers (Johnson, Johnson & Smith, 1998b). For practical and useful guidelines for using cooperative learning activities see Chapter 15.

Instructors of large classes often feel that they are unable to introduce active learning in their classes due to constraints of the physical setup as well as large numbers of students. The GAISE report (Franklin & Garfield, 2006) suggests ways that active learning can be implemented in these challenging situations. Suggestions for using cooperative learning in a large class can be found in Chapter 15 of this book.

4. Integrate the Use of Appropriate Technological Tools

It is impossible to imagine nowadays a statistics class that does not utilize technology (e.g., computers, Internet, statistical software, graphing calculators, and Web applets) in several ways. With the accessibility of computers and the prevalence of graphing calculators, students no longer have to spend time performing tedious calculations. After understanding how a formula works, they can automate this procedure using technology. This allows students to focus on the more important tasks of learning how to choose appropriate analytic methods and how to interpret results.

But, technology offers much more than a quick way to generate statistics or graph data. Technological tools also allow students to develop an understanding of abstract concepts and the interrelationships between concepts.

We view technology as an integral part of the SRLE. Technology should be used to analyze data, allowing students to focus on interpretation of results and testing of conditions rather than on computational mechanics. Technological tools should also be used to help students visualize concepts and develop an understanding of abstract ideas through simulations. Some tools offer both types of uses, while in other cases a statistical software package may be supplemented by Web applets. Technology is also used for course management with systems like Blackboard and WebCT that are playing a larger role in communication and collaboration capabilities and in assessment.

Regardless of the tools used, it is important to view the use of technology not just as a way to compute numbers, but as a way to explore conceptual ideas and enhance student learning, collaboration, and communication as well. We caution against using technology merely for the sake of using technology (e.g., entering 100 numbers in a graphing calculator and calculating statistical summaries) or for pseudo-accuracy (carrying out results to multiple decimal places). Not all technological tools will have all desired features. Moreover, new tools appear all the time and need to be carefully evaluated for if and how they may best be used to enhance student learning. We devote Chapter 5 to the subject of technology, giving examples of innovative tools and ways to use these tools to help develop students' reasoning.

5. Promote Classroom Discourse

Traditional statistics classes usually did not have much discourse, "giving" information through lectures and asking questions to "get" some answers. This is different from the kind of dialogue where students respond to each other's questions and learn to question each other as well as defend their answers and arguments. In today's statistics classes, the use of activities and technology allows for a new form of classroom discourse. Cobb & McClain (2004) describe the characteristics of effective classroom discourse in which statistical arguments explain why the way in which the data have been organized gives rise to insights into the phenomenon under investigation; students engage in sustained exchanges that focus on significant statistical ideas.

It can be very challenging to create an SRLE with classroom discourse that enables students to engage in discussions in which significant statistical issues emerge and where arguments are presented and their meaning is openly negotiated. We offer these guidelines for teachers:

- Use good questions that encourage students to speculate and think and do not necessarily have one right answer. For example, asking students what they think a distribution of Grade Point Averages would look like for students in the class, and why. Or asking students to reason about what might be compelling evidence that a student can actually distinguish between two brands of soda in a blind taste test.
- Require students to explain their reasoning and justify their answers. Then ask other students if they agree or disagree and why.
- Create a classroom climate where students feel safe expressing their views, even if they are tentative. This can be done if teachers encourage students to express their conjectures, and asking other students to comment on these conjectures,

and allowing students to test some of these conjectures using tools and software, rather than telling them whether they are right or wrong. Questions that begin with "what do you think" or "What would happen if" can lead to good class discussions. Also, if a student responds to a question posed by the teacher, and gives a wrong answer, it can be better to ask the class "what do you think?" and let them try to think about how correct or incorrect the answer is and why, rather than correcting the student.

In the lesson plans that accompany this book (fully presented in the accompanying Website), we give examples of good questions to motivate students and engage them in statistical reasoning. We also give questions that encourage students to discuss and share their reasoning, as well as questions used to help wrap-up a class that has been filled with activities and cooperative learning.

Managing discussions can be challenging and many instructors are nervous about leaving their comfortable lecture mode for an open-ended class discussion. For more information and practical advice about leading and managing discussions, see Davis (1993), and McKeachie & Svinicki (2006).

6. Use Alternative Assessment

In recent years, we have seen many alternative forms of assessment being used in statistics classes. In addition to quizzes, homework, and exams, many teachers use student's statistical projects as a form of authentic assessment. These projects vary in structure, but typically allow students to pose or select a problem, gather or access appropriate data to answer the problem, analyze the data, and write up the results in a technical report and/or presentation. In many cases, projects allow students to collaborate with peers and professionals. Other forms of alternative assessment are also used to assess students' statistical literacy (e.g., critique a graph in a newspaper), their reasoning (e.g., write a meaningful short essay), or provide feedback to the instructor (e.g., minute papers).

Students will value what you assess. Therefore, assessments need to be aligned with learning goals. Assessments need to focus on understanding key ideas and not just on skills, procedures, and computed answers. This should be done with formative assessments used during a course (e.g., quizzes, small projects, or observing and listening to students in class) as well as with summative evaluations (course grades). Useful and timely feedback is essential for assessments to lead to learning. Types of assessment may be more or less practical in different types of courses. However, it is possible, even in large classes, to implement good assessments.

Although many statistics instructors have made important changes in their class that emphasize newer learning goals such as conceptual understanding and developing students' statistical literacy, reasoning and thinking, the quizzes and exams given in most statistics classes still look very traditional. We see many examples in textbooks and on Websites of assessments that continue to focus on computation, definitions, and skills. Many instructors write their own assessments, and many of these include single number answers or forced choice options. While forced choice items can be written in ways to assess students' statistical reasoning (see Cobb & 1998), it is a difficult and challenging job to write such items. The ARTIST Website (https://app.gen.umn.edu/artist) offers an item bank of more than a thousand items that have been designed to measure students' statistical literacy, reasoning and thinking, many of which are in forced choice format but provide examples of more reasoning-based items.

We offer a separate chapter in this book that details different types of assessment methods, examples, and advice (see Chapter 4).

A Closer Look at the SRLE

We begin by describing and contrasting what is often referred to as a "traditional" class to a class that embeds the SRLE perspective. We deliberately exaggerate in the following description of the "traditional" class to illustrate the contrasting practices and underlying ideas.

A Traditional Class

The students come to class, with no anticipation of what they will learn, ready to copy down what the instructor has to say. The instructor presents a lecture that includes examples, some data analysis, perhaps some demonstrations. The students listen, take notes, and perhaps ask questions. They leave class with a homework assignment that uses information from the class they just attended. They go home, try to solve the problems by looking back at their notes or looking up worked examples in the textbook, often getting frustrated if they don't find an exact match.

Now picture a very different kind of class.

An SRLE Class

The students know that they have to prepare for class by reading a few pages in the textbook. The text has been chosen so that it is current, high quality, and readable. The students are given study questions to guide their reading and note taking. Their introduction to new content takes place outside of class, as they read the material, practice some examples, and jot down notes. Students are therefore prepared to come to class with a preliminary exposure to words and techniques, a low level of statistical literacy.

Imagine we are now going to "see" a class on comparing groups using boxplots. Class begins with a short summary of what was learned in the previous class, and students are asked if they have questions on the previous class or on the assigned reading. Students ask some questions that are answered by other students and/or the instructor. The instructor rarely answers a question directly but often asks students, "What do you think?" and if another student gives an answer asks, "Do you agree with this answer? Why?"

Now the class is ready to begin the first activity. A question is given to the students such as "Do you think that female college students spend more time on cell phones than male students?" Additional questions are asked such as "What are typical amounts of time students spend on their cell phones?" or "What type of distribution would you expect to see for cell phone usage?"

Students get into small groups to discuss these questions and sketch possible distributions, considering ideas of shape, center, and spread. The instructor asks the groups to share their conjectures and reasoning, and they listen to each other and compare their predictions. The students move to computers and access a data set containing this information that has previously been gathered about the students in

Table 3.1 Major changes between a "traditional" statistics class and an SRLE class

Aspect of the course	'Traditional" statistics class	SRLE class
Focus of course	Skills and procedures, covering content	Big ideas, developing statistical reasoning and thinking
Role of textbook	Use for examples or homework problems and to review for test	Read and take notes to prepare for class
Center	Teacher centered	Student centered
Role of the teacher	Delivers knowledge by telling and explaining	Facilitates developing of knowledge through discussion and activities
Role of technology	To compute or check answers, construct graphs	To explore data, illustrate concepts, generate simulations, test conjectures, and collaborate
Discourse	Teacher answers questions	Teacher poses questions and guides a discussion. Students present arguments. Students answer other students' questions and are asked if they agree or disagree with answers. Peer and instructor feedback.
Data	Small data sets to illustrate and practice procedures	Rich, real data sets to engage students in thinking and reasoning and making conjectures. Many data sets are generated by the students from surveys and experiments.
Assessment	Focuses on computations, definitions, formulas. Focus on short answer and multiple choice tests. Often only a midterm and final tests are given.	Uses a variety of methods, assesses reasoning and thinking. Formal and informal assessment is an integral part of learning and is aligned with learning methods and goals. Students may be asked to explain their reasoning and justify their conclusions.

the class using an online student survey. Working in pairs, students generate graphs and statistics to answer the questions on cell phone use.

Students may be guided to produce side by side boxplots, and see what these reveal about the questions being investigated. They may also compare these boxplots to two histograms or two dot plots, to see why side by side boxplots can be more effective (as well as what information is lost) when comparing groups. Students may talk about appropriate measures of center and spread for the cell phone data, revisiting those ideas from previous lessons. They may notice outliers in the data and discuss what to do: How to find out if they are legitimate values or errors, what happens to the graphs and statistics if those extreme values are removed?

The teacher's role in this class is to present the problem, guide the discussion, anticipate misconceptions or difficulties in reasoning, make sure students are engaged and on task and not experiencing any difficulties. The teacher has to know when to end discussions, when to correct mistakes, and how to provide a good summary for the activity using the work students have done, so students can appreciate what they learned from the activity. At the end of class, after the wrap-up discussion and summary, students may be asked to complete a brief assessment task, providing the teacher with feedback on their learning for that class.

It should be apparent what a major change there has been in teacher's role, student's role and responsibility, and other aspects of the course. We illustrate these differences in Table 3.1.

Challenges in Moving to an SRLE

We have presented two extremes: a "traditional" class and a class based on the SRLE. The contrast between these two approaches is large, and it is apparent that even an eager and enthusiastic teacher who wants to move from a more traditional approach to a more SRLE approach is faced with many challenges. These challenges include students, colleagues, and institution, as well as challenges inherent in instructors.

Student Challenges

Students enroll in statistics classes for a variety of reasons, but mostly to fulfill requirements for different majors. They tend to be a heterogeneous group, whose backgrounds vary from weak to strong mathematically as well as in written skills and ability to think critically. Students also vary in terms of their motivation, study habits, and learning styles, and their ability and desire to work with others.

In addition, today's teachers of statistics need to be aware that students often bring some background knowledge of statistics from their prior educational experiences and from the world around them. They are not blank slates who have never heard of an average or seen a bar graph. They also bring intuitions and misconceptions about chance and randomness. They bring an understanding of terms such as normal, random, average, sample that are not easily replaced by the formal statistical use of these terms.

Many students are used to a classroom where the teacher tells and they listen. If they miss class, they can borrow lecture notes. They are used to working on homework problems from the textbook that follow examples in the text or in class. They are used to referring to a textbook only for examples of worked out problems, but are not used to reading a textbook for building understanding or to prepare for class.

A class based on the SRLE can be quite a new experience for them and one that at first may cause them discomfort. A close look at the lesson plans and activities that accompany this book will make this evident. In our experience, students are so unfamiliar with this type of class that even when they are developing and expressing impressive statistical reasoning, they are not aware that they are learning anything. They also find that being asked to reason is difficult, and they often want the teacher to just tell them what to do to solve a problem.

What can we do to help students realize how much more they can learn and enjoy in a class such as the one we describe in our book? (And do we have experience helping students come to enjoy, value, and greatly benefit from such a course?). We think it is important to explain to students the format of the course and the rationale for teaching it in a way that may be unfamiliar to them and one that is not associated experience taking mathematics courses. This is an excellent opportunity to point out and demonstrate the difference between a statistics class and a mathematics class and how statistics is not mathematics. We have found that using a first day class activity that gets students collecting and informally looking at data and interacting with each other in small groups is effective to help them see what the course is going to look like while also illustrating important ideas about data analysis and statistics. An example of such a first day activity is given in Chapter 6 on Data.

It helps if the instructor repeatedly points out to students what content and dispositions they are learning and how their statistical reasoning is developing. It is also important that the assessment be aligned with the class experiences (the activities are not just "for fun" but have important lessons that they will be responsible for), and that quality feedback on assessments is provided on what students are learning. If class activities focus on analyzing data and discussions, but exams focus on computations and procedures, students will be unhappy with this mismatch and will focus only on the latter in their studies.

One of the biggest challenges is getting students to read their textbook. We find that by giving short assignments from engaging and readable texts, along with study questions to guide their reading and note taking, many students do learn how to read their book to prepare for class. Students quickly learn that they cannot expect the teacher to tell them what is in the text, and only by reading the textbook can they fully benefit and learn from the in-class activities. Allowing students to use their answered study questions on in class exams can also help motivate students to complete the reading and note taking assignments. We have also found that students tend to ask better questions when they have already read something about that day's topic, providing positive reinforcement to them as well.

By focusing less on computations and more on reasoning, we find that students can do well in this type of class regardless of their mathematical background. However, we do not teach a calculus-based class, but rather one for liberal arts students. We do not think that students in our course need anything more than a basic familiarity with elementary algebra to succeed in introductory statistics. In fact, students are often surprised to learn that they have to read and write in what they thought would be a "math" class, despite our efforts to convince them that "statistics is not mathematics." In fact, this requirement to think and explain can also be unsettling at first. However, as more writing across the curriculum efforts are being implemented in courses, students should be less surprised to have writing assignments in their statistics course.

Students may have difficulty accepting that there is often more than one reasonable answer to many statistics problems and that the goal of solving a statistical problem is not obtaining "The One Correct Answer." This may lead students to feel confused and uncomfortable with the fuzziness of statistical answers and claims, but with repeated experience they should learn to recognize and value the importance of justifying answers and providing supporting evidence and arguments.

As students differ in their learning styles, background, and experiences, they also differ in their ability and desire to work collaboratively in groups. Some students feel that they work better alone or are afraid other students in the group will cause them to receive a lower grade. There are many good techniques to help students learn to work together collaboratively, and we offer practical suggestions on this topic in Chapter 15.

Institutional Challenges

Many people find it difficult to change from a more traditional approach because of institutional constraints. For example, their course is a prerequisite for a second course, and they are forced to cover so many topics that they cannot spend time going into depth on the big ideas. Another constraint is class size, as larger classes can make activities and discourse more difficult. Finally, technology resources may be unavailable (e.g., only have access to Excel or graphing calculators), or students may have limited access to computers or Internet. A new challenge faced by many statistics instructors is how to design a course that embraces the principles listed above, but that is taught in a Web or hybrid learning environment that uses both Web and face to face instruction.

It is hard to offer advice in this area. One suggestion is to talk to colleagues in other departments about their desired learning goals for students and what ideas they consider really important for students to learn and retain after taking a statistics class. They might also discuss what might be removed from the course, if these learning goals are achieved, since deep learning takes more time than surface learning. Instructors can also share the GAISE guidelines that have been endorsed by the American Statistical Association as support for making recommended changes (Franklin & Mewborn, 2006). In addition, they can share the research showing that students do not learn much from traditional course (see Chapter 2), that misconceptions are stubborn and hard to overcome, and that research provides evidence that the SRLE method can be used to support desired changes. Finally, we urge statistics instructors to truly collaborate with their colleagues in working together to implement desired changes. (See more on instructor collaboration in Chapter 16.)

Instructor Challenges

In order for the recommended changes to take place, it is *crucial* for a teacher to embrace a students-centered approach to learning. For example, a teacher who is used to traditional methods of "teaching is telling" will need to develop a different perspective about his or her role as a facilitator of student learning. The instructor has to believe that students can learn and develop understanding without the teacher telling them everything they need to know. The readers of this book may be college teachers of statistics or high school teachers of statistics. While both types of teachers may have very different backgrounds, they may feel a common lack of confidence and support to teach using the methods suggested in this book.

While statisticians may know their subject matter very well and have experience analyzing different data sets, they may feel more reticent to depart from the lecture method. This is especially difficult if their experiences as students were only in the "traditional" type of course format and they have not seen or experienced other models of teaching. Moore, Cobb, Garfield, & Meeker (1995) wrote that statistics faculty are conservative and resist giving up the methods of lecturing, despite "waves of technology and of educational reform" (p. 251). However, repeating the methods that we experienced as students (lectures of new material followed by practicing problems on isolated topics) does not work well for most students enrolled in introductory statistics classes today. Hogg (1992) criticizes the reliance on traditional methods of teaching statistics that lead to poor students' outcomes. College teachers of statistics who have not taken coursework on how students learn or have a background in teaching methods may want to refer to a good reference such *How People Learn* (Bransford et al., 2000) as well as other references in this book about teaching and learning for more background on educational theory and methods.

High school teachers typically have earned their degrees in mathematics or mathematics education and have taken course work in educational psychology and teaching methods. They may have limited knowledge of statistics and experience in analyzing data and using technology. So both sets of teachers have some areas in which they can work to fill in gaps. The high school teachers can work to deepen their knowledge of statistics (see, for example, Shaughnessy J. M., & Chance, B. L., 2005) so that they may develop a deeper knowledge than they will be using to help their students to learn.

A teacher (whether or not they are in a high school or college setting) will experience many challenges. These include time constraints in class and outside of class and their own confidence, comfort, and patience. Time may be the most apparent challenge at first: time it takes to prepare for a new classroom environment; time to review and think about using lesson plans rather than lecture notes; time it takes to listen to and assess students; and time to review and reflect on what students are learning and provide useful and timely feedback. It is also important to note that tenure-track faculty who are interested using this approach may be pressured not to spend so much time on their teaching, and instead put more time into research and publication.

Another challenge is the loss of control, when the class becomes more student-centered than teacher-centered. The instructor will most likely encounter situations when data gathered (sometimes improperly) present surprises, experiments back-fire, and technology crashes or fails. It is a challenge to not (quickly) tell students answers, but to ask students what they think and to wait several seconds if no one responds to a question at first. Instructors who have not used cooperative learning can find it quite difficult to monitor groups and make sure they are working smoothly. Sometimes, interventions are needed if a group does not work well together or has gone off track.

Finally, a teacher who has been confident and comfortable lecturing may be very uncomfortable giving students an activity to work on in small groups, and walking around listening, observing, and questioning. We know teachers who have feared or experienced decreased rating in teaching evaluations when they switched from lecture classes to active learning formats, and felt that students only wanted them to lecture. Some students have been frustrated by not having an instructor tell them what then need to know, and putting more demands on the students to prepare for and participate in class.

We encourage confidence and patience. Confidence that students really can learn well from each other when working collaboratively if given good activities and guidelines, confidence that students can learn from their textbook when given study questions and a good textbook, and confidence that teaching evaluations can be quite high if active learning is used effectively. We also encourage teachers to find ways to give students structured tasks outside of class to allow more class time for activities and discourse. We also encourage collaboration (see Chapter 16) because it is always easier to make changes when you are not the only one doing this. Chapter 16 offers many suggestions for how to initiate or participate in a cooperative teaching group.

An Additional Challenge: Implementing the SRLE in an Online Course

It is important to note that more and more introductory statistics classes today are being taught in an online format or a hybrid format (part online, part face-to-face). Students appear to value the flexibility of these courses and institutions have found them to be increasingly cost-effective. There are additional challenges faced by an instructor of an online or hybrid course who wishes to move to an SRLE, such as how to find ways for students to make and test conjectures in small groups, how to develop good discussions, and how to guide students to use different technological tools. While many of the online courses that are currently taught appear to be more traditional in format, there are examples of courses that are working to make students more engaged in activities and discussions (see Everson, 2006; Everson, Zieffler & Garfield, in press; Tudor, 2006). We encourage more instructors of these courses to find creative ways to implement important aspects of the SRLE into their online learning environments.

Ways to Move to SRLE

The first step in moving towards the SRLE is to see how one's current course and materials align with the components of SRLE described in the previous sections and then pick a starting point. For example, try an activity and discussion questions included in the resources for this book.

The Website that accompanies this book provides complete lesson plans, student handouts, and in some cases data sets, for an SRLE first course in statistics, as well as other useful resources. Discussion questions, instructions, and sample students results are on the Website. While some topics are not included in our materials, you will find that others receive more emphasis in that they are introduced early (in an informal way), then revisited in more formal ways (e.g., statistical models and sampling variability). However, many aspects of these topics that can be found in most introductory textbooks are missing from our lessons, as we have chosen to emphasize the main ideas at a conceptual level and have left out many procedural aspects and applications of these ideas. A complete scope and sequence for the topics and their sequences of activities can be found at the accompanying Website.

In contrast to more traditional introductory courses, there is little focus on computation, and instead technology is used to analyze data and the students' role is to use technology wisely and to interpret their results. There is no formal study of probability and its rules, no computations involving z scores (but rather, use of a Web applet and focus on the meaning of z scores), and no use of the z test for quantitative data. Our materials may look like they are from a radically different course and not one that readers may want to instantly try. However, we encourage you to review and choose one activity or one lesson and try it out and see what happens, rather than try to adopt all the lessons as an entire course.

It is also possible to adopt the principles of SRLE in a course that covers different topics than the ones we have presented in the chapters in Part 2. We encourage instructors to see how these principles may be used to design similar activities and discussions for topics that are not addressed in our book. We also note that developing an SLRE for a course can be achieved by steady change over a period of time and that it does not have to be a radical, all at once change. In fact, success may be more likely if implemented over a period of time than if a traditional instructor tries to leap into an SLRE in a single try.

Summary

A Statistical Reasoning Learning Environment is quite different than a traditional lecture-based classroom, but it may also be quite different from many statistics classes that use some activities, real data sets, and technology tools. The six principles outlined earlier are key elements in developing a class where students are engaged in making and testing conjectures using data, discussing and explaining statistical reasoning, and focusing on the important big ideas of statistics. The subsequent chapters in this book provide guidance in moving to such a classroom. The remaining chapters in Part I offer more detailed suggestions on assessment (Chapter 4) and technology (Chapter 5), and the concluding chapters in Part III offer additional guidance on collaboration (Chapters 15 and 16). The chapters in Part II suggest lessons based on the SRLE model for each of the big ideas in an introductory statistics course. These lessons can be adopted or modified as a way to move toward the SRLE. The resources section at the end of this chapter provides additional books and articles related to the issues raised in this chapter.

Chapter 4
Assessment in Statistics Education[1]

Teachers need to design assessments backward from the task, asking at each step of the way "What's the evidence I need of students' understanding? Will this assessment get at it?"
(Wiggins[2])

Overview

How do we determine what students know and learn in our statistics classes? In recent years, much attention has been paid to assessing student learning, examining outcomes of courses, aligning assessment with learning goals, and alternative methods of assessment. This is especially true for statistics courses, where there is a growing focus on newly identified learning goals such as statistical literacy, reasoning, and thinking. Over the past two decades, dramatic changes in the learning goals of statistics has led to a corresponding rethinking of how we assess our students, and it is becoming more common to use alternative assessments such as student projects, technical reports, and oral presentations than in the past. Alternative assessment tasks can also serve as a powerful learning tool, and in fact we view assessment as an important component of the learning processes rather than only as a means for testing of students' outcomes.

In this chapter, we will consider the why and how of assessment of student outcomes in statistics, including changes to traditional assessment practices and consideration of new assessment tools, and how to utilize assessment feedback to improve course planning and instruction. This chapter encompasses both research studies about assessment in teaching and learning statistics and many resources on using assessment in statistics education.

Definitions and Goals of Assessment

Assessments are used for many different purposes. Some of these include informing students of their progress, informing teachers of individual student proficiency, and providing feedback on how effectively students are learning the desired material

[1] We gratefully acknowledge the major contributions of Beth Chance and Elsa Medina in the writing of this chapter.

[2] From: http://teacher.scholastic.com/professional/assessment/studentprogress.htm

in order to modify instruction. In addition, student assessment data may be used as a way to help instructors learn about and improve their courses, and to provide information to evaluate curriculum and programs.

Two main types of assessment are often referred to as *formative* assessment and *summative* assessment. Formative assessment is the in-process monitoring of on-going efforts in an attempt to make rapid adjustments (e.g., a cook tastes the soup and decides whether or not to add more spices). Summative assessment tends to record overall achievement, often to compare outcomes to goals, and typically occurs at the end of a process (e.g., the customers taste the soup and provide their opinion). Adjustments based on results from summative assessments can be made for the future, but are less immediately responsive. Formative and summative assessments can have different dimensions, each of which may have different goals such as promoting students' growth, improving instruction, and demonstrating students' procedural and conceptual understanding.

Three broad purposes of assessment are examined in the report, *Knowing What Students Know: The Science and Design of Educational Assessment* (Pellegrino, Chudowsky & Glaser, 2001). They are:

1. To assist learning,
2. To measure individual achievement, and
3. To evaluate programs.

The report goes on to describe the purpose of an assessment as determining priorities, and how the context of use imposes constraints on the design. Assessment is viewed as a process of reasoning from evidence. However, assessment results are only estimates of what a person knows and can do. The report stresses that every assessment, regardless of its purpose, rests on three pillars:

1. A model of how students represent knowledge and develop competence in the subject domain,
2. Tasks or situations that allow one to observe students' performance, and
3. An interpretation method for drawing inferences from the performance evidence thus obtained.

These three foundational elements – cognition, observation, and interpretation – comprise an "assessment triangle" that underlie all assessments and must be explicitly connected and designed as a coordinated whole.

Educational assessment does not exist in isolation, but must be aligned with curriculum and instruction if it is to support learning. Therefore, assessment should be carefully aligned with the important and valued learning goals. Since the focus of this chapter is on assessing student learning, the next section attempts to define some important learning goals for students. An example of using student data to assess an undergraduate statistics program is given by Peck and Chance (2005).

Defining Learning Goals: Statistical Literacy, Reasoning, and Thinking

In defining student understanding, it is important to consider different aspects or levels of understanding. Bloom's Taxonomy (1956) has long been used to categorize learning outcomes by their level of abstraction. These levels include:

- Knowledge: Can students recall basic information, terms, definitions?
- Comprehension: Can students understand, interpret, predict and make comparisons based on their knowledge?
- Application: Can students use the information to solve problems in novel situations?
- Analysis: Do students see patterns, connect and relate ideas, and make inferences?
- Synthesis: Can students combine knowledge from several areas and generate new ideas and make generalizations? Can they make predictions and draw conclusions?
- Evaluation: Can students make judgments and evaluations, recognizing strength of evidence for a theory? Can they discriminate and make choices?

This taxonomy has been utilized by assessment writers to help write items to assess a variety of levels of cognitive objectives. Despite its reputation and recognition, writers using this taxonomy are often faced with the ambiguity of figuring out exactly how to use it as they contextualize the cognitive objectives they want to assess. In addition, Bloom's taxonomy is fairly general, and several articles have pointed out problems and limitations (e.g., Stanley & Bolton, 1957; Cox, 1965; Poole, 1971, 1972; Fairbrother, 1975; Phillips & Kelly, 1975; Orlandi, 1971; Ormell, 1974; Sax, Eilenberg, & Klockars, 1972; Seddon, 1978). Many instructors have found it difficult to apply this six-level taxonomy, and some have simplified and collapsed the taxonomy into three general levels (Crooks, 1988). One way to do this is to make the first category – knowledge (recall or recognition of specific information), the second category – a combination of comprehension and application, the third category – "problem solving," while collapsing the remaining three levels.

Specific guidelines within a discipline appear to be more useful than the general categories such as those in Bloom's taxonomy (1956). We have found that using statistical literacy, reasoning, and thinking, as described in Chapter 2 helps distinguish between desired learning outcomes both in considering instructional goals as well as in writing assessment items. We now show how these outcomes might be assessed.

Statistical literacy involves understanding and using the basic language and tools of statistics: knowing what basic statistical terms mean, understanding the use of simple statistical symbols, and recognizing and being able to interpret different representations of data (Rumsey, 2002).

Statistical reasoning is the way people reason with statistical ideas and make sense of statistical information. Statistical reasoning may involve connecting one concept to another (e.g., center and spread) or combining ideas about data and chance. Statistical reasoning involves understanding concepts at a deeper level than literacy, such as understanding why a sampling distribution becomes more normal as the sample size increases. Reasoning also means understanding and being able to explain statistical processes and being able to interpret particular statistical results (e.g., why a mean is much larger or smaller than a median, given the presence of an outlier) (Garfield, 2002b).

Statistical thinking involves a higher order of thinking than does statistical reasoning. Statistical thinking has been described as the way professional statisticians think (Wild & Pfannkuch, 1999). It includes knowing how and why to use a particular method, measure, design or statistical model; deep understanding of the theories underlying statistical processes and methods; as well as understanding the constraints and limitations of statistics and statistical inference. Statistical thinking is also about understanding how statistical models are used to represent random phenomena, understanding how data are produced to estimate probabilities, recognizing how, when, and why to use inferential tools in solving a statistical problem, and being able to understand and utilize the context of a problem to plan and evaluate investigations and to draw conclusions (Chance, 2002).

Words that Characterize Assessment Items for Statistical Literacy, Reasoning, and Thinking

One way to distinguish between these related outcomes is by examining the words used in assessment of each outcome. Table 4.1 (modified from delMas, 2002) lists words associated with different assessment items or tasks.

Table 4.1 Typical words associated with different assessment items or tasks

Basic Literacy	Reasoning	Thinking
Identify	Explain why	Apply
Describe	Explain how	Critique
Translate		Evaluate
Interpret		Generalize
Read		
Compute		

Next, we present three examples to show how statistical literacy, reasoning, and thinking may be assessed.

Example of an item designed to measure statistical literacy:

A random sample of 30 first year students was selected at a public university to estimate the average score on a mathematics placement test that the state mandates for all freshmen. The average score for the sample was found to be 81.7 with a sample standard deviation of 11.45. Describe to someone who has not studied statistics what the standard deviation tells you about the variability of placement scores for this sample.

This item assesses statistical literacy because it focuses on understanding (knowing) what the term "standard deviation" means. For more examples of items assessing statistical literacy see Watson & Callingham (2003).

Example of an item designed to measure statistical reasoning:

Without doing any calculations, which of the following histograms, A or B, would you expect have a higher standard deviation and why?

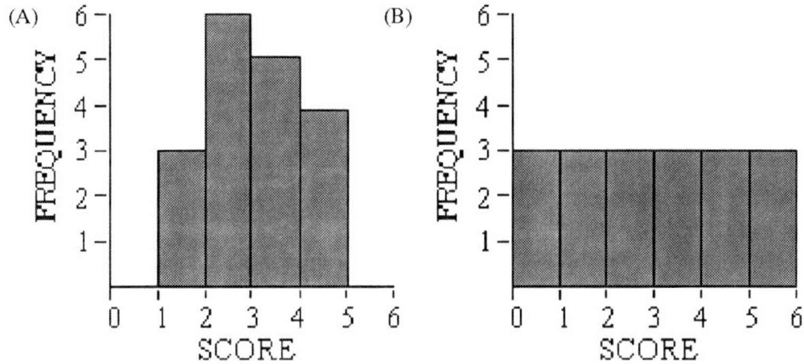

This item assesses statistical reasoning because students need to connect and reason about how the standard deviation of a distribution is affected by spread from the center (mean). They have to reason about the fact that Graph B would have a higher average deviation from the mean than Graph A, because Graph A has a higher proportion of its values clustered closer to the mean.

Example of an item designed to assess statistical thinking:

A random sample of 30 first year students was selected at a public university to estimate the average score on a mathematics placement test that the state mandates for all freshmen. The average score for the sample was found to be 81.7 with a sample standard deviation of 11.45.

A psychology professor at a state college has read the results of the university study. The professor wants to know if students at his college are similar to students at the university with respect to their mathematics placement exam scores. This professor collects information for all 53 first year students enrolled this semester in a large section (321 students) of his "Introduction to Psychology" course. Based on this sample, he calculates a 95% confidence interval for the average mathematics placement scores exam to be 69.47–75.72. Below are two possible conclusions that the psychology professor might draw. For each conclusion, state whether it is valid or invalid. Explain your choice for both statements. Note that it is possible that neither conclusion is valid.

a. The average mathematics placement exam score for first year students at the state college is lower than the average mathematics placement exam score of first year students at the university.

b. The average mathematics placement exam score for the 53 students in this section is lower than the average mathematics placement exam score of first year students at the university.

This item assesses statistical thinking because it asks students to think about the entire process involved in this research study in critiquing and justifying different possible conclusions.

Comparing Statistical Literacy, Reasoning, and Thinking to Bloom's Taxonomy

These three statistics learning outcomes also seem to coincide somewhat with Bloom's more general categories (1956) as they may be collapsed into the three levels described in the previous section. We see statistical literacy as consistent with the "knowing" category, statistical reasoning as consistent with the "comprehending" category (with perhaps some aspects of application and analysis), and statistical thinking as encompassing many elements of the top four levels of Bloom's taxonomy (application, analysis, synthesis, and evaluation).

Distinguishing among different types of desired learning outcomes can help statistics educators design assessment tasks that address the different outcomes and create classroom environments that allow multiple instructional methods and assessment opportunities. As learning goals have changed, so have uses of assessment in statistics instruction, moving away from focusing solely on the purpose of assignment grades as was common in the past (Garfield, 2000). A wide variety of instruments and methods are now availableto help instructors evaluate how statistical understanding is developing, and what changes can be made to improve student understanding. As a consequence, assessment has also been a topic of several research studies in statistics education which we discuss below.

Research on Assessment in Statistics Education

Recent research on assessment in statistics education has focused on the design, use, and evaluation of:

- instruments to assess cognitive outcomes,
- instruments to assess attitudes and dispositions, and
- alternative forms of assessment (such as, group work, portfolios, and projects) focusing on application, interpretation, communication, and understanding of the statistical process.

Assessing Cognitive Outcomes

Most assessments used in statistics courses and in statistics education research focus on learning, or cognitive outcomes of instruction. Many of the measures of learning and achievement used in courses provide only a single number summary of student performance. Due to limitations in this approach, there has been an increasing need to develop assessment tools that better measure different types of conceptual understanding to allow assessment of the development of students' reasoning, that are feasible for large-scale testing. For example, the *Statistical Reasoning Assessment* (Garfield, 2003) was developed as a multiple choice test where each correct and incorrect response was mapped to a previously identified reasoning goal (e.g., common misconceptions such as the "law of small numbers" and "equiprobability bias," Garfield, 2002b), while focusing on students' understanding of statistical concepts and ability to apply statistical reasoning.

Studies by Konold (1995) suggest that students have deeply held, often contradictory intuitions about probability and statistics prior to instruction that he maintains he would not have been able to as easily identify if he had not developed new assessment items focused on conceptual change that forced him to "articulate my major objectives and to evaluate the effectiveness of the materials I am designing." More recently, other instruments such as the *Statistics Concept Inventory* (SCI, Allen, Stone, Rhoads, & Murphy, 2004) and the *Comprehensive Assessment of Outcomes in a first Statistics course* (CAOS, Ooms, 2005) have continued this focus on developing reliable and valid instruments to assess students' conceptual understanding of statistics. The SCI is modeled after the Force Concept Inventory, which is used in assessing students' misconceptions about Newtonian physics (Halloun & Hestenes, 1985). The CAOS test focuses on statistical literacy and reasoning and has undergone extensive psychometric analyses, and baseline data have been gathered for comparisons with normative data (delMas, Ooms, Garfield, & Chance, 2006). Goals for both of these instruments include using them to collect data across student groups and settings for comparative research studies. In addition to these comprehensive tests, 11 multiple-choice tests on individual topics are available as online tests (delMas et al., 2006).

Research in science education has highlighted the effectiveness of assessment in increasing student learning. For example, Posner, Strike, Hewson, and Gertzog (1982) discuss a predict-and-test model as a way to get students to establish sufficient cognitive dissonance that subsequent instructional intervention has a stronger effect on changing common misconception. Applying this model in statistics education, delMas et al. (1999) asked students to predict (through a quiz) the behavior of sampling distributions of the mean for different populations before using technology to check their answers. Students' interaction with the technology appeared to have much more impact on their learning when this pretesting step was used.

Assessing Attitudes and Dispositions

Research has demonstrated that students' attitudes can be a powerful factor in their learning of statistics (Gal & Ginsburg, 1994), and it is worthwhile to monitor student attitudes at the beginning of the course as instruction is being planned, as well as when the course is evaluated for overall quality and effectiveness. Although this goal has received less attention in the past, partly due to the difficulty in assessing affective factors such as attitudes and beliefs involved in the learning process, it is starting to receive more attention. Many researchers in the field of mathematics education, psychology, and cognitive sciences (Mandler, 1984; McLeod, 1989, 1992; Adams, 1989; Fennema, 1989; Schoenfeld, 1983) have stressed the need to assess students' affective domain because they influence the cognitive domain. In particular, McLeod (1989) distinguishes between attitudes and beliefs and discusses why these are important for the learning of mathematics, and also defines and distinguishes between other constructs such as emotions and dispositions that are less stable and more likely to change over shorter periods of time. Preconceived notions about the subject and its usefulness appear equally important in statistics education.

Some instruments have been used to assess students' attitudes and beliefs about statistics such as *Attitudes Toward Statistics* (Wise, 1985) and *Survey of Attitudes Toward Statistics* (Dauphinee, Schau, & Stevens, 1997). However, these instruments were not designed to provide causal information (Gal, I., & Garfield, J. , Eds.). One may have to conduct interviews to find out why students have the beliefs they do about statistics. This is a limitation in trying to assess students' affective domain since it is difficult to find the time for focus groups conducting interviews. Furthermore, since attitudes and beliefs are very stable, it is difficult for an instructor to make significant changes in attitudes during one course, but these assessment instruments can at least inform a more long-term change in the teaching of statistics.

Alternative Assessment Methods

> *Even good students don't always display a good understanding of what's been taught even though conventional measures … certify success.*
>
> (Wiggins & McTighe, 1998, p. 2)

As the instructional goals shift from computational skills to deeper conceptual understanding of basic statistical ideas, the need for alternative methods of assessment is even stronger than before. As Wiggins and McTighe point out in the quote above, conventional methods often fall short of giving the full picture of what students have learned. Many innovative alternative assessment methods are available to teachers of statistics such as projects and article critiques, which are described in a later section of this chapter. While alternative assessment methods provide different types of insight into students' understanding and ability, they are also particularly conducive to promoting student learning.

Informal assessments are also an alternative to traditional quizzes and exams. During a lesson, questions can be posed orally to probe for common misconceptions and allow students to discuss their ways of thinking. Students can be asked to provide multiple solutions to a problem, and then asked to explain their reasoning. Practice in explaining reasoning and supporting judgments and conclusions will help them better perform these tasks on later assessments.

The focus on the integration of improved assessment methods and alternative forms of assessment in introductory statistics allows instructors to assess students' understanding at a deeper level, often highlighting very deeply held misconceptions about probability and statistics. Development of new assessment instruments and tasks is improving our ability to document these misconceptions and their reasoning processes, as well as track students' development of conceptual understanding as they learn new material. In particular, assessment tools can be utilized themselves as powerful instructional interventions, especially once students and faculty become more comfortable with the formative nature of this information rather than relying on them only as summative evaluations.

Designing Assessments for Statistics Classes

Assessment should support the learning of important concepts or skills and provide important information to both teachers and students National Council of Teachers of Mathematics (NCTM), 1989, 2000). The NCTM *Assessment Standards for School Mathematics* (1995) presents six assessment standards that address mathematics (the content we teach), learning, equity, openness, inferences, and coherence. Due to the multiple roles that assessment can play in statistics education, designing an assessment plan takes careful consideration. Garfield (1994) offers the following assessment framework from which to build:

- *What* is to be assessed (e.g., a concept, skill, attitude, or belief)
- The *purpose* of the assessment (e.g., to assign a grade, to help the teacher modify instruction, to help the student identify strengths or weaknesses in their understanding)
- *Who* will do the assessment (e.g., self-assessment by the student, peer-evaluation, instructor feedback)
- The *method* that will be used (e.g., project, quiz)
- The *action/feedback* that will be implemented as a result of the assessment

Each of these dimensions needs to be considered in designing an assessment to help ensure that the assessment is aligned with course goals and is as effective as possible. Because assessing students can be time-consuming, in designing assessment tasks we recommend a focus on two key tenets: (1) assess what is valued, and (2) incorporate assessment into the learning process.

Assess What Is Valued

Students pay the most attention to the material on which they are assessed. Students take their cues as to what is important in the course based on assessments. It is easy to claim that clear communication, analysis, and synthesize of topics are goals of the course. Yet even when a teacher models these skills during class, students will not focus on them if they are not held responsible for demonstrating these types of skills themselves "when it matters."

To ensure the teacher is assessing what is valued, it can be helpful to first reflect on what the teacher cares most that students should be able to do after they have taken a statistics class. Should they remember a formula? Should they be able to make decisions and provide arguments based on evidence? Should students be able to conjecture, build arguments, and pose and frame problems?

Incorporate Assessment into the Learning Process

It is also important to conceive of an assessment program as a process in which students (and faculty) can receive formative feedback during the course, rather than the traditional method of using assessments only to determine final course grades. Taking advantage of the high importance students give to assessments provides an opportunity for substantial learning and reflection (e.g., the predict-and-test model). For these reasons, it is crucial to consider student assessment *simultaneously* with the design of instructional activities and to consider how lessons learned from different assessment components can inform other parts of the course (for example, using quizzes as sample exam items to help prepare students for the types of questions they may be asked).

A K–12 teacher may find that incorporating different types of assessments into the learning process will help students develop the content knowledge and language skills necessary for success in the next school level. However, independently of what level is taught, teaching students how to learn and reflect on what they know (referred to as a *metacognitive* activity) has been shown to improve understanding in different content areas (White & Frederiksen, 1998; Schoenfeld, 1987) and would better prepare students to be life-long learners (Bransford, Brown, & Cocking, 2000).

Different Types of Assessment

Many effective types of assessment tools exist, each with different goals and purposes. Below we highlight a few common practices and their differing roles in the overall assessment process: Homework, quizzes, exams, projects, critiquing/writing assignment/lab reports, collaborative tasks, minute papers, and attitudinal measures.

1. Homework

Often students find statistical terminology and statistical way of thinking to be the most challenging aspects of their first introduction to statistics. Providing them with constant practice using terms and processes is an important component in developing their understanding and confidence with the material. Homework assignments can clearly vary in length and formality (e.g., "practice problems"), but should provide students with guidance on how to focus their studying of the topics. For example, they should include application and conceptual questions, especially if those types of questions will be asked on exams. Questions can also be included that ask students to reflect on the problem solving process. As with any assessment strategy, it is crucial to supply students with clear guidelines for the instructors' expectations (e.g., How much work needs to be shown in leading to the final answer? Should computer output be included in the write up?). Students may not be accustomed to being graded on their process (needing to supply more than just a final boxed answer at the end) or even the correctness of their approach and they will require time to adjust. Supplying guidelines for performance expectations can also help students avoid time consuming sidetracks.

In grading homework assignments, it is quite valuable to provide feedback that students can apply to later assignments. While student assistants can often help grade assignments at the university level, questions that require more interpretation might be better graded by the instructor. Being open to alternative approaches and interpretations is especially important in introductory statistics classes. We also recommend providing students with access to model solutions, whether instructor written or an anonymous student paper to help model peer work, and/or the opportunity to review and comment on faulty and correct solutions provided by other students. (For a comprehensive review of research related to the effects of homework, see Cooper, Robinson, & Patall, 2006.) Other practical considerations include how much collaboration to allow among students in doing their homework. See Chapter 15 for a discussion on the benefits of collaboration in learning statistics.

The instructor also needs to decide on the level of support or scaffolding given in response to student questions on the assignment. For example, how forthcoming should the instructor be in supplying help, answers, or hints? Will responses to students be more Socratic in nature to help students think through the process? Could additional hints be accompanied by a deduction in points available for a particular problem? Should the instructor choose to be generous in responding to questions on student homework problems, but only if questions are posed at least 24 hours prior to the due date? This type of structured support can encourage good habits in students so that they do not delay working on their homework until the last minute.

2. Quizzes and Exams

Quizzes are useful for providing timely checks of student understanding and feedback on their progress while exams can assess cumulative knowledge over a longer period of time. Low-stakes quizzes can be very helpful to students in getting a feel for what types of questions may appear on exams, especially if exam questions tend to be more conceptual in nature than textbook problems providing students with some prior knowledge of what the planned focus will be for a particular exam and the instructor's grading style. Alternatively, quizzes can be more focused mini-exams, with an accompanying higher weight in the overall course grade. This is especially effective for minimizing grading efforts, allowing homework assignments to be down-weighted or even optional.

One approach to consider is paired/collaborative quizzes. Significant student learning occurs in the debates they have with each other in developing a unified response. Students can also support each other and this approach may alleviate testing anxiety among many students. In this way, quizzes become a learning tool as well as an assessment technique. Students can be given time to think individually before the group discussion and individually graded components can also be incorporated if desired. Researchers such as Johnson and Johnson (1996) have also discussed issues in forming these groups (e.g., homogenous or heterogeneous with respect to prior performance?) and rearranging them during the term (see later comments on collaborative tasks as well).

Another practical consideration is dropping a quiz grade (or more) in the overall grade calculations. This can help ease tension if students must miss a class day, but still provide incentives to attend class. A decision may depend on whether quizzes are announced in advance and on how many are administered during the term. Use of quizzes for attendance incentives can take more prominence for early morning and Friday classes for college students.

Some instructors worry about using too much in-class time if quizzes are given every week. Short quizzes can provide meaningful insight into students' understanding. The following example shows a one-question quiz that can help the instructor to highlight and bring to class discussion a common misunderstanding:

Example (adapted from Ramsey & Schafer, 2002):

True or False? A sample histogram will have a normal distribution if the sample size is large enough. Explain why. (Sample correct responses: False, because a large sample size will give a better idea of the shape of the population from which the data have been sampled, and that population may or may not have a normal distribution).

It is clear that numerous resources now exist for accessing existing assessment tools. In selecting an assessment tool, it is important to consider what types of modification are needed to adapt this instrument to align with the instructor's learning goals.

Writing Good Items for Quizzes and Exams

Several statistics educators (e.g., Hubbard, 1997; Hayden, 1989) have suggested ways of improving exam questions such as always providing a meaningful context, using questions where the goal of the analysis has a purpose, asking more "what if," "construct a situation such that...," and comparative judgment questions, requiring explanations with calculations, identification and justification of false statements, and objective-format questions. Cobb (1998) offers ways to write good multiple choice items to assess higher order thinking and reasoning. As an alterative to writing items, the ARTIST project provides a selection of over 1000 high-quality assessment items in an online, searchable data base (see Resources section). These items are released only to instructors, and are not freely accessible on the Web by students.

According to Wild, Triggs, and Pfannkuch (1997), multiple-choice questions can still test higher level thinking skills. They provide examples of multiple-choice questions that can help students confront common misconceptions, interpret data, select appropriate techniques for data analysis, and make appropriate inferences. What multiple-choice questions cannot do is to assess open-ended thinking because it does "little to help students develop their own voices" or demonstrate students' ability to formulate questions emerging from the data.

The following guidelines can assist instructors in developing their own items for quizzes and exams:

- Include items that allow students to express their reasoning and thinking and demonstrate their use of statistical language.
- In using multiple-choice items, aim for 3–4 answer choices. Try to include responses that might match different types of student reasoning among the answer choices and try to avoid options such as "none of the above." (See Haladyna, Downing, & Rodriquez, 2002 for additional practical suggestions for writing multiple-choice items based on measurement research.)
- Add a contextual basis to existing questions that may lack a real or realistic context. For example, rather than asking students to interpret a correlation of 1.2 (an oblivious result of a calculation error), describe the problem being investigated and the variables being examined (e.g., in a study of attitudes and achievement in a statistics class, students were given assessments of math anxiety and algebra proficiency).
- Build assessment questions around existing data sources. Finding interesting data examples is often more difficult. So it may be useful to start with an interesting research study and build the statistical concepts questions around that study.

Examples of Good Assessment Items

The following items provide some examples of good assessment items, based on the guidelines in the previous section.

Example:

A study of 78 randomly selected seventh-grade students from a large Midwestern school measured students' "self-concept" (SC) through their score on Piers-Harris Children's Self-Concept Scale. Within the analysis, researchers found higher SC scores for 13 years compared to 14 year olds and constructed a 95% confidence interval for this difference to be (2.38, 29.44).

(a) Which is the best interpretation of this confidence interval?

- 95% of 13 year old and 14 years had a self concept score between 2.38 and 29.44.
- We are 95% confident that 13 year olds score between 2.38 and 29.44 points more than 14 year old scores.
- We are 95% confident that the difference in mean concept scores for the 13 year olds and 14 year olds in this study is between 2.38 and 29.44.
- 95% of intervals constructed this way will capture the difference in the population means between 2.38 and 29.44.

(b) Explain what is wrong with one of the above interpretations.
(c) Explain whether, and if so how, the above interval would differ if the confidence level had been 99% instead of 95%.

This item challenges students not only to identify a correct answer but to also identify answers that are incorrect and explain why they are wrong. It also asks students to reason about the effect of changing a confidence level to a higher value and why that would affect the interval width.

Example (adapted from Rossman & Chance, 2001):

Construct a data set of 10 possible student test scores that range between 0 and 100 so the mean is greater than median.

Students need to understand the properties of the measures to "work backwards" in solving this problem.

Example (adapted from Hubbard, 1997):

Make up a problem that could be solved with the following output.

One-Sample T
Test of mu= 25 vs <25

N	MEAN	STDEV	SE MEAN	T	PVALUE
40	23.63	3.95	0.538	–2.55	0.0054

This question requires students to not only piece together information from generic computer output, but also to generate a suitable context (i.e., one quantitative variable that is feasible for these summary statistics). This reversal in logic assesses their understanding of and ability to apply the overall process of going from a research question and data to an analysis.

Example (adapted from Chance, 2002):

Four different studies obtained data that were used in a test of the hypothesis H_0: $\mu_1 = \mu_2 = \mu_3$. Based on the information below, order these studies from smallest P-value to largest P-value. Provide an explanation for your choices. You will be graded primarily on your explanations.

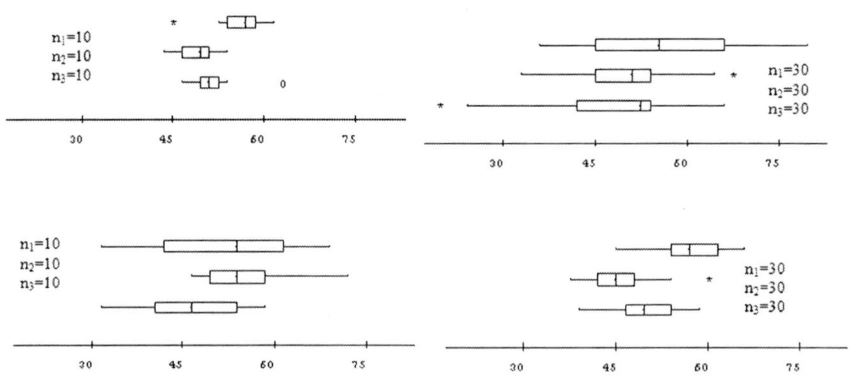

Given to students who have not studied ANOVA (Analysis of Variance), this question requires students to integrate and extend different ideas from the course (e.g., the effects of sample size, within group variation, and between group variation) on the magnitude of the p-value. A scoring rubric is less concerned with the particular ordering selected (e.g., top left vs. bottom right) but more on which of these factors are discussed by the student, the direction of the effect suggested, and the consistency of the ordering with the written explanation.

Exam questions can also be designed that require more performance and production by the students rather than memorization. Questions should also challenge students beyond traditional textbook problems, especially end of chapter questions where the solution approach is determined by which chapter is being discussed.

Example:

Your text states that "confidence intervals seek to estimate a population parameter with an interval of values calculated from an observed sample statistic." Convince me that you understand this statement by describing a situation in which one could use a sample proportion to produce a confidence interval as an estimate of a population proportion. Clearly identify the population, sample, parameter, and statistic involved in your example. Do not use any example that appears in your book.

Example (adapted from Moore & McCabe, 2005):

A university is interested in studying reasons why many of their students were failing to graduate. They found that most attrition was occurring during the first three semesters so they recorded various data on the students when they entered the school and their GPA after three semesters. [Students are given data set with numerous variables].

(a) Describe the distribution of GPA for these students.
(b) Is SAT-Math score a statistically significant predictor of GPA for students at this school?
(c) Is there a statistically significant difference between the mean GPA values among the majors at this school?

It is very important to remember that if nontraditional item formats are to be included on the exams, then students must be prepared in advance, e.g., through examples in review or earlier assignments and quizzes as discussed above. Students can also be asked to develop and submit their own example questions as part of the review process to help them anticipate what they will be asked. It is also important to think of the overall time constraints in combining these questions with more traditional type questions.

Even once the exam questions are developed, there are still many practical issues in administering and grading student responses. Considerations include whether to allow external aids (e.g., closed book vs. supplying formulas pages vs. open book), the use of technology during the exam, whether exams should be timed, approaches taken to constructing exams, how to prepare students for the exam (e.g., use of review sheets, review problems), and how to provide the best feedback to students postexam (e.g., is posting solutions online sufficient or should we always budget class time for discussion?). There are many possible views on these issues. The

ARTIST website contains an "Implementation Issues" section where leading statistics educators offer their personal views and justifications, and a panel session at the 2005 Joint Statistical Meetings also discussed these issues (Dietz, Gould, Hartlaub, & Rossman, 2005). Two suggestions we will highlight here are sharing exams with colleagues and collecting feedback in advance, and jotting down ideas for exam questions as they occur during class discussions, grading homework assignments, or reading the newspaper.

3. Projects

Projects are an important tool for both assessing many aspects of student learning as well as helping them experience different stages in posing and solving a statistical problem. Projects usually involve collecting and analyzing real data, in ways that make them a good authentic assessment (see Garfield, 1993). Projects can focus on a particular aspect of the course (e.g., Smith, 1998) or be term-long, requiring students to apply tools they learn throughout the course in one overall analysis (e.g., Chance, 2000, Fillebrown, 1994). See also the assignments John Holcomb and others have posted on the ARTIST Website (https://ore.gen.umn.edu/artist/projects.html). Holcomb provides a series of group assignments followed by individual take home exams to perform similar analyses. Projects help students realize that the optimal methods are not always practical or feasible and that compromises, with consequences, have to be made. Students also realize that data can be messy and they must consider data cleaning issues and unexpected sources of bias. Oral presentations can also be required to help students to learn to effectively communicate their analysis.

With term-long projects, periodic reports should be required, to maintain consistent progress on the project. Students can be required to work individually or in groups. Group management becomes an issue in the latter case, and it may be worth considering maintaining an individual component in assessing student performance on the project. In designing a project assessment, it is important to simultaneously consider how to assess the effectiveness of the project and student work. Students can also be asked to provide group member evaluation, and presenters can be evaluated on how well they answer questions during the question/answer period after presenting. For example, in an 8th grade class, students had to give oral presentations of their projects. It was during the question/answer period that students' misunderstandings or confusions were revealed (Lajoie, 1999). Starkings (1997) warns about the demanding nature of projects, for both instructors and students, and she provides advice to teachers planning to use projects in their classes.

4. Critiques/Writing Assignments

Most students will eventually find themselves to be consumers of statistical information rather than producers. Therefore, if a course goal is developing students' ability to process and analyze quantitative information in the media, assignments can be designed to help develop these skills.

Possible examples are:

1. Providing students with a journal article and asking them to evaluate the use of statistics and the conclusions drawn.
2. Having students "role play" and asking them to explain some conflicting information recently cited in the news to a parent, using statistical arguments, but in a way a nonstatistician could understand.
3. Writing a "meaningful paragraph" in which students are given a short list of statistical terms (e.g., sample, population, and sampling distribution) and are asked to use these terms in a short essay that explains their understanding of these terms in a particular context.

These tasks focus on requiring students to use their own words to explain their understanding, an often different experience than directly solving a given problem. Some examples are below.

Example (from Jordan, 2004):

Write a meaningful paragraph that includes the following terms: sampling distribution, population mean, sample mean, variability, normal distribution, sample size, and probability. A "meaningful paragraph" is one continuous piece of writing, which uses all of the listed words and in which the sentences "make sense and hang together." That is, the ideas in the paragraph illustrate that you understand the new terms in a way that allows you to write "meaningfully" about them. You may not simply write seven sequential sentences, for example, that merely define the terms; sentences must demonstrate relationships between the terms.

Example (from Jordan, 2004):

Suppose you receive the following letter from your dad:

Hey Kiddo,

I am worried about Grandma. Remember that she was diagnosed with high blood pressure? Well, she's currently taking the medication Makemewell to lower her blood pressure. At the time of Grandma's diagnosis, her doctor said that a randomized, double-blind experiment had been conducted and that Makemewell was shown more effective in lowering blood pressure than a placebo. To be honest, I have no idea what any of that means, but I believed and trusted the doctor. Now I've heard two stories that make me think differently. Larry, our next-door-neighbor, was taking Makemewell and he got a terrible fever that put him in the hospital. Also, my coworker, Sally, actually had her blood pressure go up while she was taking Makemewell! I'm now very suspicious of this medication.

I know that you're taking a statistics course at college. Based on the information I've given you, do you think Grandma should stop taking her medication? Whatever your opinion, will you please explain yourself thoroughly and clearly? (I will draw on your responses when I talk with the doctor.) And please don't use any statistics mumbo jumbo that I won't understand. I really appreciate your help with this.

Love, Dad

Your assignment is to type a 1–2 page letter (single-spaced, 12-pt. font) responding to your dad.

Sometimes, models or examples can be used to help students see the desired features of a finished writing assignment or project. However, students often fall into the habit of using model papers as a template instead of using their own thinking. One way around this is to give students the first model paper after they have submitted their first assignment and then allowing them to use this model as they revise or prepare the next, slightly different, assignment. Assignments that ask students to address a particular audience (e.g., your father, the soda manufacturer) seem to work well in engaging students (see Jordan, 2004).

Grading Writing Assignments

A key consideration in these types of assignments is how to grade them. Instructors may benefit from considering articles on grading more generic writing assignments (e.g., Elbow, 1997; Bean, 2001). Holistic scoring approaches can be useful in grading writing, if a rubric or guideline is used. This type of scoring is used in grading open-ended problems in the Advanced Placement Statistics Exam and works well for many assessment tasks in a statistics course. Instead of relying on an analytic scale where points are deducted for each error, holistic scoring assigns points based on a categorization of the overall quality of a student's answer and explanation. Careful design and use of scoring rubrics can produce more consistent scores, especially when only 3–5 points are assigned to each part of the rubric. The following example is adapted from an assignment developed by Jordan (2004). In this case, up to five points can be assigned to each criterion in the rubric.

Example: Rubric adapted from Jordan (2004) writing assigning.

Grading Criteria (20 points possible)

- The explanation to your dad convinces me (your teacher) that you understand the statistical concepts involved in the assignment. (5 points)
- The explanation to your dad is thorough, well organized, and clear. (5 points)
- The explanation to your dad is presented in nontechnical terms that he will understand. (5 points)
- You successfully paid attention to accepted conventions of language use (syntax, spelling, grammar, readability, etc.) (5 points)

Example: Rubric for Wetzel (2004) assignment to determine solution to a problem where they were to determine the weight of average McDonald's French fry by collecting and analyzing data.

Basics (5 points): Does the description of the problem solution include all of the required parts including accurate and appropriate use of terminology?

- restatement
- measurements – including specifics
- questions to ask – including specifics
- extra demographic question
- how to get randomize – including specifics
- other data

Organization (2 points): Is the description organized and neatly presented? Great answers also include some pluses.
Pluses:

- Does the description include any significant extras (e.g., additional detail on the data collection steps, use of terminology)?
- Significant improvements to a basic data collection design.
- Extra thought into the specifics of this context – recognizing potential problems and giving solutions.

Minuses:

- Does the description indicate that the student is mimicking a book answer and not considering the context?
- Does the description include a design that would be extremely impractical?
- Does the randomization described introduce a significant confounding variable that was not identified?

This problem requires thinking and possibly some research. A good answer will get 7 out of 10 points; in order to get full credit, you need to provide a great answer.

Such scoring rubrics provide the statistics instructor with more flexibility in allowing alternative solution approaches, and maintaining scoring consistency in a justifiable manner, and can be used to emphasize to students the equal weight given to both communication and calculation. Scoring rubrics should be provided to the students with the assignment to help clarify expectations and standards.

Teachers may consider raising their expectations as the course progresses and students build on earlier feedback. Students can also be required to submit self (and group) evaluations. With longer assignments, intermediate deadlines and/or the opportunity for revision may be possible.

5. Lab Reports

Most introductory statistics courses have some opportunities for students to also produce data, using appropriate technology tools. Having students produce technical reports that integrate computer output with discussion suitable for a nontechnical audience, helps put students in the role of the statistical analyst. Students can be asked to complete several reports during the term, with clear guidance for the technology skills and discussion questions to be addressed, as well as guidelines for quality report writing (e.g., Spurrier, Edwards, and Thombs, 1995). Chance (2000) implemented a series of lab reports, increasing in their demand of students' autonomy and integration of ideas. Several of these reports focused on a full report (introduction, data collection methods, analysis, discussion, conclusion) for analyses of data, students collected themselves (see Guidelines for Lab Reports in the Appendix). These reports were allocated 20 points for the computer output, 25 points for discussion and interpretation, and 5 points for presentation.

6. Collaborative Assessment Tasks

In a statistics course, asking students to work together to develop a common solution can be very effective as students have varied backgrounds, experiences, and mathematical sophistication. This can be done through quizzes and projects as well as in-class activities. Group work where students are responsible for jointly working together and having to agree on one answer or write up can lead to good discussions and arguments that involve explaining their statistical reasoning and enhances student learning.

One particular approach to consider is to combine group assignments with individual assessments, and even having a component of the individual assessment address what they should have learned from the group assignment (e.g., what was your project about?). For example, Harkness (2005) describes use of Readiness Assessment Quizzes that may have three components: an individual component, a group component, and an appeal process, based on reading assignments given to students before class, prior to instruction on the material. These multiple-choice questions on general concepts are first taken individually and then immediately retaken as a group of three to five students. Each student receives an individual and a group grade, and the belief is that this arrangement for the group work helps foster comprehension across the group members.

7. Minute Papers

Asking students to provide quick feedback on lessons and which topics are most and least confusing, and/or on what aspects of the course instruction are they finding the most and least helpful can help the instructor address common misconceptions or alter the course. Minute papers also convey to students that their feedback is important and can be used to help them reflect on the overall course goals and design. Students use the last five minutes of class to respond to a few questions on a half sheet of paper or index card (e.g., Angelo, T., & Cross, K. P., 1993). Minute papers can also be used to give feedback to the instructor, especially early on or in the middle of the term. This feedback may provide an instructor with ideas of things that can still be changed in the class (e.g., several students reporting that it is difficult to read the writing on the board) as well as allow the instructor to address and justify issues that would not be changed (e.g., why they would not change the software package being used).

Example:

What was the most important point of today's lecture? What did you consider to be the muddiest point of today's lecture? What would you like to learn more about?

Example:

Please provide feedback on what is working best for you in this class as well as what you would like me to change in the course structure and style.

It is important for these submissions to be anonymous if the student chooses; this allows students to feel more comfortable in giving honest responses. Many electronic course delivery systems (e.g., Blackboard®) enable anonymous surveys, while also tracking completion by individuals if credit is to be assigned. In large classes, subsections of students can be polled on different days. Discussing with the students, the feedback from the minute papers is very important. Sometimes examples can be given of comments and summaries that can be used to give the class an idea of trends in their responses. Students appreciate knowing that their opinion is important and under consideration and that they are not alone in their confusion.

8. Student Reflections and Dispositions

Many of the above tools can include self-reflection components as well as formal and informal opportunities for student feedback. Alternatively, students can be asked to complete written journals in which they are expected to contribute more reflection on the process and sharing of their reactions to the materials. They can also be asked to reflect on an experience with statistics that they have outside of the classroom (e.g., playing bingo with Grandma or compiling and explaining statistical cartoons).

Guidelines for Developing an Effective Assessment Plan

The previous section highlighted many different tools that can be used to assess students, and described how to select a particular tool for a particular assessment task. It is also important to consider how to combine the tools across the course. In this way, the teacher can make sure that, as a collection, the assessment tasks are aligned with the important goals for the course, that the action/feedback from the different components inform each other, and that no undue burden is placed on the instructor or the students. It may be useful, for example, for an instructor at the beginning of the course to map out how different assessments in the course align with learning goals of statistical literacy, reasoning, and thinking in order to make sure that the collection of items and tasks effectively evaluate each type of outcome.

It is crucial to consider the assessment plan as a course is designed and to clearly lay out the goals and purposes of the assessments as part of the course syllabus. Now, we offer some specific suggestions for developing an overall assessment plan.

- Develop an assessment blueprint that matches instructional goals to the assessment tools. This will involve analyzing the purpose and use of each assessment. For example, to assess student literacy, consider requiring student interpretations of articles and graphs. To include assessments of statistical thinking, consider requiring a student project. In fact, Wiggins & McTighe (1998) recommend de-

signing a course by beginning with the assessment goals (what outcomes are desired) and then working backwards to design the instructional component of the course.
- Provide clear guidelines ahead of time to the students (and follow them). Even if there are problems with an assignment in progress, it is often better to follow the assignment through to fruition and make notes for next time, rather than to change an assignment on students' midway.
- Use a variety of ways for students to demonstrate their understanding.
- Provide constructive and timely feedback so that there is a clear connection between the assessment and student learning.
- Challenge students to extend their knowledge (e.g., ask questions where a solution or approach is not obvious).
- Students need to feel the assessment is fair (to understand the purpose of the task and to feel that individual grades are an accurate reflection of their knowledge and effort).
- Focus on observable student behavior.
- Promote self-reflection and student responsibility.
- Consider the overall coordination of the assessment tasks throughout the course. Do not try to change all of assessment practices at once or to utilize every assessment task possible.
- Continually reflect on and refine your assessment practices and collaborate with colleagues.

Practical Considerations

We conclude with some practical advice and things to consider. We begin with the exhortation that assessment be viewed as an integral part of the learning process and not as a separate method used only to evaluate performance and assign grades. With all of the excellent assessment resources available, statistics teacher need not try to write their own tests, and in fact, are advised to utilize high-quality assessment items such as those at the ARTIST Website. We also advise instructors to think about sampling learning outcomes rather than trying to assess all topics in a particular test. It is important to remember that students' preparation for a test is part of the learning process; and one purpose of a test is to motivate study and integration of concepts.

While it may seem that every assessment of students must be read by the instructor, this is not always necessary. Students may be used as peer reviewers of papers and projects, which also helps their own learning. We also recommend that instructors take advantage of informal opportunities to assess student learning by carefully listening, asking questions, and observing students. Valuable information can be gained on how students use the statistical language, express their reasoning, and make a statistical argument as they work together in groups during class activities.

A comment made by a statistics educator brings home the point about assessing what is really valued (J. Witmer, personal communication, December 11, 2006). He said that when he examined a recent test he gave to students, that a computer would have earned a passing grade on it. That led him to reconsider the kinds of questions he asked on exams and to move from more computational and procedural questions to questions focused more on statistical literacy and reasoning and conceptual knowledge. This also relates to the issue of having students memorize material or having them able to access and use appropriate resources when solving statistical problems, as we do in real statistical work.

A final practical consideration is the use of computer-supported learning systems (such as Blackboard®), which can help in assessing students' work by tracking ongoing assessment, allowing continual revision, providing feedback, supporting collaboration, posting a paper or artifact in students' galleries, etc.

Summary

Assessment has been referred to as a way to make visible students' learning or reasoning. This is an important purpose of assessment because once reasoning is made visible, it can be "discussed, challenged, and made more robust" (Hammerman & Rubin, 2004, p. 37).

We encourage instructors to use assessment as a window to view student learning in their classes, using both formal and informal methods. We hope that teachers will move beyond using assessment solely as a mechanism for providing grades and measuring achievement, relying only on students' responses on exams for a picture of what students know and what they do not understand. We encourage instructors to use a variety of methods to better understand and measure student learning, taking notes to identify useful and not so useful components of their assessment program so that assessment tasks can be continually modified and improved, while focusing on gathering useful feedback on their teaching and on students' learning of important ideas. We end with the motto that "success builds success": We encourage instructors to make assessment a positive experience for students and one that will make them feel successful about learning statistics.

Appendix – Sample Guidelines for a Lab Report

The goal of this written lab report is to effectively communicate results in a clear and efficient manner. There is little point in getting the right answers unless it can be understood by the intended recipients. The main features of good reports are clarity, logical organization, succinctness, and clear labeling. The following guidelines should be used in writing this type of report.

Your write-up should include the following sections:

I. *Introduction*: The Introduction should be worded carefully, as that may be as far as some people get. It should provide a broad, general overview of the topic. The goal is to convey to the reader, the goals and most essential outcomes of the study. It is often easier to write this section *after* you have written the rest of the report so you know what you are trying to summarize. This summary should include your motivation for the study and why the reader should continue to read your report. Some background can be given, but leave the details to the body of the report.

II. *Data Collection Methods*: The goal of this section is to tell the reader precisely how the data was collected (what did happen, not what was supposed to happen). This should be a description of the process, not the data. This description should include details of the measuring instruments used, operational definitions of basic measurements, randomization, precautions etc. There should be enough detail so that someone could replicate your study.

III. *Results*: The goal of this section is to describe the data measured and the analyses performed. You should use tables and graphs to *summarize* the data – an effective summary is worth a thousand data points. You should also include any changes that were made to the data (scaling, transformations, missing values, points that had to be discarded). You should include a "map" telling the reader how to find the information. Tables and figures need to be labeled and numbered for easy reference. Each figure or table should be able to stand on its own and tell its own story (units!). Often Minitab can make the annotations for you, but sometimes you may want to write in a title or label by hand or with Word. Sometimes you can use computer output as is, and sometimes you will want to select out relevant information and construct your own summaries (see examples below).

   Next, you should describe the analysis methods. Be complete in your description so that the reader can assess the validity of your methods. You need to include a description of what was being tested, the statistical methods used, why they were appropriate, and any conditions (and verifications) that were made. *You do not need to include all your output, but select the output that supports your conclusions.*

IV. *Discussion*: Now you get to interpret all the above results. You should include any explanations you may have for what you have found in the data. Recall any problems you had collecting the data (e.g., is it really a random sample?) and how your interpretations may be limited. It is also possible to combine this section with the previous one, with Minitab output interspersed through the discussion.

V. *Conclusion*: Briefly summarize your report. What is your final answer to the question? What are the implications of the results? Use nonstatistical language. Include any ideas or suggestions you have for action and/or future experiments.

Appendix (Optional): If your raw data is too extensive to be included in the report, it may be placed in an appendix (again, well labeled) when necessary. Any equations or technical details can also be placed in an appendix.

Effective report writing is as laborious as it is vital. The following are some additional suggestions from Chatfield's *Problem Solving – A Statistician's Guide* book (1995):

- Document your work before your memory starts to fade
- Before you start to write, formulate your plan of attack
- After the first draft is complete, put it aside for 24 hours, then read it through as though seeing it for the first time.
- Use the spell checker
- It is often easier to write the body of the report *then* Introduction and Conclusion
- Ask another student to review it for you
- Review and revise

This type of writing takes practice; remember, we do not expect perfection on the first time! Exemplary write ups will be placed anonymously on the course Web page for your review.

Chapter 5
Using Technology to Improve Student Learning of Statistics[1]

> *Interactive software data visualization tools which allow for the creation of novel representations of data open up new possibilities for students (and teachers) to make sense of data, but also place new demands on teachers to assess the validity of the arguments that students are making with these representations, and to facilitate conversations in productive ways.*
>
> *(Hammerman & Rubin, 2004, p. 18)*

Overview

This chapter presents a broad overview of the role technological tools can play in helping high school and college introductory statistics students understand and reason about important statistical ideas. The main goal of this chapter is to provide some background of how the technology tools have evolved, a sense of the research findings and open questions on how technology impacts student learning, and concrete advice, stemming from the research literature, information on using how to use technological tools and how to avoid common pitfalls or ineffective implementations. We first summarize the impact of technology on the content, pedagogy, and even format of introductory statistics courses. Then, we highlight some of the common technological tools currently in use in statistics education and how they can be utilized to support student learning. We summarize some of the recent research insights gained with respect to using technology to aid instruction and learning in probability and statistics. While not an exhaustive literature review, the studies discussed provide additional context for a series of practical recommendations to the instructor, along with a discussion of possible obstacles and implementation issues, and questions to consider when selecting different tools.

---

[1] This chapter is based on the following journal article: Chance, B. L., Ben-Zvi, D., Garfield, J., & Medina, E. (2007, October). The role of technology in improving student learning of statistics. *Technology Innovations in Statistics Education Journal, 1*(1). Retrieved October 21, 2007 from http://repositories.cdlib.org/uclastat/cts/tise/vol1/iss1/art2/.

Technology in Today's Statistics Courses

It is hard to imagine teaching statistics today without using some form of technology. However, just 20 years ago that was very common. Today's statistics classes may be taught in a classroom with a computer projected on a screen, or may take place in a laboratory with students working at their own computers. Students commonly own a calculator more powerful than the computers of 20 years ago. Others may use a portable computer (laptop) at school, home, and on the move. An ever-growing format of teaching today is over the Internet, in the form of a Web-based course with video-taped lectures, interactive discussions, collaborative projects, and electronic text and assessment materials. The technology revolution has had a great impact on the teaching of statistics, perhaps more so than many other disciplines. This is not so surprising given that technology has changed the way statisticians work and has, therefore, been changing what and how we teach (Moore et al., 1995).

Changes in Content

Technology has led to numerous changes in statistical practice. Many analytically intractable problems that were previously inaccessible now have approximate solutions (e.g., Bayesian methods). Many assumptions that were made so that statistical models could be simplified and usable no longer need to be made. These changes in statistical practice have a direct impact on the content that should be taught, even in introductory material. For example, an entire branch of "resampling statistics" (Good, 2006) now competes with model-based inferential models in practice, while also appearing more intuitive to students. Another example is the use of statistical tables such as the z and t tables, which are no longer needed to determine rejection regions or estimate P-values when statistical software and calculators produce more accurate results much more quickly. In fact, many statistics educators now argue that previously standard topics in an introductory course (e.g., short-cut methods for calculating standard deviation) are no longer necessary to discuss in class. Finally, technology provides ways for us to visualize and explore data that have led to new methods of data analysis.

Changes in Pedagogy

While the impact of technology on the practice of statistics is irrefutable, just as powerful has been the impact of technology on statistics pedagogy and recommended practices. For example, the National Council of Teachers of Mathematics (NCTM) *Principles and Standards for School Mathematics* states that "the existence, versatility, and power of technology make it possible and necessary to reexamine *what* mathematics students should learn as well as *how* they can best learn it" (NCTM, 2000). In particular, the Guidelines for Assessment and Instruction in Statistics Education (GAISE, 2005a, 2005b) curriculum framework for PreK-12 states that "advances in technology and in modern methods of data analysis of the 1980s, coupled with the data richness of society in the information age, led to the development of curriculum materials geared toward introducing statistical concepts into the school curriculum as early as the elementary grades" (Franklin & Garfield, 2006). Similarly, the GAISE College Report directly recommends the use of technology for developing understanding of statistical concepts and analyzing data in teaching an introductory, undergraduate statistics course (Franklin & Garfield, 2006).

Moore (1997) urged a reform of statistics instruction and curriculum, based on strong synergies among content, pedagogy, and technology. However, he cautioned statisticians to remember that we are "teaching our subject and not the tool" (p. 135), and to choose appropriate technology for student learning, rather than use the software that statisticians use, which may not be pedagogical in nature. In fact, many types of technologies are available nowadays for the statistics teachers. Teachers are encouraged to view the use of technology not just as a way to compute numbers, but as a way to explore concepts and ideas and enhance student learning (Friel, 2007; Garfield, Chance, & Snell, 2000). Furthermore, technology should not be used merely for the sake of using technology (e.g., entering 100 numbers in a graphing calculator and calculating statistical summaries), or for pseudo-accuracy (carrying out results to a meaningless number of decimal places) (Franklin & Garfield, 2006). More appropriate uses of technology are accessing, analyzing, and interpreting large real data sets, automating calculations and processes, generating and modifying appropriate statistical graphics and models, performing simulations to illustrate abstract concepts and exploring "what happens if..." type questions.

Technology has also expanded the range of graphical and visualization techniques to provide powerful new ways to assist students in exploring and analyzing data and thinking about statistical ideas, allowing them to focus on interpretation of results and understanding concepts rather than on computational mechanics. Graphing calculators alone, highly valued for their ease of use, low-cost and portability, have been instrumental in bringing statistical content to lower and lower grade levels.

As the content and focus of the introductory statistics course are changing, statistics courses are looking even more different than in the past. For example, students are evaluated less on their ability to manipulate formulas and look up critical values, and more on their ability to select appropriate analysis tools (e.g., choosing techniques based on the variables involved), assess the validity of different techniques, utilize graphical tools for exploration of data, deal with messier data sets, provide appropriate interpretations of computer output, and evaluate and communicate the legitimacy of their conclusions.

Changes in Course Format

Technology has also impacted course management in the ways information is provided to students and shared among students in a class. Course management systems like Blackboard and WebCT (both can be found at http://www.webct.com)

are playing a large role, both in communication and collaboration capabilities (e.g., on-line discussion boards, video presentations and tutorials, pooling data across students, sharing instantly collected data across institutions), as well as in assessment. It is feasible to administer on-line surveys and quizzes with instant scoring and feedback provided to the students. New Web 2.0 tools such as Wiki can facilitate collaborative learning and bring about instructional change to improve student learning of statistics (Ben-Zvi, 2007). These learning systems give students and teachers more opportunities for communication, feedback, reflection, and revision.

Technology has a great potential to enhance student achievement and teacher professional development, and it will most likely continue to impact the practice and the teaching of statistics in many ways. However, technology has an impact on education only if it is used appropriately. Therefore, the focus of this chapter is on how technology can best be used to improve student learning. The following sections survey the different types of tools that are currently available and how technology can be used to support student learning. We summarize recent research results on the role of technology in the statistics classroom and then, building on this research, we suggest practical guidelines for selecting and using technology for teaching statistics. In addition, we also describe obstacles and implementation issues regarding the use of technological tools in the statistics classroom.

Technological Tools for the Teaching of Statistics and Probability

> *Using any new tool or representation necessitates change in the content and pedagogy of statistics instruction, and in many cases teachers are unprepared for these changes*
> (Hammerman & Rubin, 2004, p. 18).

The types of technology used in statistics and probability instruction can be broken into several categories: Statistical software packages, educational software, spreadsheets, applets/stand-alone applications, graphing calculators, multimedia materials, and data repositories. There is much overlap in the capabilities of the tools across these categories, yet no one tool seemingly covers all possible educational uses of technology (Ben-Zvi, 2000; Biehler, 1997a). We provide a brief summary of the types of tools available and some of their benefits and disadvantages. Other resources such as *The American Statistician* (http://www.amstat.org/PUBLICATIONS/tas) or *The Journal of Statistical Software* (http://www.jstatsoft.org) regularly provide more comprehensive software reviews. The goal of this section is to provide a flavor for the types of technological tools available, highlighting some of the more common examples of each type of tool. It is important to remember that the focus of instruction should remain on the content and not the tool, and to choose technology that is most appropriate for the student learning goals, which could involve a combination of technologies. Since new software is continually being developed for K-16 education, the following discussion does not attempt to be exhaustive.

1. Statistical Software Packages

Statistical packages are software designed for the explicit purpose of performing statistical analyses. Several packages have been used by statisticians for many years, including *SPSS* (http://www.spss.com), *S-PLUS* (http://www.insightful.com), *R* (http://www.r-project.org), *SAS* (http://www.sas.com), and *Minitab* (http://www.minitab.com). These packages have been evolving into more menu-driven packages that are more user friendly for students. The term menu-driven is used to describe a software program that is operated using file menus instead of commands. Menu-driven is commonly easier for most users as it allows the user to navigate using the mouse and to hunt and peck a bit more, which has both advantages (students do not feel as lost) and disadvantages (often using a trial and error strategy rather than real thought when choosing a command). As these packages become more user friendly, they are being increasingly used in introductory courses.

The statistical package *Minitab*, in particular, has always had a pedagogical focus and is becoming increasingly feasible as a tool that allows student exploration and construction of ideas (e.g., writing "macros" for repeated sampling, graphics that update automatically as data values are added or manipulated, ease of changing representations). *DataDesk* (Velleman, 1998; http://www.datadesk.com) is a similar package, but has focused on data exploration and interactive graphics from its initial development. *DataDesk* provides many unique tools that allow students to look for patterns, ask more detailed questions about the data, and "talk" with the program about a particular set of data. *R* (Verzani, 2005) is a language and environment for statistical computing and graphics that provides a wide variety of statistical and graphical techniques, including linear and nonlinear modeling, statistical tests, time series analysis, classification, and clustering. It is freely accessible and is being increasingly used in introductory statistics classes. Additional add-ons can be downloaded to improve the graphical interface of the program (http://socserv.mcmaster.ca/jfox/Misc/Rcmdr).

More cost-effective alternatives to these packages include student versions, which are smaller in scope (does not work for large data sets) and several stand alone statistical packages are also now available for free or at minimal cost, online. For example, *StatCrunch* (West, Wu, & Heydt, 2004; http://www.statcrunch.com), is a fully functional, very inexpensive, Web-based statistical package with an easy-to-use interface and basic statistical routines suited for educational needs.

2. Educational Software

Different kinds of statistical software programs have been developed exclusively for helping students learn statistics. *Fathom* (http://www.keypress.com/fathom), a flexible and dynamic tool was designed with the inputs of many statistics educators and educational researchers to help students understand abstract concepts and processes in statistics, and does not attempt to have the capabilities of more traditional

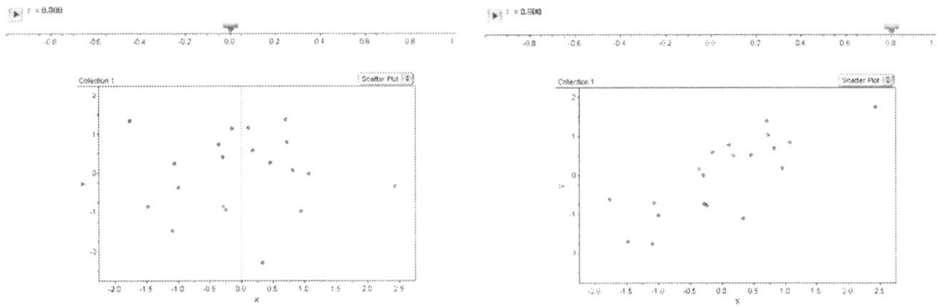

Fig. 5.1 A *Fathom* slider allows students to gradually change the value of the correlation coefficient while a scatterplot immediately updates to reflect the new strength of association between the variables

statistical software tools. Erickson (2002) described *Fathom* as a dynamic computer learning environment for teaching data analysis and statistics based on dragging, visualization, simulation, and networked collaboration. The strongest features of *Fathom* are the easy access to multiple, linked representations (see Fig. 5.5) including sliders (see Figs. 5.1 and 5.2), the ability to build and run simulations, and the many different ways of importing data from a variety of sources. One small example of the very dynamic, interactive features of *Fathom* is pointing on the edge of a histogram bar and dragging the bar, which immediately updates the graph (see Fig. 5.4).

TinkerPlots was developed to aid younger students' investigation of data and statistical concepts (Konold & Miller, 2005; http://www.keypress.com/tinkerplots). This tool has been widely field tested in math classes in grades 4–8 in both the United States and other countries (e.g., Ben-Zvi, 2006) with very positive results. Students can begin using *TinkerPlots* without knowledge of conventional graphs or different data types, without thinking in terms of variables or axes. By progressively organizing their data (ordering, stacking, and separating data icons), students gradually organize data to answer their questions and actually design their own graphs.

InspireData (http://www.inspiration.com/productinfo/inspiredata) is a commercial extended version of *TableTop* that also focuses on visual representations in

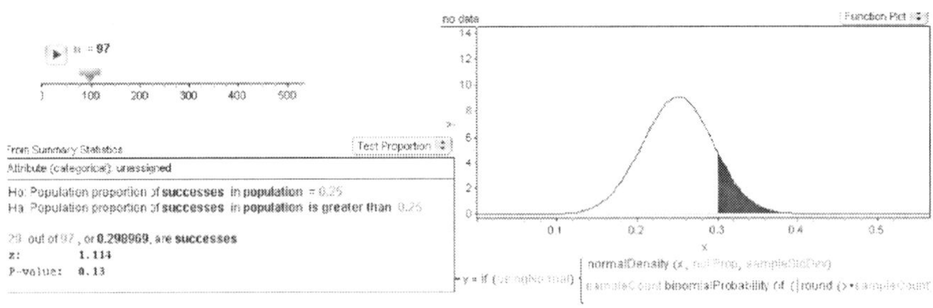

Fig. 5.2 A *Fathom* slider allows students to change the sample size in a one-sample proportion z-test and instantly see the effects on the sampling distribution, test statistic and P-value

helping grade 4–8 students "discover meaning as they collect and explore data in a dynamic inquiry process." This package also offers linked representations, animations, and easier annotation of data analyses and presentations.

Some of these educational packages are also making it easier for students to access large data sets (e.g., Census data) and for teachers to access predeveloped classroom exercises. The limited statistical capabilities may prevent their use beyond an introductory course (though they are expanding, e.g., *Fathom* now offers multiple regression), but have benefits in being less overwhelming to the students and being more geared to the point-and-click generation.

3. Spreadsheets

Spreadsheets such as *Excel* (http://office.microsoft.com/) are widely available on many personal computers. However, care must be exercised in using *Excel* as a statistical educational package. Statisticians often criticize *Excel*'s calculation algorithms and choice of graphical displays (Cryer, 2001; McCullough & Wilson, 1999; Nash & Quon, 1996). For example, it is still very difficult to make a boxplot in *Excel*. Even a histogram, one step in all statistical packages, requires a separate add-in which can take several minutes to locate and install. Creating the histogram is then still a nonintuitive process, with unappealing choices for bin limits, output location, and outcome (the default outcome is a frequency table and not a graph). Moreover, the bars in the Excel graph are not contiguous – misleading for students in understanding discrete vs. continuous data, and labeling is poor (see Fig. 5.3). *Excel* does have some strengths in helping students learn to organize data and in "automatic updating" of calculations and graphs as values are changed, and some advocate *Excel* due to its widespread use in industry and relatively easy access (Hunt, 1996). The statistical capabilities of Excel can also be expanded with additional add-ins (often accompanying statistics textbooks).

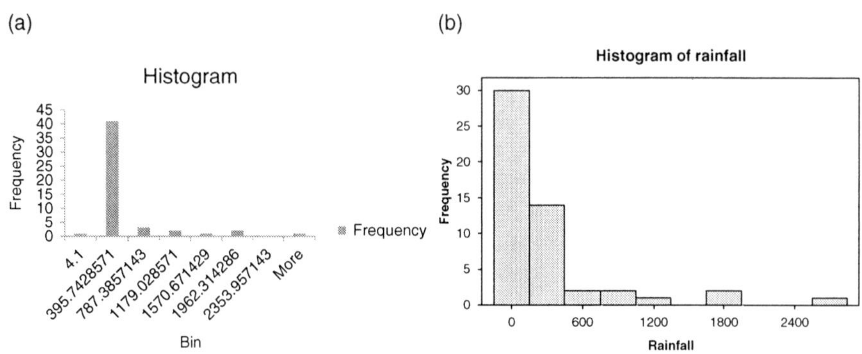

Fig. 5.3 Histograms of rainfall data using (**a**) Excel and (**b**) Minitab defaults

4. Applets/Stand-Alone Applications

Over the last decade, there has been extraordinary growth in the development of on-line applets that can help students explore concepts in a visual, interactive, and dynamic environment. An applet is a software component that usually performs a narrow function and runs typically in a Web browser. Many of the applets are easy for students to use and often capture an interesting "context" for students, e.g., the Monty Hall problem (see for example http://www.shodor.org/interactivate/activities/AdvancedMontyHall) and Sampling Reese's Pieces (see for example http://www.rossmanchance.com/applets/Reeses/ReesesPieces.html). In addition, a large number of computer programs can be downloaded from the Internet and run without an Internet connection that allow students to explore a particular concept (e.g., *Sampling SIM* allows the student to explore the nature of sampling distributions of sample means and sample proportions, freely downloadable from: http://www.tc.umn.edu/~delma001/stat_tools/software.htm).

While these tools are too numerous to list here, the *Consortium for the Advancement of Undergraduate Statistics Education* (CAUSE, http://www.causeweb.org) provides a peer-reviewed annotated list of such tools. Applets can also be found with the online *National Council of Teachers of Mathematics* (NCTM) *Principles and Standards for School Mathematics* Electronic Examples (http://standards.nctm.org/document/eexamples and http://illuminations.nctm.org/. See more applets and Web resources in the Resources section at the end of this book). What these tools often gain in visualization and interactivity, they may sometimes lose in portability. And while they can be freely and easily found on the Web, they are not often accompanied by detailed documentation and activities to guide student use. The time required for the instructor to learn a particular applet/application determines how to best focus on the statistical concepts desired, and developing detailed instructions and feedback for the students may not be as worthwhile as initially believed.

5. Graphing Calculators

Perhaps the most portable technological tool and one that is being increasingly used in 9–12 grades is the graphing calculator. A graphing calculator is a learning tool designed to help students visualize and better understand concepts in mathematics, statistics, and science (e.g., Dunham & Dick, 1994; Flores, 2006; Forster, 2007; Marlena, 2007). Advancements in technology have made the graphing calculator a powerful tool for analyzing and exploring data. Data can often be downloaded from the Web, saving students' time from keying in data. Some models provide an accessible way for students to collect and measure light, temperature, voltage, or motion data, and much more. Many statistical calculations, including inference procedures and probability distributions are now standard in most brands. Simulations can also be run in a reasonable time frame allowing students to explore concepts such as sampling distributions (Herweyers, 2002; Koehler, 2006). Student learning time is short with such technology and schools can purchase one classroom set for use at school or in a particular course. However, beyond the introductory statistics course, graphing calculators are not a reasonable substitute for statistical packages. Students also need to be wary that the output given by the graphing calculator is not sufficient communication of statistical results (e.g., "calculator-speak," graphs with no labels and scales).

Since the calculators by Texas Instruments (and several versions of Casio calculators) are the most typically used in secondary schools and colleges, what follows is a short description of some of the capabilities and limitations of these calculators for statistical analysis. The TI-73 Explorer that was designed specifically for 6–8 grades includes data collection with CBL (Calculator-Based Laboratory), descriptive statistics, and graphs for numeric and categorical data. It has a probability simulation application, which allows students to explore probability theory with interactive animations that simulate rolling dice, tossing coins, and generating random numbers as well as creating bar graphs and tables. These features are also available for other models, such as, TI-83 Plus, TI-84 Plus, and TI-89 Titanium, which also provide tests of significance and confidence intervals (one and two-sample procedures, chi-square tests, linear regression test, and ANOVA). It is possible to add more advanced applications to later models, such as, residual plots for analysis of variance, pairwise comparison in one-factor experimental design, and confidence intervals for contrasts used in one-factor experimental design (see Kuhn, 2003).

Graphing calculators, however, bear several limitations and drawbacks. Some representations and algorithms (e.g., $ax + b$ vs. $a + bx$, residuals for transformed data) are different from those in common software packages. For example, there are circumstances where students using the TI-83's random number generator for a simulation study will all obtain the same set of values (Lee, 2005), and while most statistics books provide tables based on the area under the standard normal probability curve to the left of the z-value, the TI-83 and 84 calculators provide the area between two values and the entries' syntax may be confusing to students (Lesser, 2007). There are also limitations in the level of numerical precision and speed (e.g., for simulations) of these calculators.

6. Multimedia Materials

These materials often seek to combine several different types of technology. For example, *ActivStats* (http://www.activstats.com) has been used in college classrooms, combining videos of real world uses of statistics, mini-lectures accompanied by animation, links to applet-like tools, and the ability to instantly launch a statistical software package and analyze a data set. An advantage of such an environment is that students only need to learn one type of technology. In fact, more and more, entire lessons and even textbooks are written around these types of embedded technology to make them a "living" textbook, e.g., *CyberStats* (http://www.cyberk.com; Symanzik & Vukasinovic, 2006). Many other multimedia resources are currently being developed around the world, several of which were described in the proceedings of the International Conferences on Teaching Statistics (ICOTS-5, Pereira-Mendoza, Kea, Kee, & Wong, 1998; ICOTS-6, Phillips, 2002; ICOTS-7, Rossman & Chance, 2006).

7. Data and Materials Repositories

Another popular and important use of the World Wide Web in statistics instruction is in locating and using pedagogically rich data sets and explorative activities for use with students (e.g., Schafer & Ramsey, 2003). Numerous data repositories exist. *The Data and Story Library* (DASL, http://lib.stat.cmu.edu/DASL) and the *Journal of Statistics Education* (JSE) *Dataset and Stories* feature (see http://www.amstat.org/publications/jse/jse_data_archive.html) are excellent starting places. These data sets come with "stories" outlining their background and classroom uses. CAUSE (http://www.causeweb.org) is again a good resource for datasets and peer-reviewed classroom activities.

The many types of tools and resources listed earlier offer great choices for instructors, as well as decisions about how to best use these tools, how often to use them, and for what purposes and activities. While many of the tools described are bona fide research tools, others have been developed primarily for learning purposes. We next discuss issues related to the uses of technological tools in helping students learn and reason about statistics.

How Technology Can Support Student Learning

As more and more technological tools become available, and as student facility with such tools greatly increases, it is becoming increasingly important to focus on the best ways to use such tools in the classroom. Below we provide examples of some of the effective uses of technology in the statistics classroom. It is important to keep in mind that many of these learning tools have different goals and it may be necessary to employ different tools for different learning goals and that a combination may best aid students. If using a combination of technologies, it is important to address learning curve issues for students. There is always overhead in learning to use a tool itself before students can benefit from the tool for learning statistics. Students do seem adept at learning to use different types of software in the same course, but teachers may also aim for a more consistent look and feel, at least in the instructional aids they provide to accompany the tools, and to consistently provide guidance on when to use the different tools (e.g., in assessments).

Automation of Calculations. With technology, students can carry out many calculations (and graphing tasks) in a short time, with high accuracy and few errors. For example, allowing the computer or calculator to calculate the standard deviation saves cognitive load and actual classroom time that can instead be spent on the larger concept of variation and on properties of the calculated value. Reducing the focus

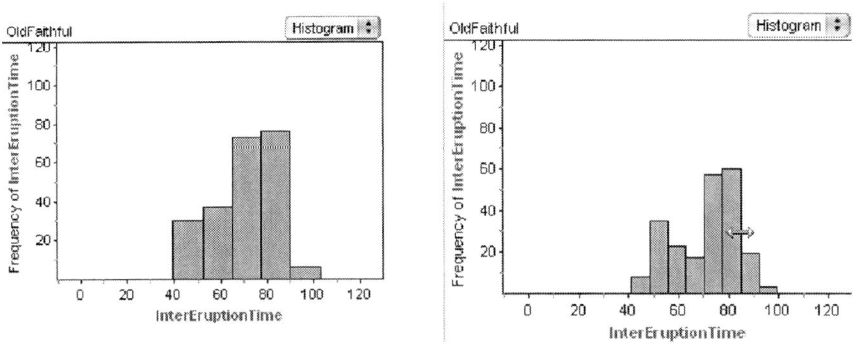

Fig. 5.4 Bin widths are changed in *Fathom* just by dragging the edges of the histogram bars

on computations frees students to spend more time focusing on understanding the concepts. There is also less focus on data entry, manipulating numbers in formulas, and on exercises using only small and/or artificial data sets. Interactivity, such as sliders, can help students focus on the effects of changing pieces in the calculation without the burden of recalculating numerous terms. Assessment tasks can evaluate student ability to explain concepts and justify conclusions rather than on how they perform rote calculations.

Emphasis on Data Exploration. The use of technology amplifies students' ability to produce many graphs quickly and easily, leaving students more likely to examine multiple graphs and different representations (Pea, 1987). For example, students can think more about the effect of bin size in a histogram (e.g., smaller bin sizes in a graph may reveal bimodal behavior that was initially hidden, see Fig. 5.4). *Fathom*, for example, allows students to click on the edge of the bar and drag, immediately updating the graph in a dynamic and interactive representation. Students may also see how different graphs of the same data provide different pieces of a story by generating different graphs on the same screen (Fig. 5.5).

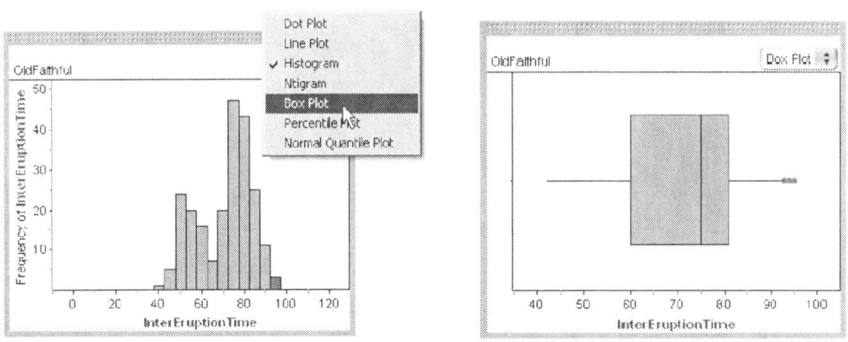

Fig. 5.5 Students easily make transitions between graph types while specific cases are highlighted in all graphs simultaneously (*Fathom*)

Technology should be utilized in the classroom to encourage students to explore data sets more in depth, to allow the data to tell a (possibly unexpected) story to the student, and to consider related conceptual issues (e.g., Erickson, 2001).

Visualization of Abstract Concepts. Technology enables visualization of statistical concepts and processes (Biehler, 1993), demonstration of complex abstract ideas, and provision of multiple examples in seconds. Students are better able to explore and "see" statistical ideas, and teachers are better able to present them to students. Such tools give students and teachers much more flexibility to ask "what if" questions. For example, we can select an individual data observation and drag it to immediately see the effects on the graph and numerical calculations (Fig. 5.6).

Another example is how *TinkerPlots* allows students to see the data values "hidden" in boxplots, as shown in Fig. 5.7.

Simulations as a Pedagogical Tool. Technology can also play a significant role in enhancing students' ability to study random processes and statistical concepts by giving them easy access to viewing and designing simulations (e.g., Chance & Rossman, 2006; Lane & Peres, 2006; Lane & Tang, 2000; Mills, 2004). These tools allow students to answer "what happens if this is repeated a large number of times" through direct observation. For middle school students, software tools

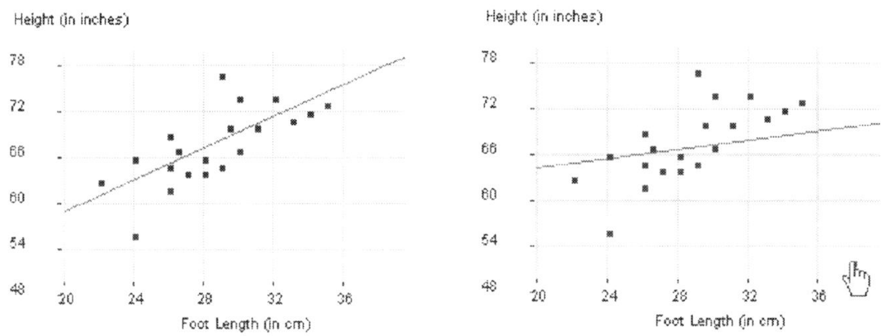

Fig. 5.6 The applet "Least Square Regression" from the Rossmanchance Applet Collection (http://www.rossmanchance.com/applets/index.html) allows student to add and drag a new observation to see how the regression line changes as the point relocates

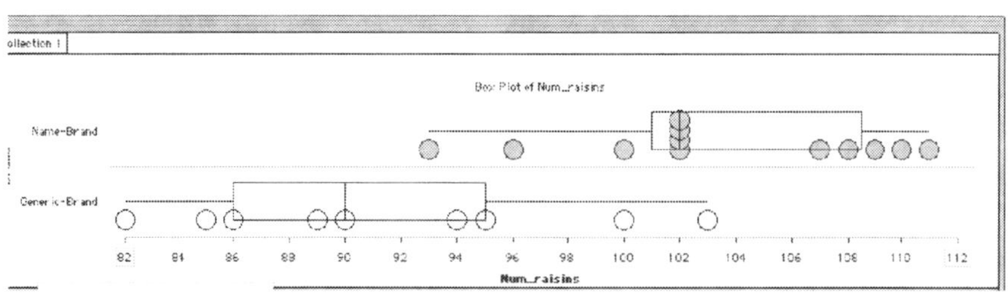

Fig. 5.7 Comparing the number of raisins between two brands using boxplots with visible case icons in *TinkerPlots*

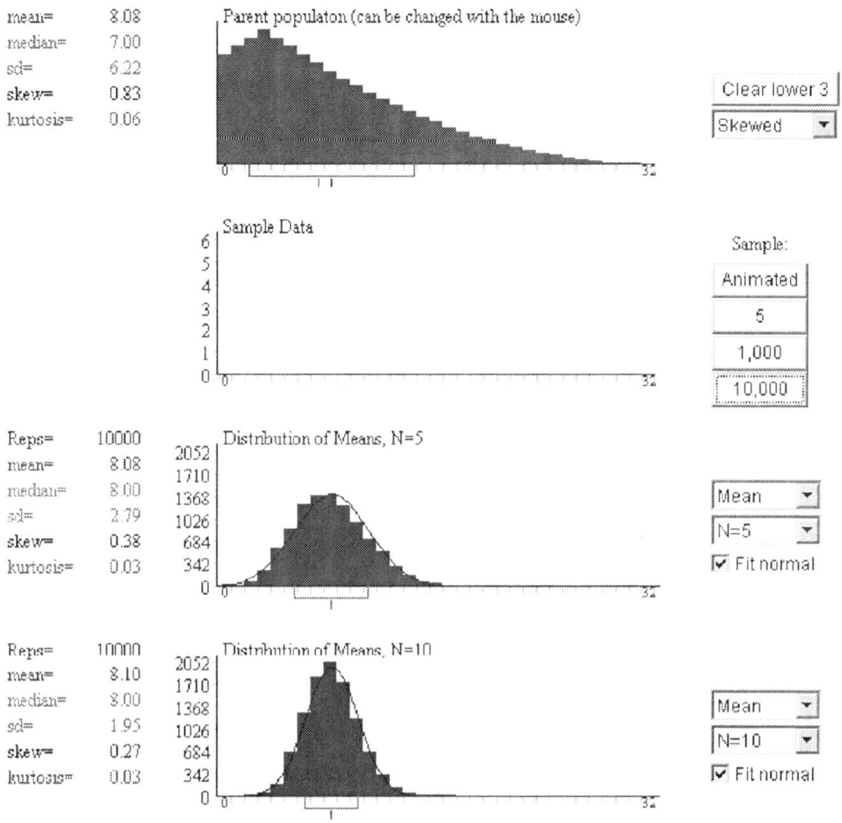

Fig. 5.8 Illustration of the Sampling Distributions applet from the "Rice Virtual Lab in Statistics" (http://onlinestatbook.com/stat_sim/sampling_dist/index.html), which allows students to specify a population shape (e.g., skewed), different sample statistics (e.g., mean), and sample sizes (e.g., n = 5 and n = 10). The above numerical and graphical displays results for 10,000 repetitions

such as *ModelChance* (http://www.umass.edu/srri/serg/projects/ModelChance, now available within *TinkerPlots*) allow students to investigate real-world applications of probability (e.g., false positives from medical screening), while helping them understand the distinction between probability and statistics. For older students, abstract concepts such as sampling distribution (Fig. 5.8) and confidence intervals (Fig. 5.9) become more concrete. Students' understanding is developed by carrying out these repetitions, controlling parameters (such as sample size, number of repetitions), and describing and explaining the behavior they observe rather than on relying exclusively on theoretical probability discussions, which can often be counterintuitive to students (delMas et al., 1999).

Students can also examine nontraditional distributions (e.g., the sampling distribution of a median) while easily analyzing the effects of different parameters (e.g., sample size, population size, and number of samples) on such conceptual

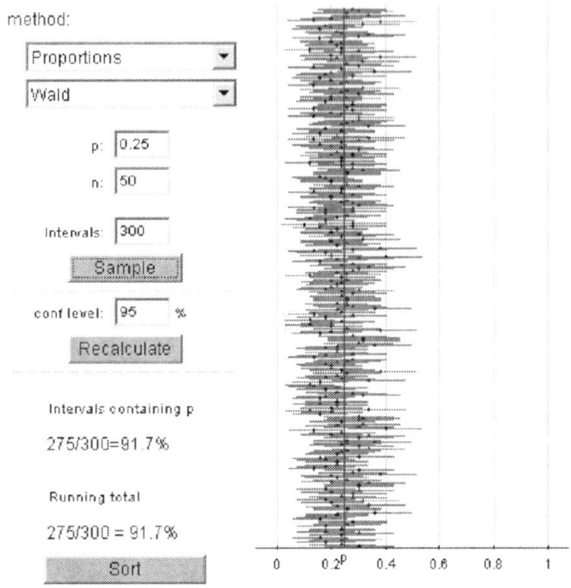

Fig. 5.9 Illustration of Simulating Confidence Intervals applet from the Rossmanchance Applet Collection (http://www.rossmanchance.com/applets/index.html) to generate 300 different random sample proportions and resulting confidence intervals, recording the percentage of intervals that succeed in capturing the value of the population proportion

ideas (see Fig. 5.10). Students can modify these parameters and initial conditions to explore and make their own conclusions.

Investigation of Real Life Problems. One of the most important uses of technology is its capacity to create opportunities for curriculum and instruction by bringing real-world problems into the classroom for students to explore and solve (Bransford et al., 2000). Technology facilitates the discussion of more interesting problems and data sets (which may be large and complicated), often accessed from the Internet. We now have the power to have students analyze real and often messy data, giving students a better idea of what statisticians do by having them go through the process of collecting, analyzing, and making conclusions to investigate their own questions. Assessment can focus on giving students data sets and having them complete a full analysis on their own, which may include "cleaning" the data first (e.g., Holcomb, 2004). Such exercises empower students as users of statistics and allow them to better understand and experience the practice of statistics (Ben-Zvi, 2004a).

Provision of Tools for Collaboration and Student Involvement. Course management systems provide communication tools (such as discussion forums, file exchange, and whiteboard), productivity tools (online student guide, searching, and progress review), student involvement tools (group work, self-assessment, student community building, and student portfolios) as well as administration tools. In these learning environments, it becomes easier for students to collaborate with other students, which can help improve the writing and communication skills needed to

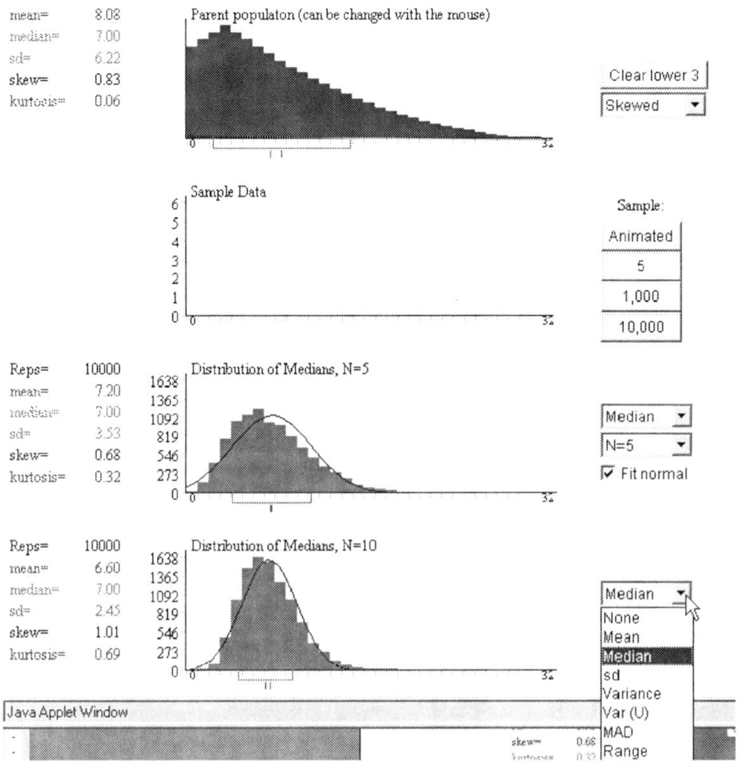

Fig. 5.10 Illustration of change to sampling distribution of a median in the "Rice Virtual Lab in Statistics" Sampling Distributions applet (http://onlinestatbook.com/stat_sim/ sampling_dist/index.html)

convey their findings. They also allow students do more learning on their own, outside of class, using Web-based or multimedia materials. This frees the instructor to minimize lecture and to spend more time on data analysis activities and group discussions.

While interest in using technology in teaching and learning statistics is great, there is not a lot of data available that provide answers to all the questions teachers may have regarding best uses of technology to support student learning. The following section provides an overview of some of the current research on using technology to teach statistics.

Overview of Research on Technology in Statistics Education

Research on the role of technology in teaching and learning statistics has been increasing over the last decade. In 1996, a special *International Association for Statistical Education* (IASE) Roundtable was convened in Granada, Spain to discuss the current state of research on the role of technology in statistics education at that time. While much of the work reported at the roundtable (Garfield & Burrill, 1997) was on the development of new tools to help students learn statistics, there was a clear call for more research on appropriate ways to use these tools to promote student learning. It was suggested that a new research agenda was needed to identify appropriate methodologies for future studies on this topic as well as to explore new ways to use technology in studying this topic (Hawkins, 1997). Given the changes in technology in the past decade, ideas about both of these aspects of technology are still emerging. In this section, we highlight some of the more recent research questions being explored and the types of studies involved, particularly with respect to developing students' statistical reasoning. The following section will suggest some implications from the research for teaching practice.

Ben-Zvi describes how technological tools are now being designed to support statistics learning in the following ways (2000, p. 128):

1. Students' active construction of knowledge, by "doing" and "seeing" statistics.
2. Opportunities for students to reflect on observed phenomena.
3. The development of students' metacognitive capabilities, that is, knowledge about their own learning and thought processes, self-regulation, and control.

In addition, technological tools can bring exciting curricula based on real-world problems into the classroom; provide scaffolds and tools to enhance learning; and give students and teachers more opportunities for feedback, reflection, and revision (Bransford et al., 2000). The types of research studies that explore technology in statistics education can be grouped into three categories:

1. Development, use, and study of particular tools (e.g., the creation and use of *Fathom* software – Biehler, 2003; *Minitools* – Cobb et al., 1997).
2. How use of particular tools help develop students' reasoning (e.g., use of *Sampling SIM* software to develop reasoning about sampling distributions – Chance et al., 2004).
3. Comparison of tools (e.g., comparing *ActivStats, CyberStats*, and *MM*Stat* multimedia – Alldredge & Som, 2002; Symanzik & Vukasinovic, 2002, 2003, 2006).

Some of the most informative studies were not designed to focus on the use of technology, but on larger teaching experiments that combined innovative instructional activities and technological tools to promote student reasoning about a particular topic, such as distribution (e.g., Bakker, 2004a, Cobb, 1999; Cobb & McClain, 2004). These studies focused on the use of a set of *Minitools*, applications created to help students move along a learning trajectory. Similarly, the studies of Makar and Confrey (2005) and Rubin, Hammerman, and Konold (2006) explore teachers' knowledge and reasoning as they use innovative software (*Fathom* or *TinkerPlots*).

Although few studies have been empirical in nature in the field of statistics education, such studies have provided valuable information on how technological tools can both improve student learning of particular concepts as well as raise new awareness of student misconceptions or difficulties (e.g., Batanero et al., 1996; Finch & Cumming, 1998; Shaughnessy, 1992). Cobb and McClain (2004) found that students were able to more easily make and test conjectures when using such tools to analyze data. While controlled experiments are usually not possible in educational settings, qualitative studies are increasingly helpful in focusing on the development of concepts and the use of skills that technology is intended to facilitate. Investigations by Miller (2000) and Lee (2000) are examples of qualitative studies of how instructors can integrate technological tools to support a student-centered learning environment for statistics education. Other examples are Biehler's (1998) use of videos and transcripts to explore students' thinking as they interacted with statistical software and research by delMas et al. (1999) that provides a model of collaborative classroom-based research to investigate the impact of simulation software on students' understanding of sampling distributions.

One research area where empirical results have been less consistent is in the use of simulation as a pedagogical tool. Research on simulation training indicates that even a well-designed simulation is unlikely to be an effective teaching tool unless students' interaction with it is carefully structured (Lane & Peres, 2006). Chance and Rossman (2006) illustrate how simulation can be a powerful tool in helping students learn statistical ideas, particularly the ideas of long-run patterns and randomness, in a concrete, interactive environment (e.g., using a simulation of sampling Reese's Pieces or shuffling playing cards to build on a tactile simulation while allowing the user to easier adjust parameters such as sample size and immediately explore the impact). Technology is also used in involving teachers in the design of computational tools that presumably encourage them as designers to reflect upon the statistical concepts incorporated in the tools under development. Healy (2006) describes how involvement of Brazilian mathematics teachers in the collaborative simulation design process helped participants come to see distributions as statistical entities, with aggregate properties that indicate how their data are centered and spread.

There have also been studies on the use of graphing calculators in teaching and learning various subjects (e.g., mathematics – Marlena, 2007; chemistry – Moore-Russo, Cortes-Figueroa, & Schuman, 2006; statistics – Collins & Mittag, 2005; Forster, 2007). Five patterns and modes of graphing calculator tool use emerged in a qualitative mathematics classroom-based study: computational tool, transformational tool, data collection and analysis tool, visualizing tool, and checking tool (Doerr & Zangor, 2000). These researchers also found that the use of the calculator as a personal device can inhibit communication in a small group setting while its use as a shared device supported mathematical learning in the whole class setting.

Although research supports the use of technology to facilitate and improve the learning of statistical concepts, Biehler (1997a) cautions that statistics educators need a system to critically evaluate existing software from the perspective of educating students and to produce future software more adequate both for learning and doing statistics in introductory courses. Thistead and Velleman (1992), in their summary of technology in teaching statistics, cite four obstacles that can cause difficulties when trying to incorporate technology into college statistics courses: equipment (e.g., adequate and updated computer labs), software (availability and costs), projection (of computer screens in classrooms), and obsolescence (of hardware, software, and projection technologies).

Nowadays, we can see increased availability of computers, access to graphing calculators and Internet, updated and more widely available software, often via CDs bundled with textbooks or on the World Wide Web, and new methods of projecting computer screens such as interactive white boards. But another obstacle is the fact that it takes time and thought to effectively incorporate new technologies. Success in the use of technology for teaching means success in placing teachers on the road to new ideas and methods of teaching, not the sudden transformation of teaching (Huffman, Goldberg, & Michlin, 2003). A first step on this road can be obtaining information on how technology can be used to support and improve students' learning in statistics courses and what technology is available to accomplish this goal.

Recommendations for Using Technology to Teach Statistics and Probability

Articles and presentations on technology in statistics education have pointed to several effective ways to use technology in the statistics classroom. Below we provide a summary of our opinions, based on the literature as well as our own experience, as to what are some important issues to consider.

- Too often in statistics courses, students become focused on the numerical calculations. This tendency can be exacerbated in a computer-rich environment, especially in using statistical packages that easily produce large amounts of (often unclear) output – students focus on the output instead of the process. Many teachers also have such expectations and may rely on drill-and-practice uses of technology rather than the student-centered, rich tasks that offer the greatest value added for the use of technology (Means, Penuel, & Padilla, 2001). Rather than let the output be the end result, we believe it is important to discuss the output and results with students and require them to provide explanations and justifications for the conclusions they draw from the output and to be able to communicate their conclusions effectively. Although students can spend time entering data (with an emphasis on ways to organize data, types of variables etc.), it is more useful to have them do only small amounts of data entry and spend more time exploring, analyzing, and interpreting data.
- While technology allows for more student-driven and open-ended explorations, this may not happen right away, as students first need to become familiar with the tool and how to use it. Sometimes students become overwhelmed or lost in the details of the instructions or programming commands and do not see the bigger statistical ideas being developed. For example, in exploring sampling distributions, they focus on how many trials to include in the simulation rather than how the sample size affects the resulting empirical distribution. Teachers therefore need to carefully structure explorations so that even while learning to use the

software, students are able to focus on the concepts rather than only paying attention to the technology or blindly following a list of commands. In this way, students may discover and construct meanings for the big ideas of statistics as they are guided through a series of investigations. For example, instructors can provide detailed instructions and ask students to interpret the results, but then also include follow-up questions asking students to explain the process in their own words or to update the steps they learned to reflect a change in the scenario or to apply the technique on their own to a new context.

Care also needs to be taken that students do not overuse the technology, searching for menus without thinking about the statistical issues, or producing computer output without interpretation. Some packages will even provide students with interpretations (e.g., *Fathom*'s "verbose" feature provides interpretations of confidence and P-values for the student), so the instructor needs to consider how much of this interpretation they want the students to generate for themselves.

Without this careful structuring and guidance, students may only be paying attention to the software and not to the statistical problem or content (Collins, Brown, & Newman, 1989). The student activities that include use of technology should embed questions that guide students in an investigation of statistical problems and encourage the students to discuss and summarize the big concepts of the lesson before they are summarized by the instructor. In this way, students should learn to conduct their own explorations with less and less structure and support from teachers. Both these points underscore that the accompanying instructions are often more crucial in impacting student learning than the specific choice of technological tool.

- Collaborative learning is often particularly helpful in statistics education, and technology can be used to facilitate and promote collaborative exploration and inquiry, allowing students to generate their own knowledge of a concept or new method in a constructivist learning environment (Huffman et al., 2003; Miller, 2000). Interactions within the groups have an important role with questioning and critiquing individual perspectives in a mutually supportive fashion so that a clearer understanding of statistical concepts emerge and knowledge of statistical ideas develops (Ben-Zvi, 2006). This type of iterative exploration of data also mirrors statistical practice and helps students develop a "habit of inquiry" (Wild & Pfannkuch, 1999).
- Statistics education is also characterized by the deep rooted misconceptions that students often hold entering the course. Technology greatly facilitates employment of a "predict-and-test" strategy that has shown to be effective in establishing the cognitive dissonance necessary for students to change their ways of thinking about a concept (e.g., Posner et al., 1982). Students can be required to predict what they will observe (e.g., expected shape of a distribution, effect of increasing number of observations) and then use the technology to obtain immediate feedback on their understanding. This is especially useful when the feedback is directly observable without the need for a lot of inference by the students, especially from numerical results that may not yet make sense to students.

We also offer the following more general reminders to consider when planning to use technology:

- Technology does not replace the teacher, but teachers need to actively observe the students, identify their difficulties, probe their thought processes and the conclusions they are forming, quickly curb any problems they are having with the technology, keep students on task, and answer questions (Feenberg, 1999). Teachers do not have downtime while students are interacting with technology.
- Technology should be chosen to facilitate student interaction and accessibility, maintaining the focus on the statistical concept rather than on the technology (Moore, 1997). This choice can depend on the learning curve, portability, and flexibility of the tool (e.g., whether the technological tool can be utilized in other places in the course such as using the same software to carry out other data analysis or simulation tasks). It is important to consider the background of the students and the goals of the course as well as the instructor's comfort level and knowledge.
- A supportive learning environment is needed that is rich in resources, aids exploration, creates an atmosphere in which ideas can be expressed freely, provides encouragement when students make an effort to understand, and allows students (and teachers) the freedom to make mistakes in order to learn (e.g., Brown & Campione, 1994; Cobb, Yackel, & Wood, 1992). Some students may still feel anxious using technology, so it can be important to integrate the use of technology gradually; it may be useful to provide examples of how the technology can save them time and minimize errors in completing calculations.

Possible Obstacles to Incorporating Technology in the Statistics Classroom

Integrating technology in the classroom has great potential to enhance teaching and learning; turning that potential into a reality can be a complex and multifaceted task. Some of the key factors for successfully integrating technology in the classroom are well-defined educational visions, curriculum design, and teacher preparation and support (Kleiman, 2004). This success comes with some "costs" and instructors and administrators need to think carefully about how to best integrate the tools into the classroom. This section presents some of the common obstacles teachers must face to create rich learning environments with the use of technology and the need for necessary support mechanisms to overcome them.

Need to Re-examine Student Learning Goals. While technology allows changes in instructional focus, these changes need to be reflected in the course goals and corresponding student assessments. At the high school level in particular, if standardized testing will be used as the one measure of students' success, this will impact how technology should be incorporated. For technology to gain the most impact on student learning, other course goals will be necessary. At the college level, changes in learning outcomes due to use of technology (e.g., doing away with the use of statistical tables and instead relying on software to compute P-values) may require buy-in from colleagues and administrators. At any level, changes resulting from incorporating technology may require endorsement from instructors, parents, and students.

Lack of Awareness of and Comfort with New Technologies. Probability and statistics are specialized subjects, and many schools may not have a faculty member whose expertise is in these areas. Since teachers' schedules are very demanding, little time is available to learn about new technologies and their capabilities. Teachers who have learned statistics decades earlier may not be comfortable using the new tools and may not believe in the value of their use. In some cases, teachers may be able to attend conferences and hear about new technologies, but there is usually not enough time for them to appreciate the benefits of the technology and fully learn how to effectively use it in the classroom. Unless teachers are provided with a long-term support for learning to use and implement technology, they are unlikely to use it in their classrooms. Internet-based communities of teachers are becoming an increasingly important tool for overcoming teacher's isolation and need for support (e.g., Levin & Waugh, 1998).

Lack of Support for Teachers. According to Ritchie (1996), schools are not yet effectively implementing instructional technologies in spite of the increase in the capacity of available educational technology. This study identified lack of administrative support as one of the most critical impediments for the integration of instructional technology. Administrative support is needed in order to provide funding for computer labs, consistent technical support for teachers, and on-going professional development for teachers to have the opportunity to learn new technologies and their uses in classrooms. Even when the technology is in place and the technical support is available, teachers need much more support and professional development in learning how to implement a new pedagogy with technology since technology alone does not make for effective teaching. To maximize the benefits of technology for students, teachers need to spend time modifying what they will teach, how they will teach it, and how they will assess it using technology (U.S. Congress, Office of Technology Assessment, 1995).

Class Time Required for Exploration. One of the largest benefits of using technology is allowing student exploration of concepts and deeper probing into large messy data sets, but such investigations can be time consuming. However, time can be saved by eliminating other components of the course such as hand calculations and replacing them with better questions or richer discussions for the benefit of more meaningful understanding by the students. Students' education will also benefit by more of these explorations occurring at lower grade levels, increasing student comfort with such explorations and perhaps leading to a less impacted curriculum at the college level.

The Fact that Technology Can Fail. It is important to realize that computers can crash, Internet sites may not be available, and so on. Therefore, teaching with technology means having a plan in case the technology fails during class. Contingency plans include making handouts of the planned lessons or transparencies revealing the expected outcomes. In fact, in probability and statistics, even working technology can lead to unexpected results. It is important for both the teacher and student to be comfortable with randomness and approximations rather than clean proofs. When technology does fail, the process of constructing the lesson often prepares teachers to pose questions; this can lead students to have a discussion that can serve as pre-amble to the next class meeting when the technology will function (Cardenas, 1998).

Time Needed to Implement Changes. Teachers should not expect inclusion of technology in their classroom to show immediate improvements in student achievement or to be a solution to all teaching difficulties (U.S. Congress, Office of Technology Assessment, 1995, p. 159). It is usually the case that effective use of technology takes refinement, trial and error, and continual improvement. Teachers should not be afraid to try something and fail; that does not mean the new teaching method is a failure, but may indicate modifications that need to occur for the implementation to be successful in an individual classroom. Educators cannot forget that part of educating students is preparing them for life outside the classroom, so even if the use of technology does not provide immediate success in the teaching and learning of mathematics or statistics concepts, students are learning how to use technologies that they may encounter in their future jobs.

Unclear Role of Distance Learning. The methods being developed for distance learning, which incorporate many innovative uses of technology, may allow schools to share resources and make a high-quality probability or statistics class at one school available to others. However, with increased distance learning courses, it is also unclear as to how much of a course can be taught exclusively using technology, what the appropriate roles of an instructor should be, and how much emphasis should still be placed on students generating calculations and graphs by hand. "Hybrid courses" that combine a distance component with less frequent face to face meetings are also being used with greater frequency (e.g., Utts, Sommer, Acredolo, Maher, & Matthews, 2003; Ward, 2004). Smith, Clark, and Blomeyer (2005) provide a synthesis of new research in K-12 online learning and Mayer (2001) provides recommendations on effective multimedia presentations.

Issues to Consider when Selecting Technology in a Statistics Class

Despite the obstacles listed above, it is still important to try to find ways to access and utilize appropriate technology to help students learn statistics. The GAISE College Report (Franklin & Garfield, 2006) lists some issues to consider when selecting technological tools to use in helping students learn statistics:

- Ease of data entry, ability to import data in multiple formats
- Interactive capabilities
- Dynamic linking between data, graphical, and numerical analyses
- Ease of use for particular audiences and availability of good instructions and materials to help learn and use the software.
- Availability to students, portability

We believe that no one tool can do it all and that there are many good tools available to use, many of which are free. Therefore, rather than thinking about one technological tool for students to use, we encourage teachers to think about what sets of tools will help student best learn statistics in each unit of the course. What is used to graphically explore data in one unit may not be the best to illustrate sampling in another.

Implementation Issues

The GAISE Report advises educators to remember that technology still needs to be used carefully. As stated earlier, instructors should first remember to worry less about the specific choice of technology and instead on *why* and *how* it will be used. Careful planning of the technology implementation is crucial to having a direct impact on student learning. We list below some of the questions to consider when planning how to use a particular technology tool with students. We do not try to provide answers to these questions, because responses will vary from class to class, topic to topic, and depending on choice of technology.

- How much time will students need to explore the technology? How can the teacher make sure that videos and simulation games do not become "play time" for students and that they are learning important ideas? How much instructor demonstration should precede student use of the technology and how much debriefing afterwards?
- How can we avoid students thinking of the computer as a "black box"? How can we make sure that students understand and trust (to a reasonable extent) what technology produces? We do not want computer visualizations to just become a black box generating data. We believe that is important for educators to use a hands-on activity with devices such as dice or M&Ms to begin an activity, and then move to the computer to simulate larger sets of results. In this way, students may better understand the simulation process and what the data actually represent. Follow-up assessment also needs to indicate to students that they will be responsible for understanding and synthesizing what they have learned from the technology.
- How consistently will the software be used? To what extent students will use the software? Is it only used in class and assignments or can it be used on exams? Are students assessed on how well they are able to use the software or is it optional to use it? How accessible will the technology be for student's use outside class? The answer to this question will determine what kind of homework assessments or group project teachers create. Teachers should keep in mind that students value what is assessed, so assessments should be aligned with learning goals. If students are expected to use a lab outside class, what are the lab hours and are these accessible for most of students?
- How many different technologies are students expected to use? For which technologies should students learn commands (i.e., software packages)? Which technologies will be demonstrations only (i.e., applets)? Incorporating many

technologies can be overwhelming for students and place conceptual understanding in second place to technological skills.

Summary

Technology has been and will continue to be a major factor in improving student learning of statistics. However, effective utilization of technology requires thoughtful and deliberate planning as well as creativity and enthusiasm. Despite the endless capabilities that technology offers, instructors should be careful about using sophisticated software packages that may result in the students spending more time learning to use the software than applying it. Even in our advanced technological society, some students are not always ready for the type of technology used in courses. Choice of a particular technology tool should be made based on ease of use, interactivity, dynamic linkages between data/graphs/analyses, and portability. Good choices if used appropriately can enhance student collaboration and student-instructor interactions, and often a combination of several different tools will be necessary.

It is also crucial to consider how best to utilize the technology (e.g., allowing predict-and-test learning situations and facilitating student interaction, not spending large amounts of time entering data). There is a need for carefully designed learning activities that guide and scaffold student interaction with technology. Set-up time should be minimal, and students often, at least initially, need to be carefully guided through the activity, with the steps building logically. As they gain more confidence with the tool and fluency with the statistical language, teachers should gradually encourage students to conduct and make sense of their own explorations, with less and less guidance and structure, while focusing on the overall larger statistical concept.

What is still lacking is more studies on the most effective ways of integrating technology into statistics courses in developing students' reasoning about particular concepts, and determining appropriate ways to assess the impact on student learning in these contexts. With an increased emphasis on statistics education at all educational levels, we hope to see more high-quality research projects in years to come that will provide information on appropriate uses of technology to improve student learning of statistics.

Part II
From Research to Practice: Developing the Big Ideas of Statistics

Introduction:
Connecting Research to Teaching Practice

In this section, we list the content of the chapters, describe the chapter structure, provide a guide to the research-based lessons described in each chapter, provide the theory behind the development of the lessons, and outline the resources in the Website that accompany each chapter.

Chapter Content

Each chapter in Part II of the book (Chapters 6 through 14) addresses one important topic or big idea in an introductory statistics course. These chapters are presented in the following order:

Chapter 6 **Learning to Reason About Data**: the nature and role of data, types of data, and methods of collecting and producing data.

Chapter 7 **Learning to Reason About Statistical Models and Modeling**: the idea of a statistical model, the uses of models in statistics, essential ideas of probability, the Normal distribution and regression as statistical models.

Chapter 8 **Learning to Reason About Distribution:** the idea of statistical distribution, understanding and interpreting graphical representations of data, introduction to ideas of shape, center, and spread.

Chapter 9 **Learning to Reason About Center**: the idea of center and representativeness; measuring center of a distribution: uses, properties, and interpretation of means and medians.

Chapter 10 **Learning to Reason About Variability**: the importance of variability in statistical thinking, the omnipresence of variation, sources of variability; measuring spread of a distribution: uses, properties and interpretation of range, standard deviation, and interquartile range.

Chapter 11 **Learning to Reason About Comparing Groups**: reasoning with center and spread in comparing groups and making informal inferences using boxplots.

Chapter 12 **Learning to Reason About Samples and Sampling Distributions**: sampling variability and sampling distributions, the effect of sample size, the implications of the Central Limit Theorem.

Chapter 13 **Learning to Reason About Statistical Inference**: tests of significance, P-values, and confidence intervals.

Chapter 14 **Learning to Reason About Covariation**: scatterplots, correlation, and simple linear regression.

These chapters present the idea and review research literature on this topic and suggest ways to build classroom activities based on the research following a progression of ideas (learning trajectory) to develop the concept.

Chapter Structure

Each chapter follows a parallel structure. This structure is outlined below:

1. Snapshot of a research-based activity

A quick glimpse of an innovative research-based classroom activity introducing the statistical topic.

2. The rationale for this activity

An explanation of how and why this activity helps build reasoning about the topic.

3. The importance of understanding the topic

Description of the topic and its importance.

4. The place of the topic in the curriculum

An analysis of the place of the topic in the curriculum of an introductory statistics course.

5. Review of the literature related to reasoning about the topic

A synthesis and summary of the most relevant, scholarly research on the topic.

6. Implications of the research: Teaching students to reason about the topic

A presentation of our views of the implications of the research about teaching the topic.

7. Progression of ideas: Connecting research to teaching

A description that outlines the progression of ideas to provide a bridge between the research and practical teaching activities. In a table format, we offer the suggested significant milestones to reach the goal of understanding and reasoning about the topic. This consists of subgoals and statistical ideas and concepts, which are listed in the left column of the table, with suggested activities for each subidea in the right column. We realize that there may be many paths to the end goal, but we offer one set that we have tried several times and feel is effective with students and appears consistent with the research.

The activities listed in the table or described in the chapter do not attempt to cover all the material that teachers may want to cover in this unit, but give examples of the types of activities that reflect some of the suggested steps of the research-based learning trajectory.

Please note that wherever we are aware of borrowing or adapting an activity, we try to credit the author, which is shown in the Table of Activities in the Appendix to this book (on page 391). There are some cases where there are so many versions of an activity or we have been using and adapting an activity for so many years, that it is hard to know exactly who created the original version.

8. Overview of the lessons

A description of the complete set of two to five lessons (and activities in these lessons) that have been designed to develop reasoning about the topic (the actual lessons are on the accompanying Website, http://www.tc.umn.edu/~aims/).

As you read through these materials, please keep in mind that *the lessons do not replace a statistics textbook*. This set of sample research-based lessons is also not exhaustive in the content it includes, but provides an overall sense of how the research implications can be integrated into an introductory course. In fact, it might be that some instructors will want to adapt or use only one lesson from this collection in a particular class or just one activity from within a lesson or topic.

In addition to the activities included in the lessons that accompany each chapter, the Website for this book provides suggestions for other types of activities instructors may want to consider.

9. Website resources for each chapter

Each chapter has a set of resources at the book's Website (http://www.tc.umn.edu/~aims/), designed to supplement the print chapters with digital resources. They include:

- The set of annotated lesson plans and activities
- Sample data files used in the lesson
- Blank student handouts for each activity
- Annotated activates that include possible students answers.

A Guide to the Research-Based Lessons

Each chapter in Part II is accompanied by a set of two to five lesson plans accessed from the Website that illustrate how class sessions might be taught building on the research reviewed in the book. A lesson plan gives the idea of what an entire class session might be. Each lesson contains one or more individual in-class activities and includes overall lesson goals, opening discussion questions, and a wrap-up discussion of the activities for the lesson. The theory behind the lessons and the components of the lesson plans are described in detail below.

The Theory Behind the Lessons

The lesson plans and activities within the lessons were developed based on the following theory:

- Learning is constructive. Students learn and build understanding through reading materials, discussions in class, and collaborative activities where they investigate statistical problems and discuss their ideas and reasoning with other students (see Chapter 15). Learning is facilitated by the making and testing of conjectures (see Chapters 2 and 3).
- Learning the big ideas (see Chapter 3) of statistics requires an "unpacking" of the ideas into components that begin with informal notions (based on and influenced by students' prior knowledge and intuitions), and build gradually toward formal notions. Moving through the sequence of informal and formal ideas takes extended time and places more importance on depth and reasoning as opposed to breadth of knowledge.
- Learning and retaining the big ideas requires the ideas to be explicitly revisited and reapplied in different contexts (and topics) after they are introduced. This also involves transferring knowledge from earlier units to applications in later units. This is based on the view of learning and development of understanding as complex, iterative, and circular (non-linear) processes.
- Learning the big ideas is facilitated by examining real data sets that engage students in statistical reasoning, although some formal aspects of the ideas may be done without a context (e.g., the Central Limit Theorem).
- Statistical reasoning requires basic statistical literacy. The role of the textbook is to help students develop basic literacy: becoming acquainted with language, terms and symbols, and some basic content before coming to class.
- Statistical reasoning is best developed by in-class collaborative, inquiry-based activities that engage students in making and testing conjectures and discussing results of data analyses.
- Statistical thinking builds on statistical literacy and reasoning. It is developed by explicit modeling of statistical thinking by the instructor throughout a course and by the student participation in real world data collection, production and analysis activities, including class activities and out of class student projects.
- Collaboration among students in the classroom facilitates and promotes student learning by providing access to multiple voices and opinions; improving the quality and completeness of student learning; and creating a positive and productive learning environment (see Chapter 15).

- Becoming knowledgeable in statistics involves both cognitive development and "enculturation", socialization of processes into the culture and values of "doing statistics." Thus, it is important to bring the practice of statistics in class closer to what it means to use statistics within the discipline.

General Comments About the Lesson Plans

The Use of Learning Goals

The student learning goals may be used to help the instructor keep in mind the goals of the lesson, to help summarize what was learned at the end of the lesson, and to plan assessment of student learning and of the effectiveness of the lesson.

Students' Prerequisite Knowledge for a Set of Lessons

Each chapter specifies some important prerequisite knowledge for the set of lessons. However, we do not specify on each lesson plan what students are expected to know before that particular lesson, as we do not believe in the linear progression of ideas and development of understanding. We view learning as a complex, iterative, and circular process. In our lessons, we emphasize making conceptual links among big statistical ideas, and we highly recommend revisiting ideas and concepts that were mentioned in previous units.

Teachers' Own Knowledge of Statistics

We assume that readers of this book are already knowledgeable about the statistical content covered in a first course. We are not trying to teach the readers (statistics teachers) content. Instead, we are suggesting ways for them to teach the content to students.

The Role of Simulations

One component of the lessons is the use of simulations and informal ideas of inference early in the course and throughout the course, so that when students reach the material of formal methods of statistical inference, they may build these formal ideas on their informal notions. To do this, we introduce the *Simulation of Samples* (SOS) *Model* shown in Chapter 6, early in the first set of lessons, when students are learning about collecting and producing data. This model, based on the research literature, presents a visual representation of three levels of data (population, samples, and sampling distributions). Many lessons on different topics involve repeated sampling of simulated data and discussions about what is a "usual" value of the sample statistic. In each case, the *SOS* Model is used to distinguish between the three levels of data and to focus this final question of what is an unusual value for a statistic on the third level.

Websites Cited

We have tried to only cite stable, well-known and reliable Websites in the lessons and activities of the following chapters. However, Website locations are likely to change. Therefore, if you find an error message when trying to locate a reference Website, we recommend using a search engine to locate the site using key words or phrases (e.g., Guessing Correlation applet). We also note that many of the referenced Web applets require Java to be installed, which can be downloaded freely from http://java.com/.

Finally, these lessons represent not only our views of what the research suggests, but they are also colored by our own experiences as teachers and our views of best practices of teaching particular topics. These lesson plans and activities are not the only ones or necessarily the best ones, but they are the ones we have developed and used and we find them to work well in promoting a Statistical Reasoning Learning Environment (SRLE, Chapter 3). Our goal is to provide examples of resources that may be reviewed to better illustrate our suggested sequences of activities, and may be used as is or modified to be used in various courses.

Chapter 6
Learning to Reason About Data

> *High on my list of elements of statistical thinking is the claim that data beat anecdotes. This is surely a learned principle, and one much neglected by public opinion.*
> *(Moore, 1998, p. 1257)*

Snapshot of a Research-Based Activity on Distribution

Students arrive in class on the first day of their introductory statistics course. After brief introductions, the instructor asks them to *Meet and Greet* each other. They are asked to stand up, to take a pad of paper with them, and to meet at least five other people by shaking hands and sharing five pieces of data about themselves. This information is to be recorded on their paper:

1. Your name.
2. The number of credits you are taking this semester.
3. Your intended major or field of study.
4. Your gut reaction when you hear the word statistics.
5. If you are a senior (yes/no).

After students have walked around and gathered enough information, they sit back down and engage in a class discussion about their data. They are asked to describe how they recorded their data. Typically, some students write down every response while others may set up a system of categories and tally marks. It is interesting to see what different organization methods were used, which leads to a discussion about what seems to be a good method to organize data and the different types of variables used to collect data (and different types of data collected, e.g., numbers, words, yes/no data; they are asked about how responses differ). Students are asked to look at their data and see which, if any variables, led to a wide variety of responses (variation in the data). There is usually more variation in credits or majors, and less variation in reactions to the word "statistics" (which sadly, are often quite negative). Students are asked about the questions used and if they could have been improved, or if they led to ambiguity of responses. For example, the question: "Are you a senior?" could be answered based on number of credits or year in school, and those might result in different answers.

Finally, the students are asked about what kinds of summaries can be made about the class by looking at the data they collected. This leads to a discussion of their sample of data, and how well their sample represents the entire class. This further leads to the idea that their samples may be too small and possibly biased and that there are better ways to take a sample of data.

Rationale for This Activity

The "Meet and Greet" activity is designed to immerse the students in data and exploratory data analysis from day 1, and to help them think about data collection as well as what you can learn from data. The lesson helps students begin to see data as a classifier, where they collect data to yield frequencies of particular values, and to also begin to see data as an aggregate. These views of data are important and are more advanced ways of thinking about data than simpler, intuitive methods of focusing on individual data values (Konold & Higgins, 2002; Konold, Higgins, Russell, & Khalil, 2003, described later in this chapter). This lesson also helps distinguish statistics from mathematics, by focusing on data and the context of collecting and interpreting data as "noisy" processes. Data that are not numerical are also examined, as three different types of data are informally handled. As we note later in the chapter, there is not much of empirical research on how students develop important ideas about data. This activity and others described in the chapter are often based on implications from studies about the nature of reasoning about data as well as more general research about student learning and effective pedagogical methods for teaching statistics, where specific research studies on learning data are not yet available.

The Importance of Understanding Data Collection and Production

Statistics is the science of learning from data. Where the data come from matters.
(Moore, 2005, p. xviii)

David Moore eloquently reminds us that the most important information about any statistical study is how the data were produced. "Before you trust the results of a statistical study," he exhorts, "ask about details of how the study was conducted." Moore goes on to state: "Data enlighten. They shed light in dark places" (Moore, 2005, p. xxiii). He explains that statistics can help guide us in using data to explore the unknown: how to produce trustworthy data, how to look at data (starting with plotting graphs), and how to reach sound conclusions that come with an indication of just how confident we can be. The three important aspects of statistical science are data production, data analysis, and statistical inference (inferring conclusions from data). This chapter deals with the first area, data production.

The Place of Data in the Curriculum

In "traditional" statistics classes, data are introduced on the first day, and distinctions are made between different types of data (categorical and quantitative). In a subsequent unit of graphs, the distinction between these two types of data is revisited as categorical data are tied to pie graphs and bar charts while quantitative data are represented in histograms and dot plots. From that point on, data are used by students as they learn different statistical summaries and procedures. Many students could leave their statistics course never questioning where data come from or realizing that how data are gathered or produced is directly related to methods of analysis and conclusions drawn.

Several years ago, this serious absence began to change as more and more textbooks included chapters on data collection (why and how to take samples and use surveys) and methods of producing data (experimental designs) (e.g., De Veaux, Velleman, & Bock, 2005; Moore, 2004; Peck, Olsen, & Devore, 2007; Rossman, Chance, & Lock, 2001; Watkins, Scheaffer, & Cobb, 2004; Utts, 2004). However, even with this added materials, many times that information was left behind and not referred to later in the course and was treated as an isolated unit. Today's curriculum recommendations (e.g., the GAISE Project, see Franklin & Garfield, 2006) encourage the study of data production and the integration of this topic throughout a class. Therefore, explorations of data include discussions about the purpose for which the data were gathered and how. While there are many suggestions about how to teach about data, so far there is little empirical research related to this topic. The next section summarizes existing research as well as important work by influential statisticians and uses this literature as a basis for suggesting a progression of ideas and activities to develop reasoning about data.

Review of the Literature Related to Reasoning About Data

In our review of the published literature, we found that the majority of research studies on reasoning about data were conducted at the primary school level, or identified *how* students reason about data rather than *how to develop* good reasoning about data. In addition, we found articles by statisticians and statistics educators expressing strong opinions about the topics related to data. For example, the statistics education literature stresses the importance of using real (and to a lesser extent, realistic) data, the importance of students planning surveys and experiments and collecting their own data, and reflecting on the processes and considerations involved in formulating a statistical question that can be answered with a survey or experimental study. In particular, the idea of where randomization plays a role in statistical study, either through random sampling or random assignment, is now regarded as an extremely important topic for students to learn (Cobb, 2007; Franklin & Garfield, 2006).

We begin this literature review with an overview of what statisticians and statistics educators have written about the nature and importance of exploratory data analysis and how we would like students to think about data, which have provided a basis for the learning goals related to this topic. Then, we summarize the literature that relates to how students understand and reason about data and methods used to collect or produce data.

The Nature of Exploratory Data Analysis

"Statistics is the science of learning from data" (Moore, 2005, p. xviii). One approach to learn from data is Exploratory Data Analysis (EDA), developed by Tukey (1977). EDA is the discipline of organizing, describing, representing, and analyzing data, with a heavy reliance on informal analysis methods, visual displays and, in many cases, technology. The goal of EDA is to make sense of data, analogous to an *explorer of unknown lands* (Cobb & Moore, 1997, p. 807). The original ideas of EDA have since been expanded by Mosteller and Tukey (1977) and Velleman and Hoaglin (1981), and others. They have become the accepted way of approaching the analysis of data (Biehler, 1990; Moore, 1990, 1992). EDA has been widely adopted by statistics educators, in large part, because it serves the need for more data and what we can learn from them, and does not focus on the underlying theory and complicated recipes (Biehler & Steinbring, 1991; Cobb & Moore, 1997; Scheaffer, 2000).

According to Graham (1987), Kader and Perry (1994), Nicholson, Ridgway and McCusker (2006), and others, data analysis is viewed as a four-stage process: (a) specify a problem, plan, pose a question, and formulate a hypothesis; (b) collect and produce data from a variety of sources (survey, experiments); (c) process, analyze, and represent data; and (d) interpret the results, discuss, and communicate conclusions. In reality, however, statisticians do not proceed linearly in this process, but rather iteratively, moving forward and backward, considering and selecting possible paths (Konold & Higgins, 2003). Thus, "data analysis is like a give-and-take conversation between the hunches researchers have about some phenomenon and what the data have to say about those hunches. What researchers find in the data changes their initial understanding, which changes how they look at the data, which changes their understanding" (Konold & Higgins, 2003, p. 194).

The focus of EDA is not on a set of techniques, but on making sense of data, how we dissect a data set, what we look for, how we look, and how we interpret. EDA postpones the classical statistical inference assumptions about what kind of model to fit to the data, with the more direct approach of "letting the data speak for themselves" (Moore, 2004, p. 1); that is, allowing the data to reveal the underlying structure and model through the translating eyes of a statistically literate viewer.

Statistical Thinking About Data

Based on in-depth interviews with six professional practicing statisticians and 16 statistics students, Wild and Pfannkuch (1999) provide a comprehensive description of the processes involved in statistical thinking, from problem formulation to conclusions. They suggest that a statistician operates (sometimes simultaneously) along four dimensions: investigative cycles, types of thinking, interrogative cycles, and dispositions. In solving statistical problems, the statisticians interviewed were particularly interested in giving prominence to grasping the dynamics of the system under investigation, problem formulation, and measurement issues. In the first stages of the investigation, statisticians also attended to data-related issues such as sampling design, data collection, data management, and data cleaning. Some types of the statisticians' thinking that emerged from the interviews are inherently statistical, such as the recognition of need for data, consideration of variation, reasoning with statistical models, and integrating the statistical and contextual.

Although statistical thinking is a synthesis of statistical knowledge, context knowledge, and the information in data to produce implications, insights, and conjectures, these researchers found that the earliest stages of a statistical investigation are driven almost entirely by context knowledge. The statistical knowledge contributes more as the thinking crystallizes. There is a back and forth shuttle between thinking in the context sphere and the statistical sphere that goes on during all phases of the statistical investigation (Wild & Pfannkuch, 1999).

Research on Students Reasoning About Data and Data Analysis

In a "teaching experiment"[1] conducted with lower secondary school students in Germany by Biehler & Steinbring (1991), data analysis was introduced as "detective" work. Teachers gradually provided students with a data "tool kit" consisting of tasks, concepts, and graphical representations. The researchers concluded that all students succeeded in acquiring the beginning tools of EDA, and that both the teaching and the learning became more difficult as the process became more open. There appears to be a tension between directive and nondirective teaching methods in this study. A study by de Lange, Burrill, Romberg, and van Reeuwijk (1993) reveals the crucial need for professional development of teachers in the teaching of EDA in light of the difficulties teachers may find in changing their teaching strategy from expository authority to guiding and from the precision of mathematics to the messiness of statistics.

Based on their observations of school students' reasoning about data, Konold, Higgins, Russell, & Khalil (2003) suggest a framework for describing increasing levels of complexity in how people understand data, focusing on how different representations and uses of these representations highlight or de-emphasize the

[1] See page 37 Chapter 2 for a description of a teaching experiment.

aggregate characteristics of data. Konold and colleagues (Konold & Higgins, 2002; Konold et al., 2003) argue that children see data in *several* simpler ways before ever noticing aggregate and emergent features of data sets. Their fourfold schema includes the following different ways of viewing data, which we consider useful for examining the thinking of adults as well as children:

1. Data as a *pointer* to the data collection event but without a focus on actual data values – in this view, data remind children of their experiences, "We looked at plants. It was fun."
2. Data as a focus on the identity of individual *cases* – these can be personally identifiable, "That's my plant! It's 18 cm tall," extreme values, "The tallest plant was 37 cm," or interesting in some other way.
3. Data as a *classifier*, which focuses on frequencies of particular attribute values, or "slices," without an overall view – "There were more plants that were 15–20 cm than 10–15 cm."
4. Data as an *aggregate*, focusing on overall and emergent characteristics of the data set as a whole, for example, seeing it as describing variability around a center, or "noise" around an underlying "signal" (Konold & Pollatsek, 2002) – "These plants typically grow to between 15 and 20 cm."

More information on the aggregate view of data and distribution is provided in Chapter 8.

Generating and Formulating Statistical Questions

In their review of the research literature on teaching students to generate questions, Rosenshine, Meister, and Chapman (1996) wrote:

> "Question generation is an important comprehension-fostering (Palincsar & Brown, 1984) and self-regulatory cognitive strategy. The act of composing questions focuses the student's attention on content. It involves concentrating on main ideas while checking to see if content is understood (Palincsar & Brown, 1984). Scardamalia and Bereiter (1985) and Garcia and Pearson (1990) suggest that question generation is one component of teaching students to carry out higher level cognitive functions for themselves" (p. 181).

Rosenshine, Meister, and Chapman (1996) found that teaching students to generate questions about the material they have read resulted in gains in comprehension, as measured by tests given at the end of the intervention. Generating and formulating a statistical question, which is the starting point of any statistical investigation, is a challenging task for school and college students. In their chapter, "Reasoning about Data", Konold and Higgins (2003) described this challenge:

> "One of the first challenges is to transform that general question about the real world into a statistical one, one that we can answer with data ... Among other things, the statistical question allows us to develop measurement instruments and data-collection procedures. By analyzing the data, we answer our statistical question, which ideally, but not always, tells us something about the real question we started with.

> In learning how to formulate questions and to collect and analyze data to answer them, students must learn to walk two fine lines. First, they must figure out how to make a statistical question specific enough so that they can collect relevant data yet make sure that in the process they do not trivialize their question. Second, they must learn to see the data they have created as separate in many ways from the real-world event they observed yet not fall prey to treating data as numbers only. They must maintain a view of data as "numbers in context" (Moore, 1992) while at the same time abstract the data from that context" (p. 195).

Studies on Students' Reasoning About Data Collection and Study Design

The guidelines for teaching the introductory college statistics course (The GAISE Project, Franklin & Garfield, 2006) suggest goals for students in an introductory course that include understanding random sampling and random assignment and the distinction between them. *Random sampling* allows results of surveys and experiments to be extended to the population from which the sample was taken. *Random assignment* in comparative experiments allows cause and effect conclusions to be drawn.

Although there is consensus among statistics educators that student data collection and analysis projects are of substantial value, the planning and piloting phases of data collection are often neglected. Short and Pigeon (1998) asked their college freshman introductory statistics, graduate statistics, and pre-service teachers to write protocols or detailed plans for how the data would be collected for a data investigation project, and to plan and conduct pilot studies before embarking on full scale data collection. They found that the protocol and pilot study assignments developed important global problem-solving and communication skills in students. One of the most important advantages of careful planning of data collections and subsequent analyses that is reported is that students keep the structure of the data they collect within the boundaries of their statistical expertise.

Another essential part of effective statistical study design is deciding when and how to conduct experimental studies rather than nonexperimental ones. This can be challenging even for college students. Heaton and Mickelson (2002) found that undergraduates had some difficulty matching appropriate data collection methods to the quantifiable questions they had posed for class projects. Derry, Levin, Osana, Jones, and Peterson (2000) described the development of undergraduates' statistical thinking ability in regard to study design and documented students' tendency to confuse the concepts of random sampling and random assignment after the course.

Given the difficulties college students have exhibited with deciding when and how to conduct experiments, one would expect experimental design to be a nontrivial matter for high school students. Groth (2003) asked high school students how they would go about designing studies to answer several different quantifiable questions. This study provides a picture of levels of thinking one might expect from high school students in regard to the design of statistical studies.

The role of "data creation" in learning about data analysis was studied by McClain and Cobb (2001). The researchers developed an approach in which the teacher talked through the data creation process with the middle school students. These conversations often involved extended discussions during which the teacher and students together framed the particular phenomenon under investigation, clarified its significance, delineated relevant aspects of the situation that should be measured, and considered how they might be measured. The teacher then introduced the data the students were to analyze as being produced by this process. The researchers found that the data creation process grounded the students' activity in the context of a problem or question under investigation and improved their ways of reasoning about data as they made statistical arguments in the course of their analyses.

Reasoning About Random Samples and Sampling

Confusion about random samples and sampling applies to school and college students as well as adults. In their seminal paper, "Belief in the Law of Small Numbers," psychologists Tversky and Kahneman (1971) wrote:

> The research suggests that people have strong intuitions about random sampling; that these intuitions are wrong in fundamental aspects; that these intuitions are shared by naïve subjects and by trained scientists, and that they are applied with unfortunate consequences in the course of scientific inquiry... People view a sample randomly drawn from a population as highly representative, that is, similar to the population in all essential characteristics. Consequently, they expect any two samples drawn from a particular population to be more similar to one another and to the population than sampling theory predicts, at least for small samples. (p. 24)

Since the publication of this article, many researchers have examined and described the difficulties students have understanding samples, sampling variability, and inevitably, sampling distributions and the Central Limit Theorem (CLT). For example, Pollatsek, Konold, Well, and Lima (1984) administered a questionnaire to 205 undergraduate psychology students in the United States. In one experiment, subjects estimated (a) the mean of a random sample of ten scores consisting of nine unknown scores and a known score that was divergent from the population mean; and (b) the mean of the nine unknown scores. The modal answer (about 40% of the responses) for both sample means was the population mean. The results extend the work of Tversky and Kahneman (1971) by demonstrating that subjects hold a passive, descriptive view of random sampling rather than an active balancing model. This result was explored further in in-depth interviews with 31 additional students, where subjects solved the problem while explaining their reasoning. The interview data replicated the first experiment and further showed (a) that subjects' solutions were fairly stable – when presented with alternative solutions including the correct one, few subjects changed their answer; (b) little evidence of a balancing mechanism; and (c) that acceptance of both means as the population mean is largely a result of the perceived unpredictability of "random samples."

In a summary of articles by psychologists on the topic of reasoning about samples, Well et al. (1990) noted that people sometimes reason correctly about sample size (e.g., that larger samples better represent populations) and sometimes do not (e.g., thinking that both large and small samples equally represent a population). To reveal the reasons for this discrepancy, they conducted a series of experiments that gave college students questions involving reasoning about samples and sampling. The researchers found that students used sample size more wisely when asked questions about which sample size is more accurate than on questions that asked them to pick which sample would produce a value in the tail of the population distribution, indicating that they do not understand the variability of sample means.

Understanding the relationship between sample and population requires grasping the representative nature of a random sample. Although studies of elementary students have found that even young children have rich informal knowledge about samples and sampling, they also have numerous difficulties in reasoning about these ideas (Jacobs, 1999; Metz, 1999; Schwartz, Goldman, Vye, & Barron, 1998; Watson and Moritz, 2000a). For example, students were reluctant to generalize from a sample to a population since they seriously doubted that any inference can be drawn beyond the sample at hand, or believed that information on all cases is necessary to draw a conclusion about a population (Metz, 1999). Studies have found that elementary students were often not able to differentiate between results produced by biased and unbiased sampling techniques, and struggled to grasp the idea of randomness and random sampling, preferring convenience sampling over random sampling (Jacobs, 1999; Schwartz, Goldman, Vye, & Barron, 1998). Although students tend to prefer convenience samples and tend to accept stratified samples when thinking about surveys, they seem comfortable with the concept of randomly generating data when considering games of chance (Konold & Higgins, 2003.)

In a rare study at the college level, Dietz (1993) reports on results of a teaching experiment in several introductory statistics courses of undergraduate mathematics education and statistics students. One activity was designed to stimulate students, who had not yet studied sampling, to think creatively about methods of selecting a representative sample from a population. The students generated possible methods for selecting a representative sample, computed various summary statistics and made plots for the variables in each sample, compared their samples statistics to the population parameters and evaluated the advantages and disadvantages of the proposed sampling methods. Dietz reported that the students have "invented" simple random sampling, systematic sampling, stratified sampling, and various combinations thereof. Students, however, had difficulties in evaluating and discussing the various "invented" sampling methods, since they based their evaluation primarily on sample and population measures of central tendency, but ignored the differences in variability. Being actively involved in their own learning and construction of sampling ideas, students better understood and longer remembered ideas related to sampling methods. Chapter 12 in this book further discusses the literature related to reasoning about samples and sampling.

Implications of the Research: Teaching Students to Reason About Data

The literature reviewed suggests that an important component of a statistics course should be on the nature of data, where data come from, how to produce or collect good data (random samples and sampling), and what types of analyses and conclusions are appropriate for data collected in different ways. In order to do this, good, rich data sets are needed. Some of these may be collected by and about the students in the course to engage them in the data collection process. Other data sets may be used as well but always considering the context and the source of data.

In trying to help students develop statistical reasoning about data, it appears helpful to model for them the kinds of questions we need to ask about data in a study, such as:

1. Was this an observational or experimental study? What types of conclusions are therefore appropriate?
2. What methods or precautions were taken to prevent biased data?
3. How and where was randomization used in the study (random sampling? random assignment?).
4. What other precautions were taken (e.g., was the study double-blinded, how were the questions phrased, was there consistency across all measurements? Are the units and measurements clear, e.g., what "operational definitions" were made along the way?).

After a unit on data collection and production is finished, these questions should be revisited throughout the course, so that students are not presented with data sets to analyze without considering where the data come from and what types of analyses are appropriate.

Role of Technology in Learning to Reason About Data

The computer has had a major impact on the use of real data in introductory statistics classes. Many rich data sets are available on the World Wide Web, and most are easily accessible by statistical software packages. For example, *Fathom* (Key Curriculum Press, 2006) and *TinkerPlots* (Konold & Miller, 2005) allow easy access to data files stored on the Internet. Data can be loaded from a Website in a variety of ways, and the software will attempt to interpret the incoming data as cases in a collection. While not 100% foolproof, this feature can greatly reduce the amount of work necessary to get data into a suitable form. Furthermore, *Fathom* can directly import samples of U.S. census microdata (data about individual people) from the Integrated Public Use Microdata Series Web site (IPUMS, http://www.ipums.umn.edu/) at the University of Minnesota, which is a coherent US census database spanning

1850–2004. This is a rich source of interesting data, which *Fathom* makes easy to access and explore.

While some Websites provide data sets that have been cleaned up and formatted to use in teaching statistics (e.g., *DASL*, http://lib.stat.cmu.edu/DASL/), others may be messy and need cleaning before their use. Some instructors are now teaching students to clean and manage data as part of their introductory course (e.g., Gelman & Nolan, 2002; Holcomb & Spalsbury, 2005). The activities allow students to develop their reasoning about what the data represent, what constitutes an outlier or an error, where and how the data were produced, and similar questions. With the abundance of data that are now available in downloadable form, it seems inappropriate to have student spend time entering data by hand into the computer or calculator, other than a few examples to learn and experience the basic process of data entry and storage.

Progression of Ideas: Connecting Research to Teaching

Introduction to the Sequence of Activities to Develop Reasoning About Data

Statistics is the science of data. We therefore begin the study of statistics by introducing the basic ideas about data: Any set of data contains information about some group of individuals, the information is organized in variables, and data represent values of a variable and show the variability of something that is measured. Data are to be viewed as numbers with a context, where the context provides meaning. There are different ways to produce data: by taking measurements, sometimes in the context of an experiment, and by asking questions, such as on a survey. Data vary based on how they are collected. The data collection method matters, because it can affect the quality of the data. Therefore, you need to know the source of the data.

Two methods of gathering data are surveys and experiments. These can be studied in any order, but we present samples and surveys first. Students learn about different types of sampling methods and the kinds of data they produce. They learn about a random sample, its characteristics, and how to take a random sample (e.g., simple random sample, stratified random sample). They learn that a random sample is needed to generalize to a larger group (population) and why this is important. They learn about characteristics of good samples and bad samples, the idea of bias, and what can lead to bias or bad data in a sample survey (e.g., voluntary response samples, poorly worded questions).

When studying about how to produce data in an experiment, students learn the importance and purpose of randomization to infer cause and effect. They learn about the basic principles of statistical design of experiments (control, randomization, and replication), and what makes it bad, i.e., *confounding* of the effect of a treatment with other influences, such as lurking variables, *lack of randomization* which causes bias, or systematic favoritism, in experiments.

After understanding basic ideas of surveys and experiments, students can learn what questions to ask about data collection when looking at data, and how to look at a statistical problem by considering the entire process. This is the beginning of statistical reasoning, which can be developed through subsequent units on data exploration and analysis. Therefore, throughout a course, students need to remember the importance of asking questions about where data come from, why they were gathered, and how that relates to the questions being investigated and the methods of analysis.

Data Sets Used in Lessons in This Book

Two main data sets were collected from students and repeatedly used in the lessons in this book. They are a student survey and a set of body measurements. Both are multivariate data sets and while in most cases, activities have students examine one variable at a time, the students also look at the variables together, getting a sense of the multivariate data and how information on one variable may inform understanding data gathered on other variables.

The survey data can be collected through an online form and contains many questions that have to do with time (see Fig. 6.1). The body measurements data were collected as part of an activity in the unit on Center and Spread (see Chapters 9 and 10). The variables measured for this data set are shown in Fig. 6.2.

Table 6.1 shows a suggested series of ideas and activities that can be used to guide the development of students' reasoning about data. While the accompanying Website (http://www.tc.umn.edu/~aims) includes some activities (organized in lessons) that illustrate these steps, many are described more generically. The activities that are not included in the Website are marked by the symbol ❖.

Introduction to the Lessons

There are four lessons on collecting and producing data. They begin with types of data and types of variables and the variability of data. The lessons provide students with experience using different methods of sampling to develop an appreciation for random sampling and help students understand different sources of variability and bias in data. Students examine surveys and consider the impact of question wording. They design and conduct an experiment to illustrate principles of random assignment as well as issues involved in designing good experiments. The importance of samples and the role of samples in making inferences are revisited throughout these lessons as a preliminary introduction to informal ideas of statistical inference.

First Day Student Survey Questions

1. Which statistics course are you in?
2. Which section are you in?
3. What is your gender? [Male, Female]
4. What is your age in years?
5. In which month of the year were you born?
6. Which day of the month were you born on?
7. How many statistics courses are you enrolled in this semester?
8. In which year did you start college?
9. Which semester did you start college? [Fall, Winter/Spring, Summer]
10. In which year do you expect to graduate from college?
11. How many credits are you registered for this semester?
12. How many college credits have you completed?
13. What is your cumulative GPA?
14. How many hours per week do you typically study, on the average?
15. How many miles do you travel (one way) from your current home to campus each day, to the nearest mile?
16. How many minutes do you estimate it will take you to travel to school each day this semester, on the average?
17. What type of transportation will you use most often to get to school this semester? [Walk, Car, Bus, Bike, Other]
18. How many minutes do you exercise each week, on the average?
19. Estimate the number of minutes you typically spend each week communicating with your parents (email, phone, in person, etc.).
20. Estimate the number of minutes you spend each day eating (meals and snacks).
21. How many minutes each day do you typically spend on the Internet?
22. How many hours of sleep do you get on a typical week night (Monday through Thursday)?
23. About how many emails do you *send* each day?
24. About how many emails do you *receive* each day?
25. How many minutes do you talk on a cell phone on a typical week day (Monday through Friday)?

Fig. 6.1 First-day student survey questions

Body data collection sheet
(Each student should complete this sheet and enter the measurements into the Instructor's computer.)

1. Head circumference -----
2. Student's head -----
3. Height -----
4. Arm span -----
5. Kneeling height -----
6. Hand length -----
7. Hand span -----

Fig. 6.2 The variables measured for the body measurement data set

Table 6.1 Sequence of activities to develop reasoning about *data*[2]

Milestones: ideas and concepts	Suggested activities
Formal ideas of data	
• Data are values of a variable	• Meet and Greet Activity (Lesson 1: "Data and Variability")
• Measurements produce data	• Meet and Greet Activity (Lesson 1)
• Data show variability	• Meet and Greet Activity (Lesson 1)
• Data are numbers with context	• Variables on Back Activity (Lesson 1)
• There are different kinds of data	• Meet and Greet Activity (Lesson 1)
• Some variability in data is due to measurement process	• Meet and Greet, Variables on Back, and Developing a Class Survey Activities (Lesson 1)
• Importance of taking good measurements by asking clear questions	• Developing a Class Survey Activity (Lesson 1)
• It is important to look at multiple variables (Multivariate data) to better understand and describe a group	• Developing a Class Survey Activity (Lesson 1)
• Sources of bias in questions	• How you Ask a Question Activity (Lesson 2: "Avoiding Bias")
• Importance of asking clear, unambiguous questions in collection survey data	• Critiquing the Student Survey Activity (Lesson 2)
• Idea, purpose and importance of random sampling	• The *Gettysburg Address* Activity (Lesson 3: "Random Sampling")
• Different methods and reasons to take samples	• Student Survey Sampling Activity (Lesson 3)
• Purpose of experiments to produce data to determine cause and effect	• Taste Test Activity (Lesson 4: "Randomized Experiments")
• Purpose of randomization in an experiment	• Taste Test Activity (Lesson 4)
• Idea of making an inference based on a result of an experiment (using simulation)	• Taste Test Activity (Lesson 4)
• Importance of randomization in drawing inferences about results of an experiment	❖ Activity involving random assignment, with introduction to permutation test to informally test if results of the experiment are surprising or due to chance. (The symbol ❖ indicates that this activity is not included in these lessons.)
• Importance of knowing sources of data: data coming from samples or from experiments	❖ Activity where students identify whether the research is a survey (observational data) or an experiment

[2] See page 391 for credit and reference to authors of activities on which these activities are based.

Table 6.1 (continued)

• Good data vs. bad data	❖ Activity where students identify potential sources of bias or confounding
• What type of conclusions can be drawn based on the type of data	❖ Activity identifying the type of conclusion given a study description
• What kinds of questions to ask about where data come from	❖ Activity where students ask appropriate questions for given sets of data
Building on formal ideas of data in subsequent topics	
• Two sources of variation in measurement data	• How Big is Your Head Activity (Lesson 1 in the Variability Unit, Chapter 10)
• Reducing variability in measurement data	• Gummy Bears Activity (Lesson 2 in the Comparing Groups Unit, Chapter 11)
• Determining cause and effect from an experiment	• Gummy Bears Revisited Activity (Lesson 4 in the Statistical Inference Unit, Chapter 13)
• Correlation does not imply causation	• Credit Questions Activity (Lesson 1 in the Covariation Unit, Chapter 14)

Lesson 1: Data and Variability

The goal of the first lesson is to help students see that there are different types of data and different ways to aggregate and display data. This lesson also helps students to see the importance of context and how statistics differs from mathematics in the emphasis of context. Student learning goals for this lesson include:

1. To get started with the statistical process of gathering and interpreting data.
2. To see that there are different types of data and that data vary.
3. To see and consider different sources of variability in data.
4. To develop a survey to use to gather data for future activities (student survey).
5. To see that statistics is different from mathematics and that context of the data is important.

Description of the Lesson

This first class of the course begins with a question about what kinds of students enroll in this class. That leads into the *Meet and Greet Activity*, which is described in detail in the beginning of this chapter. In this activity, students informally discuss many important statistical ideas, such as, methods for data recording, types of variables, types of data, variation in data, question wording, data summaries and representations, sample of data, sample represents the population, sample size, bias, and sampling processes.

The students are then asked to think about other types of information that would be interesting to gather from members of the class (*Developing a Class Survey Activity*). They get into small groups and brainstorm a set of five questions, with the restriction that these questions collect different kinds of data, so that there is at least one question that asks for numerical data, one that asks for categorical (nominal) data and one that asks for yes/no data. Students are encouraged to think of interesting ideas to ask that would produce data they would care about, so not just "year in school" or "gender." The students discuss and create questions that are turned into the instructor, who (before the next class) compiles and edits them into a class survey that they will complete online. This final version of the survey includes additional questions added by the instructor, which will be an important source of data to use in activities on other statistical topics.

A final activity of the first day is to give students an opportunity to reason about data (*Variables on Backs* Activity). Students each have a card taped to their back that has a question on it. Once again, they stand up and walk around the room, this time recording students' responses to the question on their back without knowing what the question is. All the questions are numerical and no units are allowed to be given. Examples of questions are "how many hours did you sleep last night?," "How old is our current president?," "How many counties are in this state?," and "what is the last digit of your ID number?"

After the data are collected, students sit down and look at their data and draw a graph of their choosing (any graph will do). They use this graph to make a guess about what question is taped to their back. They take turns standing up, showing their graph to the class and explaining their reasoning (e.g., "I think my question is how many pets you have because the data I got have lots of 0s, a few ones, and 2s"). After they have explained their reasoning, they can take off the card and see what question was actually taped to their back. The class discussion after this activity involves how students reasoned about their data, and what they thought about and considered as they investigated their question.

A wrap-up discussion includes comparison of statistics to mathematics, how numbers in statistics have a context, and the importance of considering the context of data.

Students are also told that this is the kind of activity they will be doing in class: gathering and analyzing data, and using samples of data to make inferences. It is stressed that *data* are the focus of the course, that we strive to collect good data, and that we are interested in studying the variability in data. They can be asked to summarize why data vary and sources of variability in data.

Lesson 2: Avoiding Bias

The focus of the second lesson is on helping students understand the idea of biased data and ways to avoid biased data in question wording and survey administration. Students suggest and discuss methods of obtaining unbiased data from a survey. Student learning goals for this lesson include:

1. To recognize common instances of bias resulting from how questions are worded or by methods or taking surveys.
2. To learn characteristics of good questions that can be answered by data.
3. To see that how you ask a question makes a difference in the quality of data collected.

Description of the Lesson

Students first see a cartoon that shows a character answering a question on a survey, which illustrates the idea of biased data from a survey and leads to a general discussion about other factors that can bias survey results, leading to bad data. After an initial discussion of how and why we take samples, and the use of surveys, students begin the first activity.

Next, in the *How you Ask a Question* activity, students respond to a set of three questions. What they do not know is that there are two different sets of these questions, worded in different ways. After the questions are answered, a show of hands is asked for the answers to each question, and it is clear that students have responded differently. One student is asked to read aloud their question 1, and then a student reads the other version of question 1. Students are then allowed to see the two sets of questions, and the data are summarized and compared for the two surveys. A discussion ensues on wording effects, how the wording of a question makes a difference in how people respond, the idea of bias in data, and different sources of bias.

This leads into the next activity *Critiquing the Student Survey*, where students read and critique the student survey they helped develop on the first day of class and determine if any of the questions was poorly worded and could lead to biased data. They suggest ways to improve question wording. (Note: they will take the revised survey online outside of class). Finally, students work in groups to discuss what kinds of questions might be posed and answered using the survey data, such as relationship and comparison type questions. They are encouraged to use the new statistical vocabulary they are learning as they talk about surveys, samples, populations, and related terms.

A wrap-up discussion summarizes what students have learned about the meaning and sources of bias in data. They begin to consider ideas of samples in taking surveys as a segue to the next day's topic on types of sampling methods.

Lesson 3: Random Sampling

Lesson 3 focuses on methods of taking samples: why they are important, how to take good samples, how samples differ from each other, and the importance of random sampling. Students take what they think is a representative sample using their judgment, and then compare this to a random sample. They see that nonrandom samples are usually biased. Student learning goals for this lesson include:

1. To understand reasons for using samples in statistical work.
2. To learn to use the basic vocabulary of sampling and surveys.
3. To understand why good samples are important and how we use samples to make inferences.
4. To understand why we rely on chance rather than our own judgment to pick a sample.
5. To learn how to take a Simple Random Sample (SRS) and why SRSs are so important.
6. To recognize and implement several kinds of probability samples (stratified random sample cluster sample, multistage sample, systematic sample).

Description of the Lesson

The lesson begins by asking students to suggest good ways to take a representative sample of five students from the class, and they discuss methods of obtaining fair and representative samples for surveys or research purposes. This leads into the *Gettysburg Address* activity. Students are shown the famous *Gettysburg Address* by Abraham Lincoln and told that statistics are often used in analysis of writing style and to identify authors of different writings. Their task is to take a good, representative sample of words from the Gettysburg Address. They do this, and then compute the average word length, being told that average word length is one statistical characteristics of a writer's style. A dot plot of the different sample averages is constructed and examined. The true population average word length is then compared to this plot, and typically, it is not anywhere the center of the graph.

Students then take Simple Random Samples of words from the Gettysburg Address, which is quickly and easily done using *Sampling Words* Java applet (http://www.rossmanchance.com/applets/index.html). They can plot distributions of sample averages and see that these samples are unbiased, and that the true population mean is in or near the center. They repeat this activity with a larger sample size and see the effect. A discussion of bias, representative samples, and the effect of sample size is followed by an activity (*Student Survey Sampling*) where students discuss how they would apply different sampling strategies to taking samples of data from students who have completed the Student Survey.

Lesson 4: Randomized Experiments

This final lesson in the Data unit involves carrying out a randomized experiment, and then considering what is a surprising result. After the randomized experiment is complete and data are gathered, students run a simulation so that they have a distribution of possible results under a chance model to compare with their sample, in order to judge if a particular result is likely to be due to chance, or is too surprising to be attributed to chance. Student learning goals for this lesson include:

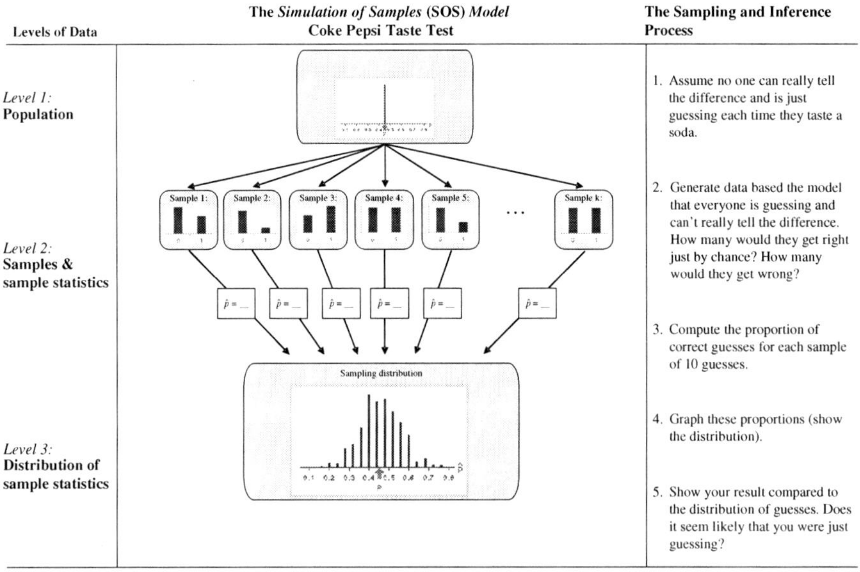

Fig. 6.3 The *Simulation of Samples (SOS) Model* for the taste test

1. To apply the characteristics of a well-defined experiment.
2. To experience the difference between an experiment and an observational study.
3. To learn to recognize instances of confounding in an experiment.
4. To learn the importance of randomization, in randomizing the assignment of treatments, and how that protects against confounding and makes cause and effect statements possible.
5. To develop an informal idea of statistical inference, as the extent to which a result is surprising given a certain claim or theory.
6. To become introduced to and familiar with the *Simulation of Samples (SOS)* Model (Fig. 6.3) as a way to represent data in a simulation.

Description of the Lesson

The lesson begins with the students being asked how many of them believe that they can correctly identify Coke and Pepsi in a blind taste test. They discuss how to determine if somebody really knows how to distinguish the two or if they are just guessing.

After a discussion about the characteristics of a good experiment and what is needed (randomization, control, and replication), students are given the details of the experiment to be performed (*Taste Test* activity). Those students who think they can identify the colas become the tasters. A group of students are assigned to pour the tastes in paper cups, with the order assigned by coin tosses. Another group of students are runners, who bring the tastes to the tasters, not knowing which is which.

The fourth group of students is recorders, who write down what the tasters think each taste is: Coke or Pepsi. The experiment has 10 trials: each taster has 10 blind tests of the soda. After the data are collected, the results are analyzed and each student receives a score for their total of correct identifications.

Next, the class discusses how high a score must be to believe that the student really was not just guessing. The *Sampling SIM* program is run to simulate what we could expect from 10 trails of this experiment if students really were guessing. They can then compare the results of the student guesses to this distribution, to see if the result is in the tails (surprising) and how far in the tails, or in the center (not surprising).

A visual model of the simulation process, adapted from Lane-Getaz (2006), the *Simulation of Samples* (SOS) Model is introduced (see Fig. 6.3). This model is used to help students understand and distinguish between the statistical model used to generate the simulation, the sample data generated, and the distribution of sample statistics for these samples (see Chapter 13 for more detail on this model). The *SOS Model* is also used to substrate the process of comparing students' experimental results to those generated by a particular model or theory (i.e., what if the student was just guessing).

Summary

The four sample lessons in the Data unit help students realize the importance of data and data collection methods in statistics. The class survey that students helped to design was used to collect data that will be analyzed in several subsequent lessons and introduces the importance of a multivariate data set. The ideas of sources of data, data collection methods, measurement issues, and variability of data will be revisited and emphasized again in many of the subsequent lessons. Finally, the importance of randomization and the use of simulation to make an inference about a surprising result are introduced.

Chapter 7
Learning to Reason About Statistical Models and Modeling

One of the most overworked words in statistics education and mathematics education is "model." Appearing in a variety of dissimilar contexts, its usage is at best unclear, and at worst, inappropriate.

(Graham, 2006, p. 194)

Snapshot of a Research-Based Activity

Class begins with a discussion about the "One-Son Policy" that was proposed for families in China to keep the birth rate down, but to allow each family to have a son. This policy allowed a family to keep having children until a son was born, at which point no more children were allowed. Students are asked to speculate about what would happen to the ratio of boys to girls if this policy was introduced and about what they would predict the average family size to be. Most students think that this policy would result in more boys or a higher ratio of boys to girls. Others think it might result in more girls, because a family might have several girls before they have a boy. Some students think that the average number of children might also increase under this policy.

After students make and explain their conjectures, they discuss how to *model* this process so that they can *simulate* data to estimate what would be the ratio of boys to girls and average family size if the One-Son Policy were implemented. First, students are introduced to the idea of making small tokens labeled "Boy" and "Girl" to model the problem. They are guided to put equal numbers of these tokens in a container, assuming boy and girl babies are equally likely. The students then draw from tokens, one at a time from the container with replacement, writing down the outcomes, for example, B, GB, B, GGGB, etc.

The students are asked to consider other ways that they might model this problem and generate data, without labeling tokens "Girl" and "Boy", and use a coin. A suggestion is made to simulate data by tossing coins, with a Head representing a Girl and a Tail representing a Boy. The students discuss their assumptions for this simulation, such as, assuming the coin tosses are equally likely to land Tails up or Heads up, and that the result is unpredictable, a *random outcome*. Preliminary data are gathered and examined, and students are surprised to see that the ratio of boys to girls is close to 1 and that the average family size is smaller than predicted. They predict what they think would happen if more data are gathered.

Students are then guided to use the *Fathom* software (Key Curriculum Press, 2006) to simulate larger data sets to answer this question. They discuss the appropriateness of the coins and *Fathom* simulations for modeling birth rates, and students bring in their own knowledge of boys being more likely to be born than girls, so that it is not exactly 50% boys and 50% girls. However, they note that the coin model was helpful in providing data that approximates real results and helps estimate an answer to the original questions about what would happen if the One-Son Policy were adopted. The students also see that it can be easier to simulate data by finding a useful *model* than to try to work out a complex probability problem.

Rationale for This Activity

The One-Son Modeling activity can take place on early in a class, even on the second day (after the first lesson plan for the topic of Data, from Chapter 6). We have introduced it this early in a course because of the importance of introducing the ideas of random outcome, model, and simulation. These ideas are interconnected and nicely illustrated in this first activity. The students can see that the results of a coin toss are random, but that repeated tosses yield predictable patterns (e.g., half Tails, half Heads). They see that coin tosses can be used as a model for birth rates, but that the model is not perfect, just useful. In other words, students see that statistical models can be useful for simulating data to answer real world questions, but they are not perfect fits to reality. They are also exposed to simulation of data as a way to examine chance phenomena, and this sets the stage for simulations as a helpful process that is used throughout the course. The research basis for this activity comes from some of the research on understanding models and modeling described in this chapter that suggests the need for students to create simple models for chance events, and to use chance devices such as coins or cards to simulate data before having a computer produce larger amounts of simulated data. We have added on the importance of using language about models and drawing students' attention to the fact that what they are doing is creating a model to represent a problem and using the model to produce data to solve the problem.

The Importance of Understanding Statistical Models and Modeling

> It is the job of statisticians to represent the data taken from the real world with theoretical models.
>
> (Graham, 2006, p. 204)

Statisticians use models in different ways, and some of these uses appear in introductory statistics courses. Two main uses of statistical models are (see schematic illustration in Fig. 7.1):

1. Select or design and use appropriate *models* to simulate data to answer a research question. Sometimes, the model is as simple as a random device, sometimes it takes the form of a statement (such as a null hypothesis) that is used to generate data to determine if a particular sample result would be surprising if due to chance. As George Cobb states: Reject any model that puts your data (the investigated sample) in its tail (2005). Sometime a data set is used to simulate (bootstrap) more data, creating a simulated population distribution to use in making statistical inferences.
2. Fit a statistical model to existing data or data that you have collected through survey or experiment in order to explain and describe the variability. This may be as simple as fitting the normal distribution to a data set, particularly when checking assumptions when using a particular procedure. It may involve fitting a linear model to a bivariate data set to help describe and explain a relationship between two variables. In advanced courses, data modeling is an essential technique used to explore relationships between multiple variables. In all of these cases, we examine the fit of a model to data by looking at deviations of the data from the model.

Figure 7.1 shows the commonalities and differences between these two uses of statistical models as well as their role in the ongoing cycle of statistical investigative work.

Why should students in introductory statistics class need to learn about statistical models? Because models are a foundational part of statistical thinking, working with models is a big element of the real work of statisticians. It is surprising; therefore, that little explicit attention is paid to the use of models in most introductory courses. The words "model" or "modeling" hardly appear in most introductory statistical textbooks. When these words are used, it is often in the context of a linear model for bivariate data or the normal distribution as a mathematical model. Sometimes, the term is used in connection with probability, as in probability models. Although many statisticians talk about the importance of data modeling and fitting models to data, most students can take an introductory statistics course and never understand what a statistical model is or how it is used.

The models statisticians use are actually mathematical models. David Moore (1990) describes the role of mathematical models in the data analysis process: "Move from graphic display to numerical measures of specific aspects of the data to compact mathematical models for the overall pattern" (Page 104). The two most commonly used mathematical models in an introductory statistics class are the normal distribution and the linear regression models. Although these two topics involve the use of mathematical models that are fitted to sample data, students rarely see these topics as related, or as examples of two different kinds of statistical models that help us analyze data.

The normal curve, which is perfectly symmetric and does not reflect the irregularities of a data set, is an idealized model that nicely fits many distributions of real data, such as measurements and test scores. Moore (1990) comments that moving from observations of data to an idealized description (model) is a substantial

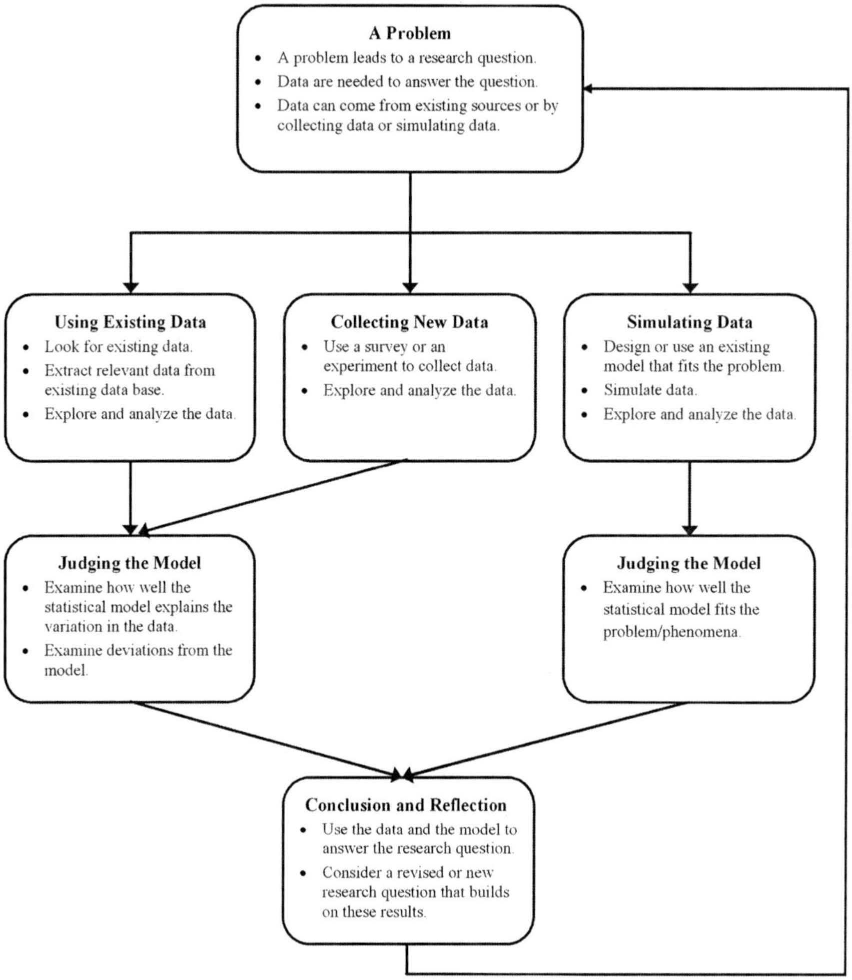

Fig. 7.1 Uses of models in statistical analysis

abstraction, and the use of models such as the normal distribution and the uniform distribution is a major step in understanding the power of statistics. Some software packages, such as *Fathom*, allow students to superimpose a model of the normal distribution on a data set. This feature helps students judge the degree of how well the model fits the data and develops students' understanding of the model fitting process.

The Place of Statistical Models and Modeling in the Curriculum

We find the lack of explicit attention to statistical models (other than probability models, in a mathematical statistics class) surprising. While there are examples of ways to model probability problems using concrete materials and or simulation tools (see Simon, 1994; Konold, 1994b), these do not appear to be part of most introductory statistics classes, and do not appear to be part of an introduction to the use of models and modeling in statistics. We have thought carefully about how to incorporate lessons on statistical models and modeling into an introductory course. Rather than treat this topic as a separate unit, we think that activities that help students develop an understanding of the idea and uses of a statistical model should be embedded throughout the course, with connections made between these activities.

The One-Son Modeling activity described earlier is a good way to introduce the related ideas of model, random outcome, and simulation. In the following unit on data (see Lesson 4, Chapter 6), we revisit the idea of modeling and simulation after students conduct a taste test and want to compare their results to what they would expect due to chance or guessing (the null model). The normal distribution is informally introduced as a model in the unit on distribution (Chapter 8) and revisited in the units on center (Chapter 9), variability (Chapter 10), and comparing groups (Chapter 11). After completing the topics in data analysis, the topic of probability distribution can be examined as a type of distribution based on a model. The normal distribution is then introduced as a formal statistical model (probability distribution) and as a precursor to the sampling unit (this activity is described in the end of this chapter). The sampling unit (Chapter 12) revisits the normal distribution as a model for sampling distributions. In the unit on statistical inference (Chapter 13), models are used to simulate data to test hypotheses and generate confidence intervals. Here a model is a theoretical population with specified parameters. Statistical models are used to find P-values if necessary conditions are met. The final model introduced after the unit on statistical inference is the regression line as a model of a linear relationship between two quantitative variables (see Chapter 14). This model is also tested by using methods of statistical inference and examining deviations (residuals). We find that the idea and use of a statistical model is explicitly linked to ideas of probability and often to the process of simulation. Therefore, we briefly discuss these related topics as well in this chapter.

Review of the Literature Related to Reasoning About Statistical Models and Modeling

> All models are wrong, but some are useful.
>
> (George Box, 1979, p. 202)

Models in Mathematics Education

Several researchers in mathematics education have applied mathematical modeling ideas to data analysis (e.g., Horvath & Lehrer, 1998). Lehrer and Schauble (2004) tracked the development of student thinking about natural variation as elementary grade students learned about distribution in the context of modeling plant growth at the population level. They found that the data-modeling approach assisted children in coordinating their understanding of particular cases with an evolving notion of data as an aggregate of cases. In another study by the same researchers, four forms of models and related "modeling practices" were identified that relate to developing model-based reasoning in young students (Lehrer & Schauble, 2000). They found that studying students' data modeling, in the sense of the inquiry cycle, provided feedback about student thinking that can guide teaching decisions, an important dividend for improving professional practice.

A related instructional design heuristic called "emergent modeling" is discussed by Gravemeijer (2002) that provides an instructional sequence on data analysis as an example. The "emergent modeling" approach was an alternative to instructional approaches that focus on teaching ready-made representations. Within the "emergent modeling" perspective, the model and the situation modeled are mutually constituted in the course of modeling activity. This gives the label "emergent" a dual meaning. It refers to both the process by which models emerge and the process by which these models support the emergence of more formal mathematical knowledge.

Models in Statistical Thinking

> Statisticians ... have a choice of whether to access their data from the real world or from a model of the real world.
>
> (Graham, 2006, p. 204)

How students understand and reason about models and modeling processes has received surprisingly little attention in statistics education literature. This is surprising since statistical models play an important part in statistical thinking. The quote by Box, "All models are wrong, but some are useful" (1979, p. 202), is a guiding principle in formulating and interpreting statistical models, acknowledging that they are ideal and rarely match precisely real life data. The usefulness of a statistical model is dependent on the extent that a model is helpful in explaining the variability in the data.

Statistical models have an important role in the foundations of statistical thinking. This is evident in a study of practicing statisticians' ways of thinking (Wild & Pfannkuch, 1999). In their proposed four-dimensional framework for statistical thinking, "reasoning with statistical models" is considered as a general type of thinking, as well as specific "statistical" type of thinking, which relates, for example, to measuring and modeling variability for the purpose of prediction, explanation, or control. The predominant statistical models are those developed for the analysis of data.

While the term "statistical models" is often interpreted as meaning regression models or time-series models, Wild and Pfannkuch (1999) consider even much simpler tools such as statistical graphs as statistical models since they are statistical ways of representing and thinking about reality. These models enable us to summarize data in multiple ways depending on the nature of the data. For example, graphs, centers, spreads, clusters, outliers, residuals, confidence intervals, and P-values are read, interpreted, and reasoned with in an attempt to find evidence on which to base a judgment.

Moore (1999) describes the role of models to describe a pattern in data analysis as the final step in a four-stage process.

> When you first examine a set of data, (1) begin by graphing the data and interpreting what you see; (2) look for overall patterns and for striking deviations from those patterns, and seek explanations in the problem context; (3) based on examination of the data, choose appropriate numerical descriptions of specific aspects; (4) if the overall pattern is sufficiently regular, seek a compact mathematical model for that pattern (p. 251).

Mallows (1998) claims that too often students studying statistics start from a particular model, assuming the model is correct, rather than learning to choose and fit models to data. Wild and Pfannkuch (1999) add that we do not teach enough of the mapping between the context and the models. Chance (2002) points out that, particularly, in courses for beginning students, these issues are quite relevant and often more of interest to the student, and the "natural inclination to question studies should be rewarded and further developed."

Reasoning About a Statistical Model: Normal Distribution

There is little research investigating students' understanding of the normal distribution, and most of these studies examine isolated aspects in the understanding of this concept. The first pioneering work was carried out by Piaget and Inhelder (1951, 1975), who studied children's spontaneous development of the idea of stochastic convergence. The authors analyzed children's perception of the progressive regularity in the pattern of sand falling through a small hole (in the Galton apparatus or in a sand clock). They considered that children need to grasp the symmetry of all the possible sand paths falling through the hole, the probability equivalence between the symmetrical trajectory, the spread and the role of replication, before they are able to predict the final regularity that produces a bell-shaped (normal) distribution. This understanding takes place in the "formal operations" stage (13- to 14-year-olds).

In a study of college students' conceptions about normal standard scores, Huck, Cross, and Clark (1986) identified two misconceptions: On the one hand, some students believe that all standard scores will always range between −3 and +3, while other students think there is no restriction on the maximum and minimum values in these scores. Others have examined people's behavior when solving problems involving the normal distribution (Wilensky, 1995, 1997). In interviews with students and professionals with statistical knowledge, Wilensky asked them to solve a problem by using computer simulation. Although most subjects in his research could solve problems related to the normal distribution, they were unable to justify the use of the normal distribution instead of another concept or distribution, and showed a high "epistemological anxiety," the feeling of confusion and indecision that students experience when faced with the different paths for solving a problem.

In recent empirical research on understanding the normal distribution, Batanero, Tauber, and Sánchez (2004) studied students' reasoning about the normal distribution in a university-level introductory computer-assisted course. While the analysis suggests that many students were able to correctly identify several elements in the meaning of normal distribution and to relate one to another, numerous difficulties understanding normal distributions were identified and described. The main conclusion in this study is that the normal distribution is a very complex idea that requires the integration and relation of many different statistical concepts and ideas. The authors recommend the use of appropriate activities and computer tools to facilitate the learning of basic notions about normal distributions (see causeweb.org for some of these resources).

Understanding Ideas Related to Probability Models

The research reviewed in Chapter 2 along with literature reviews by Falk and Konold (1998); Shaughnessy (2003); Jones (2005) and others illustrate the conceptual difficulties students have in understanding basic ideas of probability such as randomness. For example, Falk & Konold, (1994, 1997) found that people attempting to generate random number sequences usually produce more alternations of heads and tails than expected by chance. A related research result is that students tend to think that all models of random events are ones with equally likely outcomes. Lecoutre (1992) refers to this misconception as the "equiprobability bias," which she found students to use in solving different types of probability problems.

Reasoning About the Use of Models to Simulate Data

Studies on the use of simulations cover many different topics, such as how students and teachers understand statistical models in simulating data (e.g., Sánchez, 2002), the use of simulation to illustrate abstract concepts such as sampling distributions (e.g., Saldanha & Thompson, 2002) and learning to formulate and evaluate inferences by simulating data (e.g., Stohl & Tarr, 2002). While there is a strong belief that physical simulations should precede computer simulations, this has not yet been a topic of empirical study. Helpful guidelines for secondary school students on how to select models to simulate problems are included in the Art and Techniques of Simulation – a volume in the Quantitative Literacy Series (Gnanadesikan, Scheaffer, & Swift, 1987).

Biehler (1991) has presented an extensive analysis of the capabilities and limitations of simulation techniques in teaching statistics; he points out that "the different roles, goals and pedagogical perspectives for simulations have not yet been clearly analyzed and distinguished." He suggests a basic distinction between "the use of simulating as a method for solving problems, similar to the professional use outside school and the use of simulation to provide model environments to explore, which compensate for the 'lack of experience" (p. 183).

To study teachers' opinions about the instructional use of models in simulating data, Sánchez (2002) interviewed six high school teachers who participated in a workshop of simulation activities using *Fathom*. The analysis of their responses included four general aspects: the role of simulation in teaching; the different steps to follow in a simulation; the complexity of starting situations; and the statistical concepts that take part in simulation activities. The results show that teachers deem as important only certain aspects of simulation, but neglect the fundamental concepts of randomness and distribution. Sánchez noted that the teachers have centered their attention in certain modeling aspects like formulation of a model and its simulation, but neglected other aspects like the analysis of results and the validation of the model.

Implications of the Research: Teaching Students to Reason About Statistical Models and Modeling

Other than some literature on reasoning about the normal distribution or a linear relationship in bivariate data, there is little research illuminating how students come to learn and use statistical models. Therefore, we are speculative in putting together a research-based sequence of activities for this topic.

The literature reviewed implies that it is important to make models an explicit topic in the introductory statistics course, and to help students develop this idea and its multiple meanings and uses through experiences with real statistical problems and data, rather than through a formal study of probability. The literature also suggests that understanding the idea of randomness is difficult and that carefully designed activities should be used to help students understand the ideas of random outcomes, random variables, random sampling, and randomization.

There are also implications from the literature about the role of probability in an introductory college statistics course. It is suggested that in order for students to develop a basic understanding of statistics in a first college course, they only need the basic ideas of probability that were introduced above. As Garfield and Ahlgren wrote in 1988 "useful ideas of inference can be taught independently of technically correct probability." Therefore, we propose helping students understand the ideas of random outcomes and a probability distribution. They do not need to learn the language and laws of probability in such a course. Indeed, research shows that even if students encounter these topics in probability in an introductory statistics course, few students understand and can reason about this topic (see reviews by Garfield & Ahlgren, 1988; Hawkins & Kapadia, 1984; Shaughnessy, 2003).

We agree with David Moore's (1997) claim that mathematical probability is a "noble and useful subject" and should be part of the advanced coursework in statistics, and instead of being part of the introductory course, should be learned in a separate course that is devoted to this topic. The literature also implies that in order to develop a deep understanding of the basic ideas outlined in Chapter 3, the traditional course should be streamlined. Following the lead of Moore (1997), our candidate for the guillotine is formal probability. Moore recommends that an informal introduction to probability is all that is needed and that this begins with experience with chance behavior, usually starting with physical devices and moving to computer simulations that help demonstrate the fundamental ideas such as the Law of Large Numbers.

Technological Tools to Help Students Develop Reasoning About Models

Many Web applets are available to help students see and use the normal distribution (e.g., http://www.rossmanchance.com/applets/NormalCalcs/ NormalCalculations.html) and fit a line to data (e.g., http://www.math.csusb.edu/faculty/stanton/m262/regress/regress.html). Applets can also be used to simulate data for a probability problem (see rossmanchance.com). The simulation tool *Probability Explorer* (Stohl, 1999–2005; http://www.probexplorer.com/) enables school students and teachers to design, simulate, and analyze a variety of probabilistic situations. Model Chance is a new program that is now being developed as part of the *TinkerPlots* project, to help student model problems and simulate data to estimate answers to these problems (http://cts.stat.ucla.edu/projects/info.php?id=4).

Progression of Ideas: Teaching Students to Reason About Statistical Models and Modeling

Introduction to a Sequence of Activities to Develop Reasoning about Statistical Models and Modeling

While most textbooks do not introduce the term "model" until the normal distribution and may not use it again until regression, we believe that the term should be introduced early in a course and used frequently, to demonstrate and explain how models are used in statistical work. Since there is no empirical research on an optimal approach for helping students develop the idea of model in an introductory statistics class, we offer one of several possible sequences of activities that we have found useful in our courses. These activities develop the idea of a model to simulate data, and the idea of a model to fit to a given set of data.

We believe that the idea of a model can be presented informally in the first few days of class using a fairly simple context. Physical simulations using devices such as coins can be used to model a random variable as part of solving a problem, and once data have been simulated using coin tosses, the computer can be used to quickly produce large amounts of simulated data. It is important that students be made aware of the use of a model, to represent key features of the event and to produce simulated data.

In the Taste Test activity in the Data topic (see Chapter 6), when the notion of an experiment is introduced, the idea of a null model (what would happen if results were only due to chance) can be reintroduced. In this context, the model is used to simulate data to informally determine a P-value, indicating whether or not an individual's identification of soda brand in a blind taste test may not just be due to chance. The formal idea of model may be encountered in the context of random variables. Students can model random outcomes such as coin tosses, dice throwing, and drawing of cards. Each time students can describe the model, e.g., for a standard deck of Poker cards, where the outcome is a Heart, this can be modeled as a binomial variable with a probability of $1/4$. These models can be used to simulate data, which leads to examining probability distributions. Different probability problems can also be modeled first using coins or dice, and then using applets or software to simulate data. In this case, the problem is modeled by the coins or dice, and then this model is replicated by a technology tool. After generating empirical probability distributions for binomial random variables, the probability distribution for the normal distribution can be introduced as a model that is often used to describe and interpret real data. This model is seen again as students begin to examine distributions of sample means and the Central Limit Theorem. Null models are encountered again in the unit of inference to generate distributions of sample statistics to run tests of significance. In our proposed sequence of activities, the final model presented is that of the linear model, used in the unit on covariation, to model the relationship of two quantitative variables.

What is unique about the lessons in this chapter is that unlike all the other chapters in part II of this book, we are not offering a unit on models but rather providing several activities that illustrate and use the idea of statistical models throughout an introductory course. While the informal ideas of model are introduced at the beginning of a course, the formal ideas are encountered midway through the course, and then revisited at additional times in other topic areas, rather than in a set of sequential lessons in a unit of their own. Table 7.1 shows a suggested series of ideas and activities that can be used to guide the development of students' reasoning about models and modeling.

Introduction to the Lessons

Three lessons are designed to help students develop an understanding of the importance and use of statistical models. Other activities involving models are interspersed throughout other topics in this book. The first lesson has students model birth outcomes in a situation where there is a "One-Son Policy." A probability model is used to simulate data that is summarized to answer some research questions. The second activity in this lesson has students use first a physical and then a computer simulation to model the "Let's Make a Deal" game, using the simulated data to find the best strategy for playing the game. The second lesson has students create binomial models for different random devices (coins, dice, and cards) and use these

Table 7.1 Sequence of activities to develop reasoning about statistical models and modeling[1]

Milestones: ideas and concepts	Suggested activities
Informal ideas prior to formal study of statistical models	
• Models can be used to portray simple random outcomes. Random devices and computers can be used to simulate data to answer a question about this context	• One-Son Modeling Activity (Lesson 1: "Using Models to Simulate Data")
• A random outcome, unpredictable, but giving a predictable pattern over the long run. The more data there is, the more stable is the pattern	• Let's Make a Deal Simulation (Lesson 1)
• Designing and using a model can help to answer a statistical question	• Let's Make a Deal Simulation (Lesson 1)
• The idea and importance of random samples (revealing the predictable pattern of random outcomes)	• The Gettysburg Address Activity (Lesson 3, Data Unit, Chapter 6)
• Models can be used to generate data to informally test an experimental result to provide evidence about whether or not this result is due to chance	• Taste Test Activity (Lesson 4, Data Unit, Chapter 6)
• Distinguish between the model, the simulated data, and the sample data	• Taste Test Activity (Lesson 4, Data Unit, Chapter 6)
• The normal distribution as a model for some distributions of real world data	• Sorting Histograms Activity (Lesson 2, Distribution Unit, Chapter 8)
• The mean is a good summary of the center of a normal distribution	• Choosing an Appropriate Measure of Center Activity (Lesson 2, Center Unit, Chapter 9)
• The mean and standard deviation are good summaries for a normal distribution	• How do Students Spend their Time Activity (Lesson 4, Comparing Groups Unit, Chapter 11)
Formal ideas of statistical models	
• Random variables and random outcomes	• Coins, Cards, and Dice Activity (Lesson 2: "Modeling Random Variables")
• Equally likely model does not fit all random outcomes	• Coins, Cards, and Dice Activity (Lesson 2)
• A probability distribution as a model	• Coins, Cards, and Dice Activity (Lesson 2)
• Probability problems can be modeled using random devices and simulation tools	❖ Activity where cards are used to model a problem, such as Random Babies activity in Chance and Rossman (2006) (The symbol ❖ indicates that this activity is not included in these lessons.)

[1] See page 391 for credit and reference to authors of activities on which these activities are based.

Table 7.1 (continued)

• Characteristics of normal distribution as a model	• What is Normal? (Lesson 3: "The Normal Distribution as a Model")
• What does normal data look like?	• What is Normal? (Lesson 3)
• Using the normal distribution as a Model	• Normal Distribution Applications (Lesson 3)
Building on formal ideas of models in subsequent topics	
• How and why the sampling distribution of means can be modeled by the normal distribution	• Central Limit Theorem Activity (Lesson 3, Samples and Sampling Unit, Chapter 12)
• The null hypothesis as model to which we compare sample data	• Balancing Coins Activity (Lesson 1, Statistical Inference Unit, Chapter 13)
• When testing a hypothesis, it is often important to check the condition of normality of the sampling distribution	• Research Questions Involving Statistical Methods (Lesson 5, Statistical Inference Unit, Chapter 13)
• The regression line is a useful model of bivariate relationships between quantitative variables	• Diamond Rings Activity (Lesson 2, Covariation Unit, Chapter 14)
• Checking the fit of a model to data, by examining residuals from a regression line	• da Vinci and Body Measurements Activity (Lesson 2, Covariation Unit, Chapter 14)

to generate and then simulate data (on the computer) to examine and compare probability distributions (each having a different shape and expected value). The third lesson introduces the normal distribution as a model of a probability distribution. This model is used to fit to samples of data (e.g., do the data sets appear to have a normal distribution?) and to demonstrate when it is appropriate to use this model in analyzing data.

Lesson 1: Choosing Models to Simulate Data

The first lesson begins with the *One-Son Modeling* activity described at the beginning of this chapter. Next, students consider the chances of wining on the game show *Let's Make a Deal* to determine whether one strategy has a higher chance of wining than another, modeling the game first with cards and then with a Web applet. The student learning goals for this lesson include:

1. Be able to use simulation as a tool for answering statistical questions
2. Be able to develop models to simulate data
3. Examine probability as an indication of how likely is an event to happen.
4. Realize that their intuitions about probabilities may be misleading.

Description of the Lesson

The students are told about the "One-Son policy" that was proposed for China as a way to decrease their birthrate, but still allow families to produce a son.

> In the early 1990's China considered adopting a "One-Son Policy", to help reduce their birthrate by allowing families to keep having children until they had a son. Under this plan a family has a child. If it is a son, they stop having children. If it is a daughter, they can try again. They can keep trying until they have a son and then they stop having children.

The following questions are posed to students to engage them in reasoning:

1. *If a country adopted a policy that let families have children only until they had a boy, and then they had to stop, what would you expect to happen?*
2. *What would the average number of children per family be? Would you expect more boys or more girls overall?*

Students present their answers and reasoning. Simulation is introduced as a tool statisticians use to generate data to estimate an answer to this type of question.

Working in pairs, students take a yogurt container that contains two slips of paper. One is labeled "B" for boys. One is labeled "G" for Girls. They randomly draw one slip of paper from the container and that will be the first child born in a simulated family. If they draw a "B", then they are told to stop; the family is done having children. They enter the data on a chart indicating the result for the first family. If the result is a "G", they draw again. They keep drawing until they draw a "B", then they stop and enter the data on the chart for that family. Students repeat this process for five simulated families, getting results such as: GB, B, GGB, B, GB, etc.

After this first round of collecting simulated data, the number of children in each family is counted along with the number of girls and the number of boys. Students examine this small set of data and are asked if the results confirm their original predictions. Next, they consider what they would get if they did not have slips of paper, but instead only had a coin. Students figure out how to do this same simulation using only one coin that is tossed. They discuss and write out the process so that another group could run the simulation by following their directions. Students usually determine that one outcome (heads) will represent a Boy and the other outcome will represent a Girl. They describe how to repeatedly toss a coin until it lands heads, recording the data each time to simulate families. Next they simulate this data using the coin, generating data for five more families.

Now student groups have data for 10 simulated families based on the two sets of simulations. They tally the total number of girls, the total number of boys, find the ratio of boys to girls, and find the average number of children per family. Again, they compare these results to their initial conjectures. These results are also shared with the class and graphed on the board.

Next, students are asked what they would expect to find if they repeated this simulation many more times. They are shown how to use the *Fathom* to run this simulation and to gather data for more simulated families, as shown in Fig. 7.2.

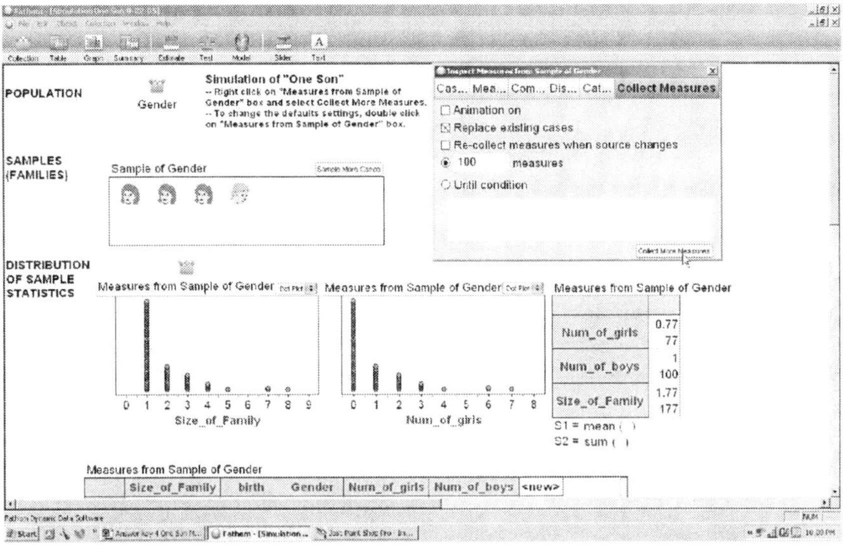

Fig. 7.2 Running the One-Son simulation in *Fathom*

A discussion of these data reveals that the results tend to stabilize as more data are collected. Students are asked what they would expect if more data were generated, and this leads to an informal discussion of the Law of Large Numbers. Next, a discussion of choosing models to simulate data leads students to consider real life factors to consider using a coin or an equally likely binomial model to simulate births. They suggest ways in which the model of equally likely outcomes may not perfectly fit the data, such as the fact that the probability of having a boy is actually more than .5. They consider how these factors could affect simulated results. A wrap-up discussion includes issues involved in selecting and using models (first the coin, and then the computer simulation of a binomial model with $p = .5$) to simulate a real world phenomenon (birth rates of boys and girls).

Students are next introduced to an activity (*Let's Make a Deal Simulation*), where they will check their intuitions about chance events by using simulation to determine probabilities in a game show setting. *Let's Make A Deal* is introduced as a popular TV game show from the 1970s.

> The task: *Suppose you are given three doors to choose from. Behind one door there is a big prize (a car) and behind the other two, there are goats. Only Monty Hall knows which door has the prize. You are asked to select a door, and then Monty opens a different door, showing you a goat behind it. Then you are asked the big question: Do you want to stay with your original door, or switch to a different door? What would you do?*

Students discuss in small groups whether they think there is a higher chance of winning the prize if they stay with their first door selection or should they switch to the remaining door and why. A show of hands in the class indicates that most students think that either the contestant should stay with the first choice or that it does not make a difference; staying or switching are equally likely to result in winning a prize.

Next, students think about how they could test these conjecture. They design a simulated game using index cards. Each group is given three index cards, and they write lightly on each one of the following outcomes: "Goat", "Goat", and "Car". One person is the game host (Monty Hall) and knows where the car is when the cards placed face down in order. The other group members take turns being contestants. A recording sheet is used to write down what strategy was used each time (stay or switch doors) and what the outcome was ("Win" or "Lose"). After a few rounds, the data are pooled for the class, and it appears that "switch" is resulting in more wins. But is that just due to chance or a stable result? Students then use a Web applet that simulates this problem to quickly simulate large amounts of data, revealing that the chance of winning when they switch doors is about 2/3, and the chance of winning when they stay with their first choice about 1/3. Students are asked about the correctness of their original intuitions and why they were incorrect. A quick explanation may be given about why this happens, or students may be given a written explanation about why it pays to switch doors when playing this game.

The *Let's Make a Deal Simulation* activity concludes with a discussion about the use of models and simulations to easily generate data to estimate answers to statistical questions, and when to trust results of simulations (e.g., when the model is a good fit to the problem and when there are enough simulations to generate a stable result).

Lesson 2: Modeling Random Variables

This lesson engages students with modeling random variables using coins, cards, and dice. Students construct probability distributions to represent each of the scenarios and make predictions on probabilities of other events based on these histograms. They first generate data with concrete materials and then move to *Fathom* to simulate larger data sets and see the stable trend emerge. Student learning goals for this lesson include:

1. Understand use of models to represent random variables and simulate data.
2. Understand how to interpret visual representations for probability.
3. Use a simulation to generate data to estimate probabilities
4. Gain an informal understanding of probability distributions as a distribution with shape, center, and spread.

Description of the Lesson

Students begin by considering different random devices they have encountered, such as coins, cards, and dice (the *Coins, Cards, and Dice* activity). Then they make conjectures about the expected number of:

- *Heads on five tosses of a fair coin?*
- *Hearts in five draws from a poker deck?*
- *Twos in five rolls of a fair die?*

They then reason about what is the same and what is different about these three experiments. Next, students are led to define three random variables as follows:

> X: Number of heads in five tosses
> Y: Number of hearts in five cards dealt
> Z: Number of 2's in five rolls

They now create models of each variable to generate (by using the actual devices) and then simulate data using the computer, which can be graphed and summarized to compare the distributions for each random variable.

Students begin with the Coin variable. They toss a coin five times and count the number of heads and then repeat this 10 times, making a frequency distribution for the number of heads that show up on each toss of five coins. Each time they toss the five coins, they check the value of X (Number of heads in five tosses) in the table, then find the relative frequency probability for each value of X. Next, students open a *Fathom* file that generates data based on this model of equally likely outcomes as shown in Fig. 7.3.

Students are asked how the simulated data compare to the data they generated by physically tossing coins, and if they expect these results to be similar for all students in the class. This leads to a discussion of the idea of two aspects of a random outcome: (1) that an individual outcome is unpredictable but (2) that you can predict patterns of outcomes for many repetitions, such as the proportion of

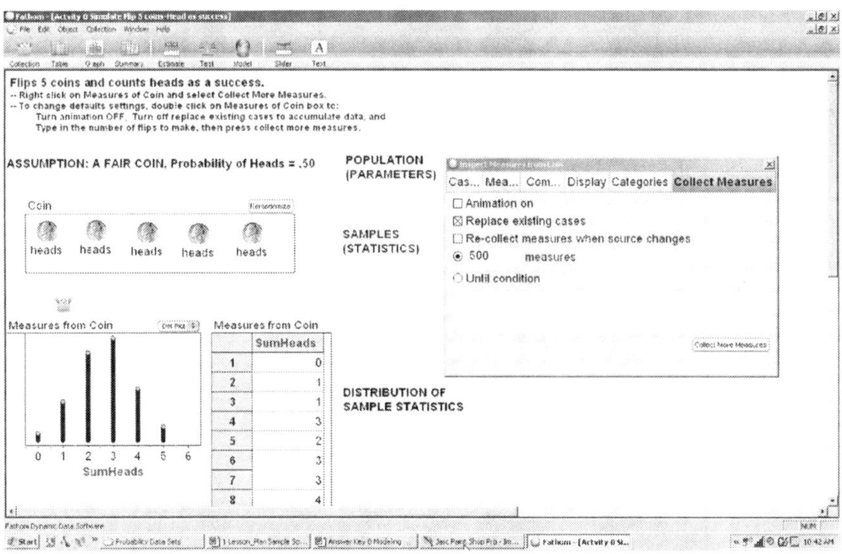

Fig. 7.3 Simulating the number of heads in five tosses of a fair coin in *Fathom*

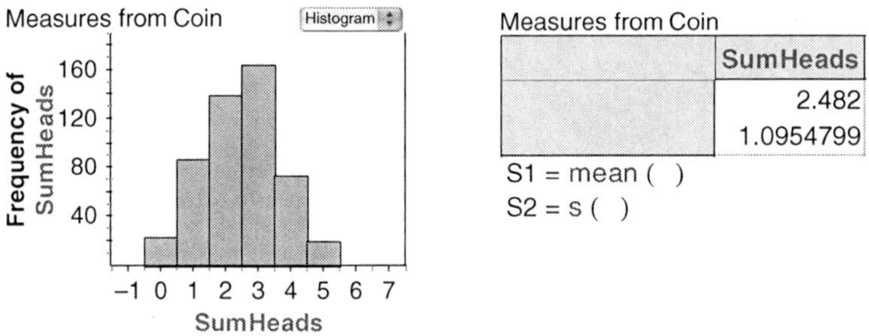

Fig. 7.4 The results of simulating the number of heads in five tosses of a fair coin in *Fathom*

heads for many tosses of a fair coin. The simulation is repeated for 500 trials, and relative frequencies are found for the six possible outcomes (0–5 heads). Students again discuss whether they think the results will vary from student to student and see that the larger sample size has much less variability. They graph the data and informally describe shape, center, and spread. They see that the expected number of heads for five tosses would be 2–3 heads, but that results can vary from 0 to 5 heads. They can generate the mean and standard deviation and interpret these values for the distribution (see Fig. 7.4).

The next part of the *Coins, Cards, and Dice* activity has students take a deck of cards, shuffle, draw a card, replace it, draw, and replace, five times. They count the number of hearts that showed up in their sample of five draws (with replacement). A discussion of the term "replacement" challenges students to reason about why they would replace the card drawn each time and how that would affect the results of the experiment so it is not similar to the coin tosses. This is an informal introduction to the idea of *independent events* without going into probability theory. Next, students use *Fathom* to simulate this experiment. They discuss what the model will be, that it is no longer one of equally likely outcomes, but that the chance of getting one heart when randomly drawn from a deck is now one-fourth. The students run a new simulation based on a binomial model with $p = .25$, graph the results, and find measures of center and spread, which are different from those in the Coin example (Fig. 7.5).

This activity is repeated a third time for the Dice example. First, students roll a dice five times counting the number of 2s that show up. They record the data and compare it with the class. Then they simulate the data on *Fathom*, first discussing how the model must be changed, based on the new probability of getting a two, which is one-sixth.

The results of the three experiments are compared. Students consider what was the same and what was different across the three experiments and how the differences in the experiments reveal themselves in the histograms of data. The expected value is also compared for each experiment, as well as where the same value (e.g., 4) appears in each histogram. This can lead to a discussion on how likely this outcome is (or is not) for each experiment, based on whether it is in the tail of the distribution, a precursor of the P-value concept.

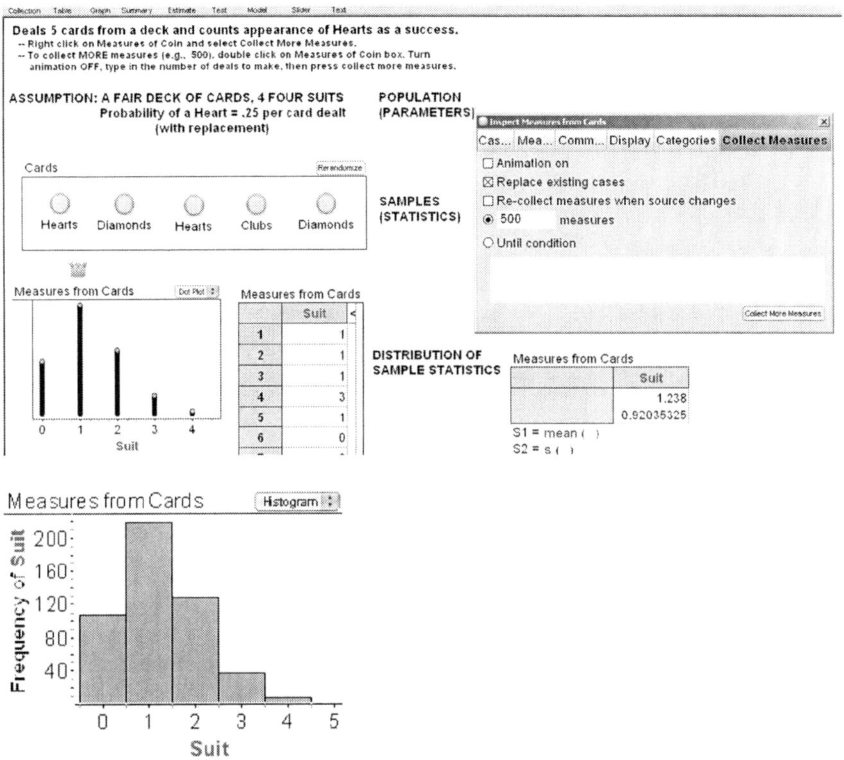

Fig. 7.5 The *Fathom* simulation of the number of hearts in cards

Lesson 3: The Normal Distribution as a Model

This lesson introduces students to the formal model of the normal distribution. The characteristics are examined, as students make and test conjectures about whether data gathered on different variables have a normal distribution. The unique characteristics of the mean and standard deviation in a normal distribution are used to examine percentages of data within one, two, and three standard deviations of the mean. The idea of the standard normal z-score is introduced and used to locate different areas of the distribution, using a Web applet. Student learning goals for this lesson include:

1. Understand and reason about the normal (and standard normal) distribution as a *model*
2. Understand and reason about the important characteristics of this distribution, e.g., the percentages of data within 1, 2 and 3 standard deviations of the mean.
3. Use standard deviations and z-scores to measure variation from the mean.

Description of the Lesson

Students are asked what they think when they hear the word "normal" and how they could tell if something is not normal. They contrast a normal data value (e.g., body temp of 98.6 F) in different contexts. They discuss how they think statisticians use this word and how their use is different from everyday use. Note, they have used the term *normal curve* informally in the unit on distribution (Chapter 8) when describing the shape of a data set.

In the *What is Normal* activity, students consider again the body measurements gathered earlier in the course (for which some may show a symmetric, bell-shaped pattern and others do not).

- Height
- Hand span
- Hand length
- Kneeling height
- Arm span
- Head circumference

The students make conjectures about which of these variables would have data sets that when graphed appear to have a normal distribution and why they predict those variables to have a normal distribution. They consider how to test their conjectures, and use the computer to generate graphs in *Fathom*. Next they select one distribution that looks approximately normal, to draw a picture of the graph on their handout and label the axes, the mean, and two standard deviations in each direction. Then they mark their own data value for this measure (e.g., their height) on the graph. They describe the location of their value in the overall graph as follows: *Are you close to center? In the tails? An outlier?* Next students find the z-score for their body measurement for that variable and explain what this z-score tells about the location of their body measurement relative to the class mean.

Students are instructed to open a Web applet that gives areas under the curve for a normal distribution. They use this applet to find the proportion of the distribution that is *more* than their value (e.g., what is the proportion of the curve representing values higher than their height) and then *less* than their value. Students discuss whether the obtained results from the applet makes sense to them and why or why not.

Next, students find the value that is one standard deviation above the mean and one standard deviation below the mean. They use the Applet to find the proportion of the distribution that is between these two values, and then repeat this for two and three standard deviations above and below the mean.

A class discussion focuses on which of the body measurements seemed to be normal and how they can tell how well the *normal model* fits a data set. Statisticians fit models to data, and this is illustrated by drawing a curve over the plot of a data set. Students consider and discuss how good a fit there is of the model to the data.

Next, students re-examine the use of the Web applet. They see that when they found the proportions of the distribution that were above and below their own data value, it was from a normal distribution that has the class mean and standard deviation as the class data. They used the model of the normal curve to estimate proportions. Students see that it depends on how well the data fit the model and that when they use z-scores to find percentages (probabilities) using the normal curve, they are using a statistical model. The use of models to explain, describe, estimate, or predict is revisited, recalling the earlier use of models to simulate births of boys and girls.

In the *Normal Distribution Applications* activity, students apply the use of intervals used with the normal distribution (i.e., the middle 68, 95, 99.7%, referred to as the Empirical Rule) to real data sets. They explore and discuss when it is appropriate to use this rule (a model) to describe or make inferences about a particular set of data. They explore when and how to use the model to solve the problems in context. A last part of the activity is to see a *Fathom* demo on "What are Normal Data?"

In a final wrap-up discussion, students are given this popular quote by Box, "all models are wrong, but some are useful" (1979, p. 202) and discuss how this quotation applies to the normal distribution as a model.

Summary

Models are one of the most important and yet least understood ideas in an introductory statistics course. This chapter has tried to make the case that the idea of statistical model should be made explicit and used repeatedly in an introductory statistics course, so that students become familiar with the importance of models and modeling in statistical work. We believe that ideas of probability are best introduced in this context, without having to go into the formal rules and vocabulary that are better saved for a course in mathematical statistics or probability. We also encourage the explicit discussion of how models are used to simulate data, from informal uses early in the course to formal uses as part of tests of significance later in the course. When introducing and using the normal distribution as a model of certain univariate data sets or the regression line as a model of certain bivariate data sets, we hope instructors will describe the importance and use of these modes, and fitting models to data, modeling important aspects of both statistical practice and statistical thinking for the students to see.

Chapter 8
Learning to Reason About Distribution

Statisticians look at variation through a lens which is "distribution".

(Wild, 2006, p. 11)

Snapshot of a Research-Based Activity on Distribution

Groups of three to four students are each given an envelope containing 21 different pieces of paper, each with a different histogram printed on it. Students sort the graphs into piles, so that the graphs in each pile have a similar shape. After sorting them into the piles (e.g., normal/bell-shaped, left-skewed, right-skewed, bimodal, and uniform), students choose one histogram from each pile that best represents that category, and these selections are shared and discussed as a class. Students use their own informal language to "name" each shape: bell-shaped, bunched to one side, like a ski slope, camel humps, flat, etc. These informal names are matched to the formal statistical terms such as normal, skewed, bimodal, and uniform. Finally, students consider which terms are characteristics (e.g., skewness) that can apply to graphs in more than one category (e.g., a distribution that is skewed and bimodal) vs. those that can only be labeled by one name (e.g., uniform or normal distribution).

Rationale for This Activity

Although this activity may seem like a game for elementary school students, the activity involves some important challenges and learning outcomes for high school and college students in a first statistics course. First of all, this activity helps students look at histograms as an entity, rather than as a set of data values and cases, which research has shown to be a key problem in reasoning about distributions. Secondly, students often fail to see the general shape of distributions, because of the effects of randomness (the "noise"); and expect to see perfect shapes like the models given in their textbooks. This activity helps them see that there are many types of "normal" distributions or skewed distributions. They learn to look beyond the individual features of the graph and see the more general or global characteristics. Finally, this activity focuses on the language used to describe distributions, which can often be confusing to students. The word "normal" in statistics refers to a bell-shaped curve

J.B. Garfield, D. Ben-Zvi, *Developing Students' Statistical Reasoning*,

that has certain characteristics while in everyday life it means typical or not unusual. Students can take an informal name of a shape (e.g., ski slope for a right-skewed distribution) and map them to the correct statistical labels, which can then be used to help remind them of the statistical term (e.g., a teacher talking about a skewed curve can say, "remember it is like a ski slope").

The Importance of Understanding Distribution

We begin this section with a poignant illustration, offered by Bill Finzer to participants at the Fourth International Research Forum on Statistical Reasoning, Thinking, and Literacy, the focus of which was on "Reasoning about Distribution" (Makar, 2005)

The Little Prince, by de Saint-Exupéry (2000) begins with this drawing.

To adults, the drawing looked exactly like a hat.
To the child artist who drew it and to the little prince, it was a drawing of a boa that had eaten an elephant.

If the little prince showed this picture to a statistician, he would say: "This represents a distribution of data."

We like this example because it shows how statisticians look at irregular shapes of data sets and look beyond the details to see a general shape and structure. This is usually a first step in any data analysis and leads to important questions about the data to be analyzed, such as: What mechanism or process might have led to this shape? Are there any values that need to be investigated (e.g., possible outliers)?

A graph of a distribution reveals variation of a quantitative variable. According to Wild (2006), statisticians respond to the "omnipresence of variability" in data (Cobb & Moore, 1997) by investigating, disentangling, and modeling patterns of variation represented by distributions of data. He suggests that statisticians look at variation through a lens that is "distribution."

Students encounter two main types of "distribution" in an introductory statistics class. The first type is distributions of sample data that students learn to graph, describe, and interpret. These are *empirical* distributions of some particular measured quantity. The second type of distribution encountered is a *theoretical* one, e.g., normal or binomial distributions, which are actually probability models (Wild, 2006).

Although the two types share many common features (e.g., they can be described in terms of shape, center, and spread), it is important to help students distinguish between them because of the way in which we use them. The distinction that underlies *empirical* versus *theoretical* distributions relates to variation. When examining an empirical distribution, the focus is on description and interpretation of the message in the data, and thinking about what model may fit or explain the variation of the data. Theoretical distributions are models to fit to data, to help explain, estimate, or make predictions about the variability of empirical data. Yet, a third type of distribution, students encounter in a statistics course is a distribution made up of sample statistics, which again has both empirical and theoretical versions. These *sampling distributions* are discussed in detail in Chapter 12.

The Place of Distribution in the Curriculum

Empirical distributions are the foundation of students' work in an introductory statistics course, either beginning a course or following a unit on collecting and producing data (experiments and surveys, Chapter 6). This chapter focuses mainly on teaching and learning issues related to *empirical* distributions, while Chapters 7 and 12, respectively, also discuss *theoretical* distributions. In addition, this chapter focuses on understanding a single distribution, primarily in the form of dotplots and histograms, while Chapter 11 examines these graphs along with boxplots in the comparison of two or more distributions.

The methods of Exploratory Data Analysis introduced by Tukey (1977) have had a big impact on the way distributions are taught in today's courses. Students use many ideas and tools to explore data and learn to think of data analysis as detective work. Students usually learn multiple ways to graph data sets by hand and on the computer. These methods include dotplots (also called line plots), stem and leaf plots, histograms, and boxplots. Students learn that different graphs of a data set reveal different characteristics of the data. For example, a histogram or dotplot gives a better idea of the shape of a data set, while a boxplot is often better at revealing an outlier. A stem-and-leaf plot or dotplot may give a better idea of where there are clumps or gaps in the distribution.

Current statistical software programs (e.g., *TinkerPlots, Fathom*) allow students to easily manipulate data representations, for example, to transform one graph of a data set to another, display several interlinked graphs of the same data set on one screen (a change in one will show in the others), and some allow students to highlight particular data values and see where they are located in each graph. These explorations are used to ask questions about the data: What causes gaps and clusters? Are outliers real data values or errors in data collection or coding? What factors may help explain the features revealed in a graph of a distribution? In most introductory courses along with learning how to graph distributions of data, students are taught to look for specific features of distributions and begin to describe them informally (e.g., estimate center and range) and then more formally (e.g., shape, center, and spread).

Distribution is one of the most important "big ideas" in a statistics class. Rather than introduce this idea early in a class and then leave it behind, today's more innovative curriculum and courses have students constantly revisit and discuss graphical representations of data, before any data analysis or inferential procedure. In a similar vein, the ideas of distributions having characteristics of shape, center, and spread can be revisited when students encounter theoretical distributions and sampling distributions later in the statistics course.

Review of the Literature Related to Reasoning About Distribution

The research literature provides a strong case that understanding of distributions, even in the simplest forms, is much more complex and difficult than many statistics teachers believe. Although little of the research includes college students, the results of studies on precollege level students and precollege level teachers demonstrate the difficulty of learning this concept, some common misconceptions, and incomplete or shallow understandings that we believe also apply to college students.

Much of the research on distribution emerged because of the consensus in the statistics education community that it is a basic building block for a web of key statistical ideas, such as variability, sampling, and inference (e.g., Garfield & Ben-Zvi, 2004; Pfannkuch & Reading, 2006). Other studies (e.g., Reading & Shaughnessy, 2004; Watson, 2004) focused on broader questions than how students reason about distribution, but yielded relevant results. For example, Chance et al. (2004) assert that the knowledge of distribution and understanding of histograms are necessary prerequisites to learning and understanding sampling distributions.

Developing an Aggregate View of Distribution

A major outcome of several studies on how students solve statistical problems is that they tend not to see a data set (statistical distribution) as aggregate, but rather as individual values (e.g., Hancock, Kaput, & Goldsmith, 1992). Konold & Higgins (2003) claimed that, "students need to make a conceptual leap to move from seeing data as an amalgam of individuals each with its own characteristics to seeing the data as an aggregate, a group with emergent properties that often are not evident in any individual member" (p. 202). They explained this challenging transition in the following way:

> With the individuals as the foci, it is difficult to see the forest for the trees. If the data values students are considering vary, however, why should they regard or think about those values as a whole? Furthermore, the answers to many of the questions that interested students—for instance, Who is tallest? Who has the most? Who else is like me?—require locating individuals, especially themselves, within the group. We should not expect students to begin focusing on group characteristics until they have a reason to do so, until they have a question whose answer requires describing features of the distribution. (Konold and Higgins, 2003, p. 203)

To explore the emergence of second graders' informal reasoning about distribution, Ben-Zvi and Amir (2005) studied the ways in which three second grade students (age 7) started to develop informal views of distributions while investigating real data sets. They described what it may mean to begin reasoning about distribution by young students, including two contrasting distributional conceptions: "flat distribution" and "distributional sense". In the "flat distribution" students focused just on the values of distribution and did not refer at all to their frequencies, while students who started acquiring a "distribution sense" showed an appreciation and understanding that a distribution of a variable tells us what values it takes and how often it takes these values. The gradual transfer from the incomplete perception of a distribution towards the more formal sense of distribution presented an immense challenge to these students.

In a teaching experiment with older students (seventh grade students in Israel), Ben-Zvi and Arcavi (2001) show how students were able to make a transition from *local* to *global* reasoning, from *individual-based* to *aggregate-based reasoning*. The researchers found that carefully designed tasks (e.g., comparing distributions, handling outliers), teachers' guidance and challenging questions, along with motivating data sets and appropriate technological tools helped students to make this transition.

Konold, Pollatsek, Well, and Gagnon (1997) interviewed two pairs of high-school students who had just completed a year-long course in probability and statistics. Using software and a large data set students had used as part of the course, these students were asked to explore the data and respond to different questions about the data and to support their answers with data summaries and graphs. The results suggest that students had difficulty in thinking about distributions and instead focused on individual cases. They did not use the methods and statistics learned in the course when comparing two distributions, but instead relied on more intuitive methods involving comparisons of individual cases or homogeneous groups of cases in each group. Results were re-analyzed along with results from two other studies (Konold, Higgins, Russell, & Khalil, 2003) and the following types of responses were suggested as ways students reason about a distribution of data.

1. Seeing data as Pointers (to the larger event from which the data came).
2. Seeing data as Case-values (values of an attribute for each case in the data set).
3. Seeing data as Classifiers (giving frequency of cases for a particular value).
4. Seeing data as an Aggregate (the distribution as an entity with characteristics such as shape, center, and spread).

The authors note that although an important goal in statistics is to help students see a distribution as an aggregate, they feel it is important to pay attention to students' initial views of data and to carefully help them gradually develop the aggregate view (Konold et al., 2003).

Understanding the Characteristics of a Distribution

Several studies focused on how students come to conceive of shape, center, and spread as characteristics of a distribution and look at data with a notion of distribution as an organizing structure or a conceptual entity. For example, based on their analysis of students' responses on the National Assessment of Educational Progress (NAEP) over the past 15 years, Zawojewski and Shaughnessy (2000) suggest that students have some difficulty finding the mean and the median as well as difficulty selecting appropriate statistics. They explain that one of the reasons that students do not find the concepts of mean and median easy may be that they have not had sufficient opportunities to make connections between centers and spreads; that is, they have not made the link between the measures of central tendency and the distribution of the data set. Mokros and Russell (1995) claim that students need a notion of distribution before they can sensibly choose between measures of center and perceive them as "representatives" of a distribution.

Reasoning About Graphical Representations of Distributions

One of the difficulties in learning about graphical representations of distributions is confusion with bar graphs. In elementary school, students may use bars to represent the value of an individual case (e.g., number of family pets), or a bar can represent the frequency of a value (e.g., number of families with one pet). Today, some statistics educators distinguish between these two types of representations, referring to case-value plots as the graphs where a line or bar represents the value of an individual case, or student. In contrast, the bars of a histogram represent a set of data points in an interval of values. While case-value and bar graphs can be arranged in any order (e.g., from smallest to largest or alphabetical by label), bars in a histogram have a fixed order, based on the numerical (horizontal) scale. Furthermore, while the vertical scale of a histogram is used to indicate frequency or proportion of values in a bar (interval), the vertical scale for a bar graph may represent either a frequency or proportion for a category of categorical data, or it may represent magnitude (value of a case presented by that bar). These differences can cause confusion in students, leading them to try to describe shape, center, and spread of bar graphs or to think that bars in a histogram indicate the magnitude of single values (Bright & Friel, 1998).

Establishing connections among data representations is critical for developing understanding of graphs; however, students cannot make these connections easily and quickly. To find instructional strategies that help learners understand the important features of data representations and the connections among them, Bright and Friel (1998) studied ways that students in grades 6, 7, and 8 make sense of information in graphs and connections between pairs of graphs. They report that students benefited from these activities by recognizing the importance of "the changing roles of plot elements and axes across representations", and, therefore, suggest that teachers need to "provide learners with opportunities to compare multiple representations of the same data set" (p. 87). They also suggest to promote rich discourse about distributions of data in the classroom to help students understand the important aspects of each representation.

Students' recognition of graphical aspects of a distribution as an entity was studied by Ainley, Nardi, and Pratt (2000). They observed young students (8–12 years) who collected data during ongoing simple experiments and entered them in spreadsheets. They noted that despite limited knowledge about graphs, students were able to recognize abnormalities (such as measurement errors) in graphs and to take remedial action by adjusting the graphs toward some perceived norm. The researchers have labeled this behavior, "normalizing," an activity in which children construct meanings for a trend in data and in graphs. Ainley and her colleagues claim that children gained this intuitive sense of regularity from everyday experience, experience gained during the activity, their sense of pattern, or from an emerging perception of an underlying mathematical model. The researchers recommend the use of computer-rich pedagogical settings to change the way in which knowledge about data graphs is constructed.

Helping Students to Reason with Graphs of Distributions

Students often see and use graphs as illustrations rather than as reasoning tools to learn something about a data set or gain new information about a particular problem or context (Wild & Pfannkuch, 1999; Konold & Pollatsek, 2002). Current research on students' statistical understanding of distribution (e.g., Pfannkuch, 2005a; Watson, 2005) recommends a shift of instructional focus from drawing various kinds of graphs and learning graphing skills to making sense of the data, for detecting and discovering patterns, for confirming or generating hypotheses, for noticing the unexpected, and for unlocking the stories in the data. It has been suggested that reasoning with shapes forms the basis of reasoning about distributions (Bakker, 2004a; Bakker and Gravemeijer, 2004).

Others refer to developing skills of visual decoding, judgment, and context as three critical factors in helping students derive meaning from graphs (Friel, Curcio and Bright, 2001). Reasoning about distributions is more than reasoning about shapes. It is about decoding the shapes by using deliberate strategies to comprehend the distributions and by being cognizant of the many referents, which are bound within the distributions. Furthermore, students have to weigh the evidence to form an opinion on and inference from the information contained in the distributions (Friel et al., 2001). Such informal decision-making under uncertainty requires qualitative judgments, which are much harder than the quantitative judgments made by statistical tests (Pfannkuch, 2005a).

In one of the rare studies at the college level, delMas, Garfield, and Ooms (2005) analyzed student performance on a series of multiple-choice items assessing students' statistical literacy and reasoning about graphical representations of distribution. They found that college students, like younger peers in middle school described above, confused bar graphs and histograms, thinking that a bar graph of individual cases, with categories on the horizontal scale, could be used to estimate shape, center, and spread. They also thought that such a bar graph might look like the normal distribution. They tended to view flat, rectangular-shaped histograms as a time series plot showing no variation, when these graphs typically show much variation in values. The researchers also identified errors students make in reading and interpreting horizontal and vertical axes. Based on the difficulties students appeared to have reading and interpreting histograms, the authors questioned whether students should be taught to use only dotplots and boxplots to represent data sets. After questioning colleagues, they concluded that there were important reasons to keep histograms in the curriculum as a way of representing distributions of data, because of the need for students to understand the ideas of area and density required for understanding theoretical distributions, and because dotplots are not feasible for very large data sets.

Technological Tools to Develop the Concept of Distribution

Technology can play an important role in developing distributional reasoning by providing easy access to multiple representations and endless opportunities to interactively manipulate and compare representations of the same data set. However, this is not a simple task. Biehler (1997b) reports that despite using an innovative software tool to generate and move between different graphs of data, interpreting and verbally describing these graphs were profoundly difficult for high school and college students, unless they had a conceptual understanding of the foundational concepts.

To study the impact of technology on distributional understanding, Cobb (1999), McClain & Cobb (2001), and Bakker & Gravemeijer (2004) examined how a hypothetical learning trajectory, translated into a particular instructional sequence, involving the use of *Minitools* (Cobb et al., 1997) supported the development of students' statistical reasoning about distribution. *Minitools* are simple but innovative Web applets that were designed and used to assist students to develop the concept of distribution. Results of these teaching experiments suggest that students' development of relatively deep understandings of univariate distribution are feasible goals at the middle school level, when activities, discussion, and tools are used in particular ways.

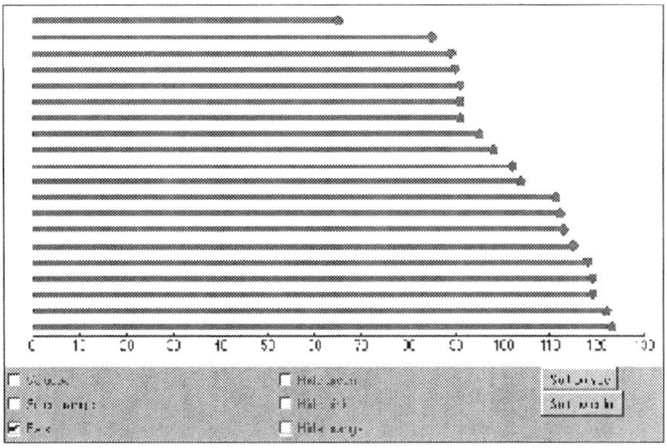

Fig. 8.1 Life span of batteries displayed by a case-value bar graph in *Minitool 1* (sorted by size). Each horizontal bar represents the life span in hours for a particular battery

For example, one aspect of distribution, shape, can be seen by looking at histograms or dotplots. To understand what dots in a dotplot represent, students need to realize that a dot corresponds to a value of a particular variable, and each dot represents one case that has that particular value. To help students develop this insight, a tool shows them case-value bars (*Minitool 1*, see Fig. 8.1). These bars seem to correspond to students' intuitive ways of organizing and displaying a set of data. Students then are helped to make a transition to a second *Minitool* (Fig. 8.2), which takes the end points of the case-value bars and stacks them in a dotplot. While each case in *Minitool 1* is represented by a bar whose relative length corresponds to the value of the case, each case in *Minitool 2* is represented by a dot in a dotplot. Figures 8.1 and 8.2 display just one data set at a time; however, all given data sets in *Minitools* include two groups (e.g., comparing two brands of batteries), which better help students develop distributional reasoning.

Minitool 2 has options to organize data in ways that can help students develop their understanding of distributions. For example, the dotplot can be divided into equal groups or into equal intervals, which support the development of an understanding of the median and quartiles, boxplot, density, and histogram, respectively

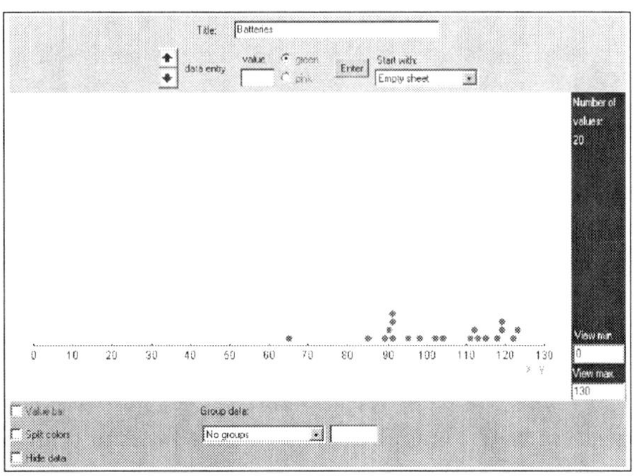

Fig. 8.2 The same data set of life span of batteries displayed by a dotplot in *Minitool 2*

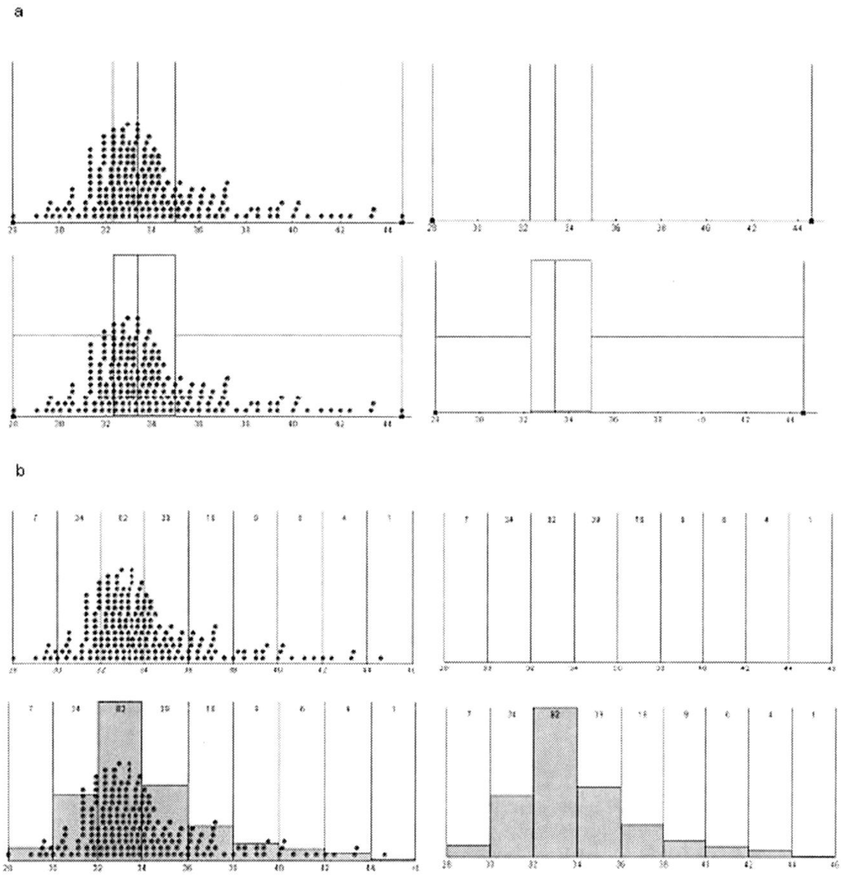

Fig. 8.3 (**a**) Four equal group and boxplot overlay option with and without data. (**b**) Fixed interval width and histogram overlay options with and without data

(see Fig. 8.3). Many of the features of the *Minitools* have been incorporated into the recently published *TinkerPlots* software (Konold & Miller, 2005).

The combination of the two *Minitools* graphs was found to be useful in helping students develop the idea of distribution. Bakker and Gravemeijer (2004) identified patterns of student answers and categorized an evolving learning trajectory that had three stages: Working with graphs in which data were represented by horizontal bars (*Minitool 1*, Fig. 8.1), working with dotplots (*Minitool 2*, Fig. 8.2), and focusing on characteristics of the data set, such as bumps, clusters, and outliers using both *Minitools*.

Based on their research, Bakker and Gravemeijer (2004) suggest several promising instructional heuristics to support students' aggregate reasoning of distributions: (1) Letting students invent their own data sets could stimulate them to think of a data set as a whole instead of individual data points. (2) *Growing samples*, i.e., letting students reason with stable features of variable processes, and compare their conjectured graphs with those generated from real graphs of data. (3) Predictions about the *shape* and location of distributions in hypothetical situations. All these methods can help students to look at global features of distributions and foster a more global view.

Implications of the Research: Teaching Students to Reason About Distribution

The results of these studies suggest in general that it takes time for students to develop the idea of distribution as an entity, and that they need repeated practicing in examining, interpreting, discussing, and comparing graphs of data. It is important to provide opportunities for students to build on their own intuitive ideas about ways to graph distributions of data. Some of the research suggests that students use their own informal language (e.g., talking about 'bumps' and 'clumps' of data) before learning more formal ones (e.g., mode, skewness). The research also suggests that teachers begin having students use graphical representations of data that show all the data values (e.g., dotplots or stem-and-leaf plots) and carefully move from these to more abstract and complex graphs that hide the data (e.g., histograms and boxplots), showing how different graphs represent the same data. Several studies suggest a sequence of activities that leads students from individual cases (case-value bars) to dotplots to groups of data points (clusters in intervals) to histograms. This sequence can later be used to develop the idea of a boxplot (see Chapter 11). New computer tools (e.g., *Minitools*, *TinkerPlots*, and *Fathom*) show promise for helping to guide students through this process and to allow them to connect different graphical representations of distributions.

Teaching Students the Concept of Distribution

The strongest message in the research on understanding graphs and distributions is that statistics teachers need to be aware of the difficulties students have developing the concept of distribution as an entity, with characteristics such as shape, center, and spread. While most textbooks begin a unit on descriptive statistics with graphs of data, when to use them, how to construct them and how to determine shape, estimate center and spread, we believe that there are some important steps to precede this. We think it is best to begin data explorations with case-value bars that represent individual cases, a type of graph students are very familiar with and that is more intuitive for them to understand and interpret. Then, this type of graph can be transformed to a dotplot using diagrams or a tool such as *TinkerPlots*. For example, a diagram of case-value bars such as the one shown below (Fig. 8.4), for a set of students test scores, can be converted by the students to a dotplot (Fig. 8.5), taking the end point of each case-value and plotting it on a dotplot.

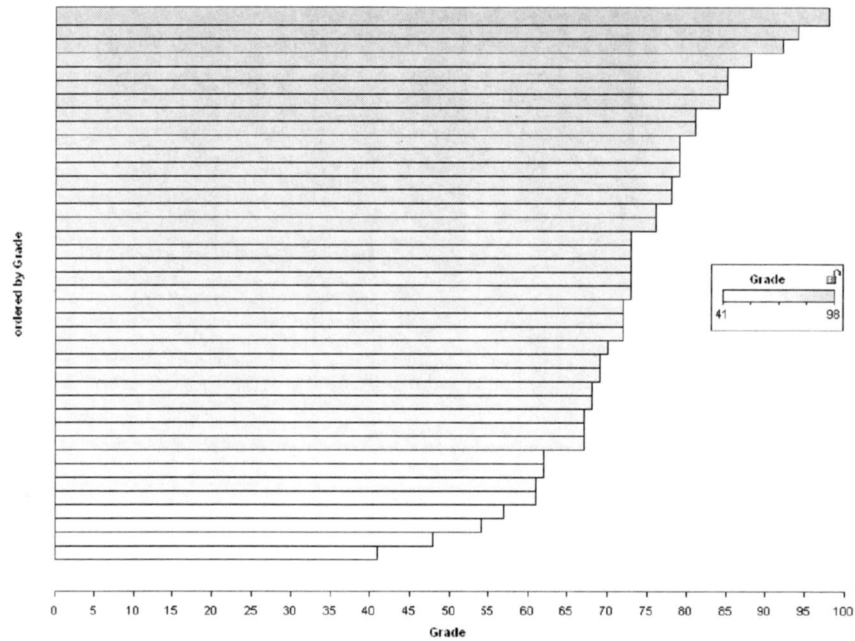

Fig. 8.4 Exam scores displayed by a case-value bar plot in *TinkerPlots*

We then suggest having students talk about categories that are useful for grouping the data, such as students who scored in the 40s, 50s, 60s, etc. (Fig. 8.6). These groupings could lead to bars of a histogram (Fig. 8.7). Students should be encouraged to compare the three types of graphs of the same data, discussing what each graph does and does not show them, how they compare, and how they are different. We also suggest moving from small samples of data to large samples, to continuous curves drawn over these graphs to help students see that plots often have some common shapes. This can also be done by giving students sets of graphs to sort and classify, as described in the snapshot of an activity in the beginning of this chapter, so that students can abstract general shapes for a category of graphs of distributions.

Another way to help students develop ideas of distribution is to help them discover characteristics of distribution. Giving them sets of graphs to compare can help

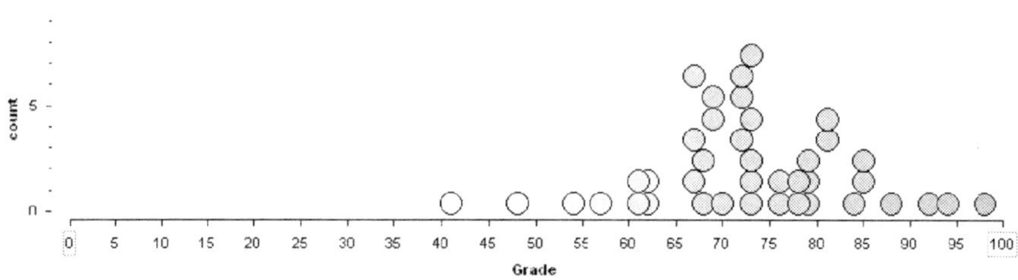

Fig. 8.5 The same exam scores displayed by a dotplot in *TinkerPlots*

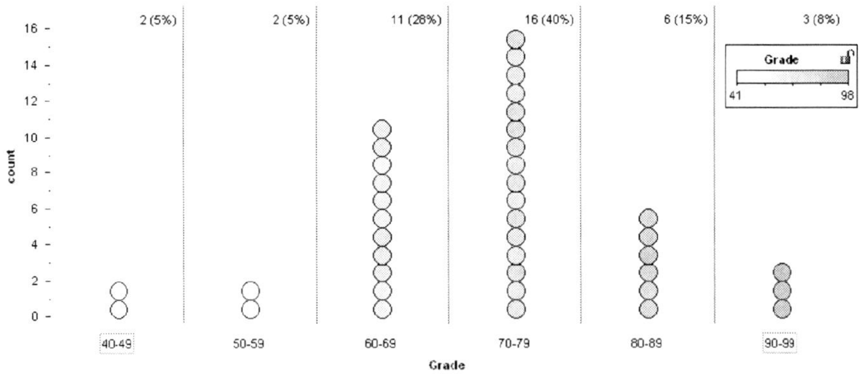

Fig. 8.6 Exam scores grouped by bins of 10 in *TinkerPlots*

do this. For example, an activity originally developed by Rossman et al. (2001) giving students a set of graphs that are similar in shape and spread, but have different centers can allow students to discover these characteristics. This can be repeated with distributions that have similar shapes and centers but differ in spread. We provide examples of these activities in Lesson 1. We focus on characteristics of distribution first using dotplots, which are easier for students to read and interpret than histograms.

The research suggests that having students make conjectures about what a set of data might look like for a particular variable and sample, can help students develop their reasoning about distribution. Rossman and Chance (2005) have also built on these ideas in their activities that have students match different dotplots to variables, forcing students to reason about what shape to expect for a particular variable, (e.g., a rectangular distribution for a set of random numbers or a skewed distribution for a set of scores on an easy test). We think it is important to have students do both activities: Draw conjectured distributions for variables and match distributions to variable descriptions. After students have studied measures of center and spread,

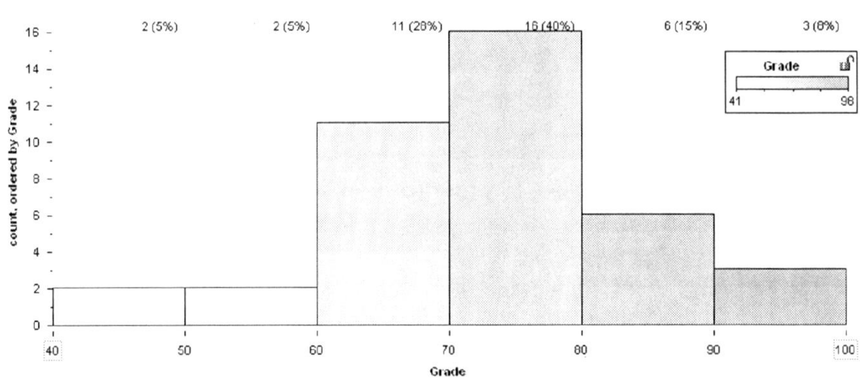

Fig. 8.7 Exam scores displayed by a histogram in *TinkerPlots*

it is suggested that they try to match graphs of distributions to a set of summary statistics, as in the histogram matching activity by Scheaffer, Watkins, Witmer and Gnanadesikan (2004b, pp. 19–21). An even more challenging version of this activity is to have students match two different representations of the same data set, such as histograms to boxplots (also found in Scheaffer et al., 2004b, pp. 21–22), which we include in Chapter 11 on Comparing Groups.

Finally, the research suggests that technology can be used to help students see the connections between different graphical representations of data, helping students to build the idea of distribution as an entity. We see three important uses of technology:

1. To visualize the transition from case-value graphs to dotplots to histograms, all based on the same data set.
2. To illustrate the ways that different graphs of the same data reveal different aspects of the data, by flexibly having multiple representations on the screen at the same time, allowing students to identify where one or more cases is in a graph.
3. To flexibly change a graph (e.g., making bins wider or narrower for histograms) so that a pattern or shape is more distinct, or to add and remove values to see the effect on the resulting graph.

Fortunately, today's software tools and Web applets readily provide each of these different types of functions.

Progression of Ideas: Connecting Research to Teaching

Introduction to the Sequence of Activities to Develop Reasoning about Distribution

Although most statistics textbooks introduce the idea of distribution very quickly, amidst learning how to construct graphs such as dotplots, stem and leaf plots, and histograms, the research literature suggests a learning trajectory with many steps that students go through in order to develop a conceptual understanding of distribution as an entity. Students need to begin with an understanding of the concept of variable, and that measurements of variables yield data values that usually vary. A set of data values can be visually displayed in different ways, the most intuitive way being individual cases, e.g., case-value graphs. After students understand how to interpret case-value graphs, they can be guided to understand dotplots, and then histograms, as one representation is mapped or transformed to the next, showing their correspondence. In this process, it is important to emphasize the role of grouping the data into different intervals (bin size), which has an effect on the graph shape, and the possible interpretations that can be drawn from it.

When first looking at dotplots and histograms, students should be encouraged to use their own language to describe characteristics, and to move from small samples of data to larger ones, as they move from empirical distributions (dotplots) to theoretical distributions (density curves). Throughout each of these phases, students should make conjectures about data sets and graphs for particular sets of data where they can reason about the context, and then should be allowed to test these conjectures by comparing their graphs to real graphs of data. The ideas of distribution are explicitly revisited in every subsequent unit in the course: in studying measures of center and spread, in reasoning about boxplots when comparing groups, when studying the normal distribution as a model for a univariate data set, when studying sampling distributions, understanding P-values, and reasoning about bivariate distributions.

Table 8.1 shows the steps of this learning trajectory and what corresponding activities may be used for each step.

Table 8.1 Sequence of activities to develop reasoning about distribution[1]

Milestones: ideas and concepts	Suggested activities
Informal ideas prior to formal study of distribution	
• Understand that variables have values that vary and can be represented with graphs of data	• Variables on Back Activity (Lesson 1, Data unit, Chapter 6)
• Understand simple graphs of data where each case is represented with a bar (e.g., case-value graphs)	❖ An activity where students summarize and interpret data sets that are of interest to them, such as class survey data given in case-value plots. Have students arranged the points on the horizontal scale in different orders. (The symbol ❖ indicates that this activity is not included in these lessons.)
• A distribution is a way to collect and examine statistics from samples	• Gettysburg Address Activity (Lesson 3, Data unit, Chapter 6)
• A distribution can be generated by simulating data	• Taste Test Activity (Lesson 4, Data unit, Chapter 6)
• Understanding a dotplot	❖ An activity where students see how the data can be represented in a dotplot, and how this plot gives a different picture than a case value plot
Formal ideas of distribution	
• Characteristics of shape, center, and spread for a distribution	• Distinguishing Distributions Activity (Lesson 1: "Distributions")
• Features of graphs, clustering, gaps, and outliers of data	• Distinguishing Distributions Activity (Lesson 1)
• A continuous curve as representing a distribution of a large population of data	• Growing a Distribution Activity (Lesson 1)
• An understanding of histogram by changing one data set from a dotplot to a histogram, by forming equal intervals of data. Recognizing the difference between these two types of graphs	• What is a Histogram Activity (Lesson 2: "Reasoning about Histograms")
• The abstract idea of shape of histogram and recognition of some typical shapes	• Sorting Histograms Activity (Lesson 2)

[1] See page 391 for credit and reference to authors of activities on which these activities are based.

Table 8.1 (continued)

Milestones: ideas and concepts	Suggested activities
• Understand that histograms may be manipulated to reveal different aspects of a data set	• Stretching Histograms Activity (Lesson 2)
• Recognize where majority of data are, and middle half of data	❖ An activity where students describe graphs in terms of middle half of data and overall spread
• Recognize the difference between bar graphs of categorical data, case value graphs, and histograms of quantitative data	❖ An activity where students examine and compare these three types of graphs that all use bars in different ways
• Only certain types of graphs (e.g., dotplots and histograms) reveal the shape of a distribution	• Exploring Different Representations of the Same Data Activity (Lesson 2)
• Reason about what a histogram/dotplot would look like for a variable (integrate ideas of shape, center, and spread) given a verbal description or sample statistics	• Matching Histograms to Variable Descriptions Activity (Lesson 2)
Building on formal ideas of distribution in subsequent topics	
• Idea of center of a distribution and how appropriate measures of center depend on characteristics of the distribution	• Activities in Lessons 2 (Center Unit, Chapter 9)
• Idea of variability of a distribution and how appropriate measures of variability depend on characteristics of the distribution	• Activities in Lessons 1 and 2 (Variability Unit, Chapter 10)
• How a boxplot provides a graphical representation of a distribution	• Activities in Lessons 1, 2, 3, and 4 (Comparing Groups unit, Chapter 11)
• How boxplots and histograms reveal different aspects of the same distribution	• Matching Histograms to Boxplots Activity (Lesson 3, Comparing Groups Unit, Chapter 11)
• Probability distribution as a distribution of a random variable that has characteristics of shape, center, and spread	• Coins, Cards, and Dice Activity (Lesson 2, Modeling Unit, Chapter 7)
• The normal distribution as a model of univariate data that has specific characteristics of shape, center, and spread	• Activities in Lesson 3, The Normal Distribution as Model (Models Unit, Chapter 7)
• The idea of sampling distribution as distributions of sample statistics that can be described in terms of shape, center, and spread	• Activities in Lessons 1, 2, and 3 (Samples and Sampling Unit, Chapter 12)
• How statistical inferences may involve comparing an observed sample statistic to a sampling distribution	• Activities in Lessons 1 and 2, (Inference Unit, Chapter 13)
• Bivariate distribution as represented in a scatterplot	• Activities in Lesson 1 (Covariation Unit, Chapter 14)
• Characteristics of a bivariate distribution such as linearity, clusters, and outliers	• Activities in Lesson 1 (Covariation Unit, Chapter 14)

Introduction to the Lessons

The two lessons on distribution contain many small activities, which together lead students from exploring one set of data in a simple, intuitive form, to more sophisticated activities that involve comparing and matching graphs. While it is not at all "traditional" to spend two full class sessions on the idea of distribution and basic graphs, we feel strongly that unless students understand this idea early on, they will not understand most of the subsequent course material at a deep level.

Lesson 1: Distinguishing Distributions

In the first activity of this lesson, students are given several different groups of dotplots and asked to determine the distinguishing feature that distinguishes each of the dotplots in a group. In this way, the students discover the characteristics of shape, center, and spread, and features such as clusters, gaps, and outliers. The activity also helps students see distributions as a single entity with identifiable characteristics. The second activity has students make predictions about graphs for a variable measured on their class survey and then make and test predictions about what would happen if the sample size were increased. Student learning goals for this lesson include:

1. To develop the idea of a distribution as a single entity rather than individual points.
2. To recognize different characteristics of a distribution and understand these characteristics in an intuitive, informal way.
3. To recognize differences between graphs of small and large samples, and how graphs of distributions stabilize as more data from the same population is added.
4. To develop an understanding of a density curve as it represents a population.

Description of the Lesson

The lesson begins with a discussion of the term "distribution" and how this differs in everyday usage and in statistics. Students reason about and discuss pattern, what we need to know to draw a reasonable graph without knowing the data values.

Next, in the *Distinguishing Distributions* activity, students are given a series of dotplots that depict the distributions of hypothetical exam scores in various classes. For example, they are asked:

> For classes A, B, and C, what is the main characteristic that distinguishes these three graphs from each other? What might explain this difference? (Fig. 8.8)

Each set of graphs reveals a different characteristic of distribution: e.g., center, spread, shape, and outliers. Students are then asked what features are important to examine when describing distributions, what to look for, features that are always present such as shape, center and spread, and ones that might be present or absent, depending on the data set.

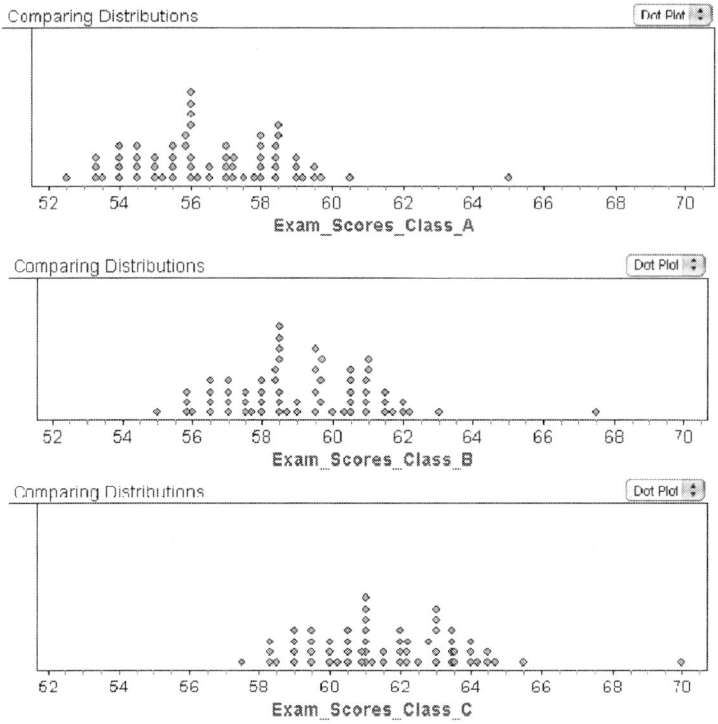

Fig. 8.8 Distributions of hypothetical exam scores in various classes

The following activity (*Growing a Distribution*) has students making predictions about number of hours students in their class study each week, both in terms of a typical value and a range of values. Next, they gather a sample of five data values from students in the class and graph this using a dotplot. More data are gathered (e.g., the entire class) and new dotplots are generated and described. Then students are asked to imagine if four classes of students were combined, and to draw their imagined plot of study hours per week. Finally, they consider all students at the university, and draw a smooth curve to represent this population. The students repeat this by looking at dotplots of three other variables based on student survey data, and then draw a smooth curve to represent the distribution of all students at their school for these variables.

A wrap-up discussion focuses on differences between dotplots and smooth curves. Students discuss what we mean by the term distribution, what are some of the common characteristics of a distribution of quantitative data, what information a graph of a distribution provides and what information a histogram provides that is not revealed by looking at a bar graph or case value graph (e.g., shape, center, and spread). The students consider when a histogram might be a better representation of data than a dot plot, what information can be determined by looking at a dotplot more easily than in a histogram, and what information is lost or not shown by a histogram.

Lesson 2: Exploring and Sorting Distributions

This lesson begins with a data set of body measurements gathered from a set of students that includes kneeling heights. After students make some predictions about this variable, they examine and describe a dotplot of class data on kneeling heights. The students are led through a transformation of this dot plot into a histogram, using *TinkerPlots* software. They compare different representations of the same data to see how different features are hidden or revealed in different types of graphs.

In the third activity, students sort a set of histograms into different piles according to general shape, which leads students to recognize and label typical shapes, guiding them to see these distributions as entities, rather than as sets of individual values. The students further develop the idea of shape by changing bin widths on histograms using *Fathom* software to manipulate and reveal how the size of the bins used affects the stability of the shape. In the fifth and final activity, students match histograms to variable descriptions, reasoning about the connections between visual characteristics of distributions and variable contexts.

Student learning goals for this lesson include:

1. To understand how a distribution is represented by a histogram and that a histogram (or dotplot) allows us to describe shape, center, and spread of a quantitative variable in contrast to a bar graph (bar chart).
2. To understand the differences between case value graphs, bar graphs (case-value bars) of individual data values and graphs displaying distributions of data such as histograms.
3. To understand how graphical representations of data reveal the variability and shape of a data set.
4. To recognize and label typical shapes of distribution, using common statistical terms (normal, skewed, bimodal, and uniform).
5. To understand that the shape of a graph may seem different depending on the graphing technique used, so it may be important to manipulate a graph to see what the shape seems to be.

Description of the Lesson

In the *What is a Histogram* activity, students begin by making conjectures about what they would expect to see in the distribution of kneeling height data. This data set is then used to help students develop an understanding of a histogram, by using *TinkerPlots* to sort the data into sequential intervals, and then fuse the intervals into bars.

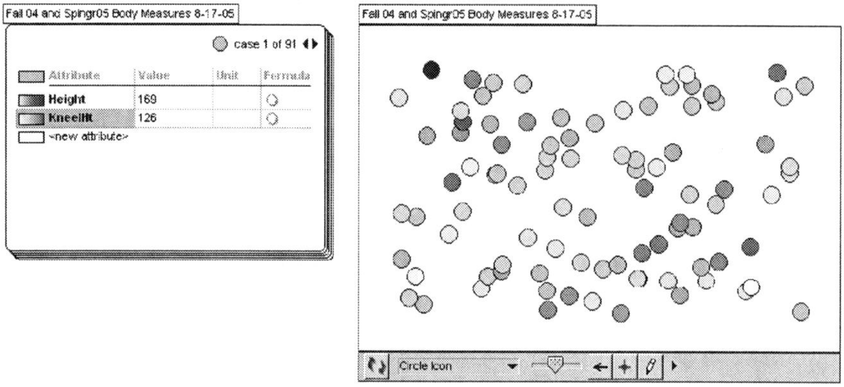

Fig. 8.9 Kneeling heights scattered dots in *TinkerPlots*

They begin with a graph of the data values, scattered as shown in Fig. 8.9, and after playing with and examining the data, are guided to arrange the dots into case-value bars, to sort the smaller values of kneeling heights as shown in Fig. 8.10.

Students are then guided to use the *Separate* operation in *TinkerPlots* to separate the cases into intervals that are then fused into a histogram, as shown in Figs. 8.11 and 8.12.

In the *Stretching Histograms* activity, students use a histogram applet to examine the effect of bin width on shape, seeing that larger and smaller bin widths may obscure shape and details (such as gaps, clusters, and outliers). Students are then encouraged to think about the difference between the different types of plots they have seen (*Exploring Different Representations of Data* activity). These plots

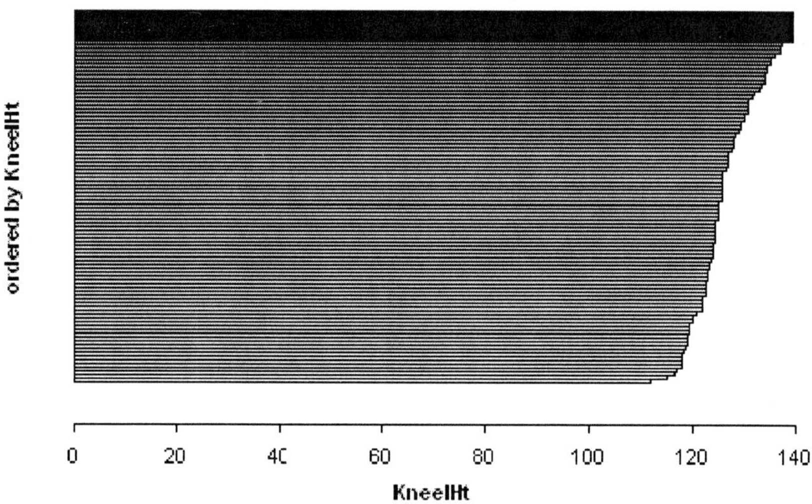

Fig. 8.10 A case-value graph ordered by kneeling heights

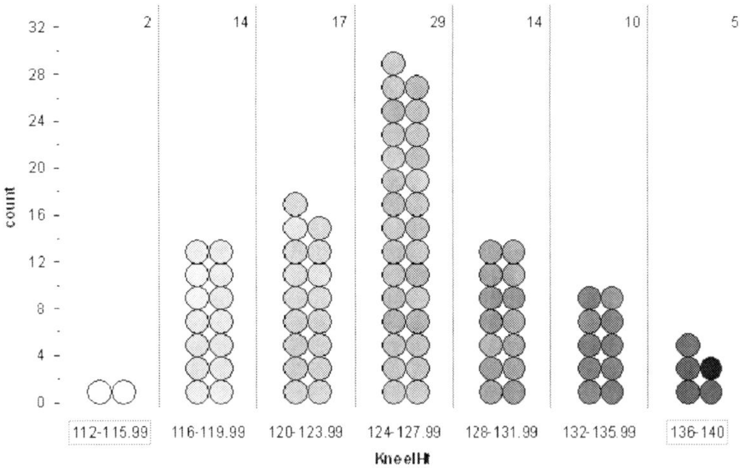

Fig. 8.11 Stacked kneeling height data with frequencies noted in each interval

include the dotplot, the case-value graph (value bar graph), and the histogram. They are then asked to consider which graphs help them better estimate the lowest five values, where "most" of the kneeling heights are clustered, the middle or typical value of kneeling heights, the spread, and the shape of the data. Students discuss how to determine the best representation to answer a particular question and why one representation may be better than another.

In the next activity (*Sorting Histogram*), students work in small groups to sort a set of 21 different graphs, representing different data sets. Students sort the stack of graphs into piles, according to those that look the same or similar, then describe what is similar about the graphs in each group, pick one representative graph for

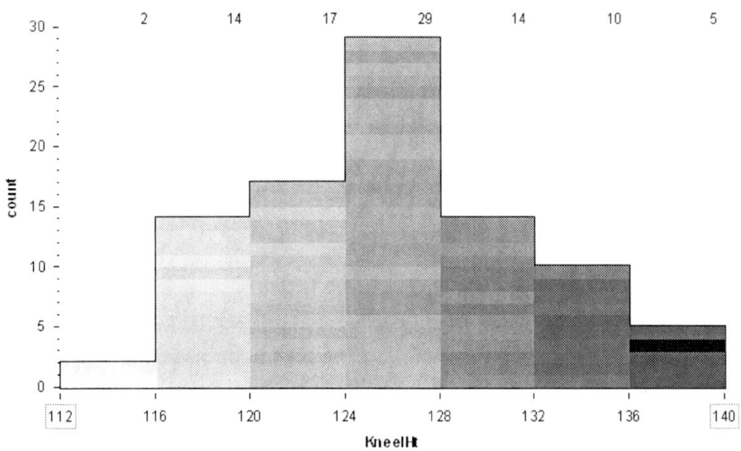

Fig. 8.12 A histogram presenting the kneeling height in seven bins

each group, and come up with their own term to describe the general shape of each group. A whole class discussion follows where group results are compared and then the correct statistical terms for the graphs (uniform, normal, right- and left-skewed, bimodal) can be introduced or reinforced. Models (uniform, normal) are described in terms of symmetry and shape (bell-shape or rectangular). Other distributions that do not fit these models can be described in terms of their characteristics (skewness, bimodality or unimodality, etc.). A discussion of which descriptors can and cannot go together may follow. For example, normal and skewed cannot go together.

During this discussion, some important points are addressed:

- Ideal shapes: density curves vs. histograms (theoretical vs. empirical distributions)
- Different versions of ideal shapes
- Idea of models, characteristics of distributions
- Statistical words vs. informal descriptors
- Other ways to describe a distribution
- Why is it important to describe a distribution?
- Normal, skewed, uniform, bimodal, and symmetric: which can be used together? How well do they fit the graphs? Which fit best?

Next, students work in groups to discuss and sketch what they expect for the general shape for some new data sets, and use the statistical terms to describe the shape of each. For example, the salaries of all persons employed at the University or grades on an easy test. There is another whole class discussion to compare answers and explanations.

The final activity (*Matching Histograms to Variable Descriptions*) has students match descriptions of variables to graphs of distributions, helping to develop students' reasoning about behavior of data in different contexts and how this is related to different types of shapes of graphs. A wrap-up discussion focuses on the characteristics of distributions and how they are revealed in different types of graphs.

Summary

Understanding the idea of distribution is an important first step for students who will encounter distributions of data and later distributions of sample statistics as they proceed through their statistics course. While most textbooks ask students to look at a histogram or stem plot and describe the shape, center, and spread, many students never understand the idea of distribution as an entity with characteristics that reveal important aspects of the variation of the data. The focus of the activities described in this chapter is on developing a conceptual understanding of distribution, and we have not included activities where students learn to construct different graphs, a topic well covered in most textbooks. We encourage instructors to repeatedly have students interpret and describe distributions as they move through the course, whether plots of sample data or distributions of sample statistics.

Chapter 9
Learning to Reason About Center

> *A statistician sees group features such as the mean and median as indicators of stable properties of a variable system—properties that become evident only in the aggregate.*
> *(Konold & Pollatsek, 2002, p. 262)*

Snapshot of a Research-Based Activity on Center

Pairs of students are given a set of 10 small Post-It® notes and a number line that goes from 17 to 25. They are asked to think about a group of college students whose mean age is 21, and construct a dot plot, using the Post-It® notes, of an age distribution with a mean of 21. Most of the students quickly figure out that they can stack all 10 Post-It® notes at the 21 point. But then, they are told that one of the values is 24, so they have to figure out where the other values are on the number line (e.g., move one note to 18, to balance out the 24, being 3 units above the average of 21). Students are instructed to move one Post-It® note to 17, and arrange the rest of the Post-It® notes so the mean is still 21. Finally, they move all of the Post-It® notes so that none are at 21 but the mean age is 21. Students use different strategies to do this, such as making sure that every note above the mean has a value equally placed below the mean. Other students may balance one note that is four units above the mean with two notes that are two units below the mean, etc. Different results are compared and discussed.

Students are then instructed to draw dot plots of 10 points that have the same mean of 21, and then draw deviations from the mean for their graphs. They consider how the deviations balance each other out, so that if one value is moved producing a deviation of −3, another value must be moved to have a deviation of +3. This leads to a discussion of what a mean "means" in terms of these deviations all cancelling each other out to be zero.

Rationale for This Activity

Although most students have learned how to calculate means before entering their statistics course, research studies reveal that few understand what a mean really is or what it tells us about data. Research also shows that students often have difficulty understanding how the mean and median differ and why they behave differently (e.g., in the presence of outliers). This activity helps students to build a conceptual

J.B. Garfield, D. Ben-Zvi, *Developing Students' Statistical Reasoning*,

understanding of what the mean and median actually mean and how they are affected by the different values in a data set. This lesson also introduces the idea of deviation early in the course, as a way to understand the idea of a mean and what it represents. This deviation idea, which begins to connect ideas of center and variability, is revisited when learning about variability and the standard deviation, and again when considering residuals in a regression analysis. This activity helps students develop a conceptual and a procedural understanding of the mean. By having the students physically manipulate data values on a number line, they are better able to see and reason about the idea of deviation from the mean and the balancing of these deviations.

The Importance of Understanding Center

> The idea of data as a mixture of signal and noise is perhaps the most fundamental concept in statistics.
>
> (Konold & Pollatsek, 2002, p. 259)

Understanding the idea of center of a distribution of data as a signal amidst noise (variation) is a key component in understanding the concept of distribution, and is essential for interpreting data graphs and analyses. While students develop informal ideas of center in the earlier units as they graph and describe distributions of data, they later encounter the idea of center more formally as they learn about different measures of center, how to compute them, what information they provide, and how we use them. However, it is impossible to consider center without also considering spread, as both ideas are needed to find meaning in analyzing data.

The Place of Center in the Curriculum

Traditional textbooks first introduce center, thenintroduce spread, and then move on to the next topic. However, it may be more helpful to study these topics together because they are so interrelated (Konold & Pollatsek, 2002; Shaughnessy, 1997).

It is hard to imagine a situation where one would summarize a data set using only a measure of center without talking about the spread of the data or how much variability there is around that measure of center. However, there are many instances, particularly in the media, in which only measures of center are provided for a data set, in some cases, leading to incorrect conclusions. When comparing groups or making inferences we need to examine center and spread together: the signal, and the noise around the signal. While these ideas are introduced in early units on exploring data, these concepts re-appear when looking at theoretical models such as the normal distribution and sampling distributions. Later on, the ideas of center (and spread) are revisited when making statistical inferences about samples of data.

Review of the Literature Related to Reasoning About Center

> (The knowledge of) computational rules not only does not imply any real understanding of the basic underlying concept, but also may actually inhibit the acquisition of more adequate (relational) understanding.
>
> (Pollatsek, Lima, & Well, 1981, p. 202)

Understanding Means

How students understand ideas of center has been of central interest in the research literature. Research on the concept of average or mean was at first the most common topic studied on learning statistics in the school level (see Konold & Higgins, 2003; Shaughnessy, 1992, 2003). The studies suggested that the concept of the average is quite difficult to understand by children, college students, and even elementary school preservice and in-service teachers (Russell, 1990; Groth & Bergner, 2006).

Early research typically focused on the single idea of center rather than looking at the interrelated concepts of center and spread, and on procedural understanding. These studies focused primarily on the mean, either as a simple average of a single data set or a weighted mean. An early interest was on students' understanding of the mean as a *balance model* (e.g., Hardiman, Well, & Pollatsek, 1984; Strauss, 1987), which is a common method taught in a statistics course. A balance model illustrates how values are placed on a balance beam at distances from the mean so that the deviations from the mean are equal. Hardiman et al. (1984) tested whether improving students' knowledge of such balance rules through experience with a balance beam promoted deep understanding of the mean. Forty-eight college students enrolled in psychology classes participated in the study which involved pretest, training, and posttest of paper and pencil items. Students who were given the balance training performed significantly better on the posttest problems.

Other studies identified several characteristics of the mean and then examined students' understanding of these characteristics (Goodchild, 1988; Mevarech, 1983; Strauss & Bichler, 1988; Leon & Zawojewski, 1993). Mevarech (1983), for example, found that high school students made mistakes in solving problems about means because they believed that means have the same properties as simple numbers, and that it is helpful to provide students corrective-feedback instruction as they solve problems involving reasoning about the mean. Strauss and Bichler (1988) found that fourth- through eighth-grade students had a difficult time understanding seven properties of the mean. Leon and Zawojewski (1993) looked at school and college students' understanding of four components of the mean. Using different testing formats, they found that some properties of the mean are better understood than others. The two properties, "the mean is a data point located between the extreme values of a distribution," and "the sum of the deviations about the mean equals zero" were better understood by students in this study than the two properties, "when the mean is calculated, any value of zero must be taken into account," and "the average value is representative of the values that were averaged."

Gal, Rothschild, and Wagner (1989, 1990) found that sixth-grade students are generally unable to use the mean to compare two different-sized sets of data. Later work showed that students had difficulty working backward from a mean to a data set that could produce such a mean (Cai, 1998). Study by Mokros and Russell (1995) expanded on this task by having students manipulate data values to produce a given mean and studying how students reasoned during this process.

Earlier research has also concentrated on understanding weighted means. Pollatsek et al. (1981) reported data from interviews of college students indicating difficulties they had in understanding the need to weight data in computing a mean. While mathematically sophisticated college students can easily compute the mean of a group of numbers, this study indicated that a surprisingly large proportion of these students do not understand the concept of the weighted mean, which is a concept that they often encounter (e.g., grade point averages). When asked to calculate a mean in a context that required a weighted mean, most subjects answered with the simple or unweighted mean of the two means given in the problem, even though these two means were from different-sized groups of scores. Callingham (1997) found that the same problem in a study of preservice and in-service teachers. As a result of their study, Pollatsek et al. (1981) wrote that "for many students dealing with the mean is a computational rather than a conceptual act" (p. 191). They concluded that knowledge of "computational rules not only does not imply any real understanding of the basic underlying concept, but may actually inhibit the acquisition of more adequate (relational) understanding" (p. 202).

What students remember about the mean? In general, it appears that many students who complete college statistics classes are unable to understand the idea of the mean. Mathews and Clark (2003) analyzed audio-taped clinical interviews with eight college freshmen immediately after they completed an elementary statistics course with a grade of "A." The point of these interviews was neither to see how quickly isolated facts could be recalled, nor was the point to see how little students remember. Rather, the goal was to determine as precisely as possible the conceptions of mean, standard deviation, and the Central Limit Theorem, which the most successful students had shortly after having completed a statistics course. The results are alarming since these top students demonstrated a lack of understanding of the mean, and could only state how to find it, arithmetically. Interviewing along the same lines, a larger (n=17) and more diverse sample of college students from four distinct campuses, Clark et al. (2003) found overall the same disappointing results. These researchers call, therefore, for pedagogical reform that will dis-equilibrate the process image of statistical concepts that students bring with them to college in order to enable them to encapsulate the process of statistical concepts into objects that are workable entities (Sfard, 1991). For example, they recommend creating situations in which students have to determine and reflect which measure of center is more appropriate.

Understanding Medians

Difficulties in determining the medians of data sets have also been documented by research. Elementary school teachers have difficulty determining the medians of data sets presented graphically (Bright & Friel, 1998). Only about one-third of twelfth grade students in the United States who took the NAEP test were able to determine the median when given a set of unordered data (Zawojewski & Shaughnessy, 2000).

Measures of Center as Typical Values

The *typical value* interpretation of the arithmetic mean has received a great deal of attention in curriculum materials and in research literature (Konold & Pollatsek, 2002). The following is an example of a problem set in a *typical value* context:

> The numbers of comments made by 8 students during a class period were 0, 5, 2, 22, 3, 2, 1, and 2. What was the typical number of comments made that day? (Konold & Pollatsek, 2002, p. 268)

Several studies have provided insights about students' thinking in regard to typical value problems. Mokros and Russell (1995) studied the characteristics of fourth through eighth graders' constructions of "average" as a *representative* number summarizing a data set. Twenty-one students were interviewed, using a series of open-ended problems that called on children to construct their own notion of mathematical representativeness. They reported that students may respond to *typical value* problems by: (i) locating the most frequently occurring value; (ii) executing an algorithm; (iii) examining the data and giving a reasonable estimate; (iv) locating the midpoint of the data; or (v) looking for a point of balance within the data set. These approaches illustrate the ways in which school students are (or are not) developing useful, general definitions for the statistical concept of average, even after they have mastered the algorithm for the mean.

Levels of Reasoning About Measures of Center

In an investigation of the development of school students' thinking in regard to measures of center, Watson and Moritz (1999) placed a structure on the categories of thinking documented by Mokros and Russell (1995). Continuing this line of research, Watson and Moritz (2000c, 2000d) also used the SOLO taxonomy (Structure of Observed Learning Outcomes, see Biggs and Collis, 1982) to rank students' responses to interview questions about averages. They observed movement from Unistructural, to Multistructural, and finally to Relational levels of reasoning as students developed from thinking about centers first from "mosts and middles" and finally to the mean as "representative" of a data set. Jones, Thornton, Langrall,

Mooney, Perry, and Putt (2000) and Mooney (2002) found that the ability to be thoughtful and critical about applying formal measures to typical value problems marks a relatively sophisticated level of statistical reasoning.

Measures of Center as "Signals in Noisy Processes"

A rich spectrum of student reasoning about center is identified by Konold and Pollatsek (2002): mean as typical value, mean as fair share, mean as data reducer, and finally, mean as signal amid noise. These researchers suggest that students should be given more opportunities to work with statistical problems set in contexts that involve searching for "signals in noisy processes." The following item is an example of a data analysis problem that involves detecting a signal in a noisy process:

> A small object was weighed on the same scale separately by nine students in a science class. The weights (in grams) recorded by each student were 6.2, 6.0, 6.0, 15.3, 6.1, 6.3, 6.2, 6.15, 6.2. What would you give as the best estimate of the actual weight of this object? (Konold & Pollatsek, 2002, p. 268)

In the case of the repeated measures problem above, the arithmetic mean of the weights that are bunched closely together could be viewed as a signal that estimates the true weight of the object. The measurement of the object can be viewed as a noisy process that contains variation stemming from various possible sources. Konold and Pollatsek (2002) acknowledge the possible cognitive complexity in using repeated measurement problems with students, pointing out that the mean as a reliable indicator of signal was not universally accepted by scientists during the early development of the discipline of statistics (Stigler, 1986). Hence, they call for more research on students' thinking in such contexts in order to help advise instruction.

Patterns of thinking about average in different contexts were investigated by Groth (2005) who studied fifteen high school students. He used problems set in two different contexts: determining the typical value within a set of incomes and determining an average set in a signal-versus-noise context. Analysis of the problem-solving clinical interview sessions showed that some students attempted to incorporate formal measures, while others used informal estimating strategies. Students displayed varying amounts of attention to both data and context in formulating responses to both problems. Groth pointed out the need for teachers to be conscious of building students' statistical intuitions about data and context and informal estimates of center, and connecting them to formal measures without implying that the formal measures should replace intuition.

Selecting an Appropriate Measure of Center

Another focus of research has been on the challenge of choosing an appropriate measure of center to represent a data set. The National Assessment of Education

Progress (NAEP) data confirm that school students frequently make poor choices in selecting measures of center to describe data sets (Zawojewski & Shaughnessy, 2000). Choosing an appropriate measure of center was also a challenge for students enrolled in an Advanced Placement high school statistics course (Groth, 2002). Similar results were found by Callingham (1997) who administered an item containing a data set structured, so that the median would be a better indicator of center than the mean, to a group of preservice and in-service teachers. Callingham reports that most of them calculated the mean instead of the more appropriate median.

In a study on statistical reasoning about comparing and contrasting mean, median, and mode of preservice elementary school teachers, Groth and Bergner (2006) described four levels, basing these on the SOLO Model:

- *Unistructural*-level: responses did not contain any strategy other than definition-telling when asked to compare and contrast the three measures.
- *Multistructural*-level: responses included definition-telling along with a vague notion that the mean, median, and the mode are all tools that can be used to analyze a set of data; responses did not reflect an understanding that each of the measures is intended to measure what is central or typical to data sets.
- *Relational*-level: responses differ from Multistructural responses in that they included recognition of the fact that the mean, median, and mode all measure the center of the data or what is typical about the data in some manner.
- *Extended abstract*-level: responses include all of the characteristics of those classified at the relational level, but go beyond relational-level responses in that they included discussions of when one measure of center might be more useful than another.

Groth and Bergner's (2006) study illustrated that attaining a deep understanding of these seemingly easy statistical concepts is a nontrivial matter, and that there are complex conceptual and procedural ideas that need to be carefully developed.

Using the History of Measures of Center to Suggest Instruction

The history of statistics can be a source of inspiration for instructional design. Bakker and Gravemeijer (2006) systematically examined examples of how measures of center were described and used, starting in ancient historical periods, and from countries such as India and Greece, in contexts involving mathematics and science. Based on their analysis of these examples, Bakker and Gravemeijer (2006) formulate hypotheses about how students in grades 7 and 8 (12–14-years old) could be supported in learning to reason with mean and median. The following ideas stemming from the historical phenomenology were found to be most fruitful for helping young children understand center.

1. Estimation of large numbers can challenge students to develop and use intuitive notions of mean.
2. Students may use the midrange as a precursor to more advanced notions of average.
3. Repeated measurement may be a useful instructional activity for developing understanding of the mean (cf., Petrosino, Lehrer, & Schauble, 2003).
4. To support students' understanding of the median, it is helpful to let them visually estimate the median in a dot plot and look for a value for which the areas on the left and right are the same.
5. Skewed distributions can be used to make the usefulness of the median a topic of discussion.

Such a historical study can help to "unpack" and distinguish different aspects and levels of understanding of statistical concepts and help instructional designers to look through the eyes of students. Note that some of these hypotheses are in accordance with the results of the research studies described above.

Implications of the Research: Teaching Students to Reason About Center

What has been striking over 25 years of research is the difficulty encountered by students of all ages and teachers in understanding concepts related to center. Although students may be able to compute simple arithmetic means, they need help in understanding what means actually mean. Activities can help students develop meaningful models such as balancing of data values by manipulating deviations from the mean to sum to zero.

The research suggests that careful attention be paid to developing the concepts of measures of center, focusing on mean and median rather than mode. These ideas should be first introduced informally as students estimate and reason about typical value for data sets, both large and small, prior to formally studying these topics in a unit on measure of center. Students may be asked to make and test conjectures about typical values using real data sets. The research also suggests that students have opportunities to explore the characteristics of the mean and median and how they are affected by different types of data sets and distributions. Developing an understanding of deviation may be an important part of understanding the mean as a balance point, so activities helping students to see and reason about deviations may help them better understand the mean. The literature suggests both visual, interactive activities as well as explorations with real data utilizing technology to produce measures of center, especially for data sets where values are changed (e.g., outliers are added or removed). Finally, the idea of the center as a signal in a noisy process should be developed, examining trends in repeated measurements. This also suggests that ideas of center be introduced along with ideas of spread or variability, and that these ideas are repeatedly connected as students explore and interpret data.

Progression of Ideas: Connecting Research to Teaching

Introduction to the Sequence of Activities to Develop Reasoning About Center

The idea of typical value as a summary measure of data set, shown graphically, is first introduced in early units whenever students make or examine graphs of numerical data. While students may intuitively look at the mode or peak of a graph or look at the middle value on the horizontal axis, they can be guided to think about the mean and median as typical values by looking at different graphs where mode or middle scale value do not seem to represent good "typical" values. This will help motivate the need for examining different measures of center and when to use them. These informal examinations and estimates should include estimates of spread of the data as well, as students respond to questions such "what is a typical value for these data" and "how spread out are the data?," learning to connect ideas of center and variability from the beginning of the course.

When formal measures of center are introduced, students are guided to explore their properties using physical and then computer manipulations of data. Properties of the mean and median can be explored and examined in this way. It is helpful for students to actually work backwards, starting with a given value of mean or median to reason about how different data sets may be constructed and altered to produce those given values. This can be done first for mean and then for median. Students then conjecture what typical values they might find for different types of variable, taking into account the shapes and characteristics of graphs of these variables. These conjectures can then be tested using real data and technology, and discussions can examine which measures are more useful summaries for each variable and why.

When students begin to study formal measures of variability, they see the relationship between mean and standard deviation, and between medial and Interquartile Range, and how it makes sense to use these pairings when summarizing different types of distributions (e.g., means and standard deviations for symmetric distributions, medians and IQR for very skewed distributions, and distributions with outliers). The idea of examining center at the same time as variability as a way to compare groups is then encountered as students learn about and compare boxplots. When the Normal distribution is introduced, students will see that the mean has special properties and use in relation with stand deviations and z-scores.

The mean is again examined when learning about samples and how the mean stabilizes as sample size gets very large, and the role of the mean in the Central Limit Theorem. As students move from sampling to statistical inference, they again encounter the mean, distinguishing between using the mean in an inference based on a large sample from using the mean as a summary measure of a single data set (when a median might be a better typical value given the shape of the distribution). The measures of center are also encountered in the unit on covariation when students look at trends by examining medians of sequential boxplots, and later as centers of distributions of the two variables. Table 9.1 shows a suggested series of ideas and activities that can be used to guide the development of students' reasoning about models and modeling.

Table 9.1 Sequence of activities to develop reasoning about center[1]

Milestones: ideas and concepts	Suggested activities
Informal ideas prior to formal study of center	
• Idea of center as a typical or representative value for a graph of a variable (e.g., dot)	• Distinguishing Distributions (Lesson 1, Distribution Unit, Chapter 8)
• The mean as somewhere in between the highest and lowest value, but not necessarily the middle value or the midpoint of the horizontal scale	• What does the Mean Mean Activity (Lesson 1: "Reasoning about Measures of Center")
Formal ideas of center	
• Properties of the mean as a balance point and the value for which all deviations from that value sum to zero	• What does the Mean Mean Activity (Lesson 1)
• How the mean is affected by extreme values	• What does the Mean Mean Activity (Lesson 1)
• The median as the middle value in a data set	• What does the Median Mean Activity (Lesson 1)
• Properties of the median: under what conditions it changes or stays the same	• What does the Median Mean Activity (Lesson 1)
• Comparing properties of the mean and median	• Means and Medians Activity (Lesson 1)
• The idea of a typical value	• What is Typical Activity (Lesson 2: "Choosing Appropriate Measures")
• Understanding why and how to use appropriate measures of center for a sample of data for a particular variable	• Choosing an Appropriate Measure of Center Activity (Lesson 2)
Building on formal ideas of center in subsequent topics	
• How center and spread are used together to compare groups	• Activities in Lessons 1, 2, 3 and 4, Comparing Groups Unit (Chapter 11)
• The idea and role of mean in normal distribution	• What is Normal Activity (Lesson 3, Statistical Models Unit, Chapter 7)
• Recognize stability of measures of center as sample size increases. When sample grows, see how measures of center predict center of larger population, and how it stabilizes (varies less)	• Sampling activities in Lessons 1, 2, and 3, Samples and Sampling Unit (Chapter 12)
• Role of mean in making inferences	• Activities in Lessons 1, 2, 3, 4, and 5 (Statistical Inference Unit, Chapter 13)
• Role of mean in bivariate distribution	❖ An activity involving fitting and interpreting the regression line to bivariate data. (The symbol ❖ indicates that this activity is not included in these lessons.)

[1] See page 391 for credit and reference to authors of activities on which these activities are based.

Introduction to the Lessons

There are two lessons on reasoning about measures of center. They begin with the physical activity described earlier where students manipulate Post-It® notes on a number line to develop an understanding of mean, and then median. Students use a *Fathom* demo to contrast how the mean and median behave for different types of data sets. Students make and test conjectures about typical values, testing these using software to generate graphs and statistics. The last activity has students compare features and uses of different measures of center when summarizing sample data.

Lesson 1: Reasoning About Measures of Center

While students have heard of means and medians before they enter an introductory high school or college statistics course, this lesson helps them develop a conceptual understanding of the mean and median. There are three parts to the lesson: an activity where students move dots on a plot to explore properties of the mean, a similar activity with the median, and then *Fathom* demos to further illustrate the properties of these concepts. Student learning goals for this lesson include:

1. Develop a conceptual understanding of the mean.
2. Understand the idea of deviations (differences from the mean) and how they balance out to zero.
3. Understand how these deviations cause the mean to be influenced by extreme values.
4. Develop an understanding of the median as a middle value that is resistant to extreme values.
5. Understand the differences between mean and median in their interpretation and properties.
6. Understand how to select appropriate measures of center to represent a sample of data.

Description of the Lesson

In the first activity described at the beginning of the chapter, (*What does the Mean Mean* activity) students are told that the average age (mean) for students in the class is 21 years and consider what we know about the distribution of students' ages for this class (e.g., "Are they all about 21 years old?"), and explain their answer first in a small group and then to the class. They also explain where this value of 21 came from and how it was produced. Students make conjectures about the ages of these 10 students and use 10 Post-It® notes to form a series of dot plots on a given number line so that the average is 21 years. Students move one Post-It® note to 24 years, and later one to 17, and figure out how to move *one or more* of the other Post-It® notes to keep the mean at 21 years, discussing their strategies and reasoning with their group and then the class. The term *deviation* is introduced to represent the distance of each data value from the mean and students examine deviations for their plots under different conditions, seeing how they have to balance to zero.

In the second activity (*What does the Median Mean?*), students reduce their Post-It® notes to 9 and arrange them on the same number line used earlier so that the *median* is 21 years. Again, they are given different constraints (e.g., change one of the values that is currently higher than 21 years) and they determine if and how the median is affected. Finally, students discuss and summarize what would they have to do with a data value in the plot in order to change the median.

In the final activity of this lesson (*Means and Medians*), students observe and discuss two *Fathom* Demos: *The Meaning of Mean* and *Mean and Median* to further understand properties of these measures. The lesson ends with a wrap-up discussion about use interpretation, and properties of the mean and median.

Lesson 2: Choosing Appropriate Measures

This lesson introduces the idea of choosing an appropriate measure of center to describe a distribution. It has students predict typical values for variables that have different distributions. The lesson then has students find the actual mean and median for those variables using computer software and examine the distributional features that made their prediction closer to either the mean or median. It also introduces the idea of outlier influence on these measures of center. Student learning goals for this lesson include:

1. Deeper conceptual understanding of mean and median.
2. Understand when it is better to use each as a summary measure for a distribution of data.
3. How to generate these statistics using *Fathom* Software.

Description of the Lesson

Students are first asked how we can describe the typical college students taking an introductory statistics course, and in what ways do students in this class differ? They discuss how people use the words: typical, average, and normal in an everyday sense and how these words are used as statistical terms: mean, median, center, and average.

In the *What is Typical* activity, students consider a set of variables that were measured on their first day of class Student Survey. Working in pairs, they predict what they might expect as a *typical value* for all students enrolled in their statistics class this term. They are reminded that a typical value is a single number that summarizes the class data for each variable. They write their prediction in the "First Prediction" column of the table shown below (Table 9.2).

Next, they generate dot plots of the data using *Fathom* software to see if their original predictions seemed reasonable. Based on the graphs, they are allowed to make revised predictions for the typical value for each of the variables, which are written in the table above in the "Revised Prediction" column.

Table 9.2 Predicting and verifying typical values in the *What is Typical Activity*

Attribute from Student Survey	First Prediction	Revised Prediction	Statistics from Fathom	
			Mean	Median
Age				
Number of statistics courses you are taking this semester				
Credits registered for this semester				
Total college credits completed				
Cumulative GPA				
Hours a week you study				
Number of emails you send each day				
Number of emails you receive each day				

Students use *Fathom* to find the mean and median for each of these variables and complete the last two columns of the table above. They discuss how close were their revised predictions to the "typical" values produced by *Fathom* and for which attributes were their predictions most accurate. They also consider what results were most surprising to them and why, and whether in general, were their revised predictions closer to the means or medians of these variables.

Students are asked to discuss:

- Which measures of center were closest to their intuitive ideas of "typical" values?
- What information do means and medians provide about a distribution?
- How to decide whether to use the mean or median to summarize a data set?
- In statistics, what do they think is meant by the word "typical"?

In *Choosing an Appropriate Measure of Center* activity, students suggest conditions where the mean and median provide similar information and when they give different information for the same data set. This leads to a discussion of which measure is more appropriate for each variable and why, and how to choose the best measure of center for a data set.

Students are asked if it is all right to compute a mean or median without first looking at a graph of data and then why that is not a good idea. They reason about what information is missing if all they are given is a measure of center, including what they know and not know if all they were given were measures of center. This provides a segue to discussion on spread (the next unit) and reinforces the connection between center and spread. In a wrap-up discussion, students are asked to imagine a variable that could be measured in two different settings that might yield data sets that have the same mean and different amounts of spread, one with a little spread and one with a lot of spread, and explain their reasoning.

Summary

The two lessons in the unit on measures of center are closely connected to ideas of distribution and variability, so that the ideas of mean and median are always connected to these concepts and contexts. The intent of the lessons is to help students build a conceptual understanding of mean and median as well as the idea of center of a distribution, through physical manipulations of data values, making and testing conjectures about typical values, and discussing the use and properties of these two measures. While the concepts may seem simple, and not worth two full lessons, we believe that that these lessons provide important foundations for and connections to subsequent units in the course.

Chapter 10
Learning to Reason About Variability

Variation is the reason why people have had to develop sophisticated statistical methods to filter out any messages from the surrounding noise.
(Wild & Pfannkuch, 1999, pp. 235–236)

Snapshot of a Research-Based Activity on Variability

Students are given pairs of histograms with the assignment to discuss and determine which graph in each pair would have a higher standard deviation and why. First, students are encouraged to actually draw deviations on the histograms and draw lines from the mean to represent the number of deviations for each bar of the histogram (e.g., a bar representing five values would correspond to five lines of the same length, showing deviations from the center). Some of the histograms are easier to compare than others, such as those that have a few bars close to the center versus a histogram with most bars far from the center (see examples in Fig. 10.1). The difficult comparisons are for graphs that have the same range, the same frequencies for each bar, but different numbers of bars (representing different possible values of the variables).

Thinking about the size and number of deviations from the mean helps the students reason about which graph would have a larger standard deviation. It also helps them confront some misleading intuitions such as looking at graph A in the second example below, and initially describing this graph as having no variability, because it is confused with a time series graph or a bar graph where the height of the bars indicates that they all have the same measurement. Drawing deviations from the center helps students to realize that this histogram is different than a bar graph or time series, and that there are different deviations from the center that can be "averaged" to produce an estimate of the standard deviation.

After students compare and discuss their answers, they enter the data for each graph into *Fathom* (Key Curriculum Press, 2006) and have the standard deviations computed. They use *Fathom* to calculate the actual squared deviations from the mean for each graph. After the students have checked their answers in this way, the class discusses each pair of histograms and why the standard deviations were larger or smaller in each pair. As they do this, they construct a set of factors that appear to influence the size of the standard deviation (e.g., more bars farther from the mean) and those that do not seem to affect the size (e.g., "bumpiness" of the graph, or different heights of the bars).

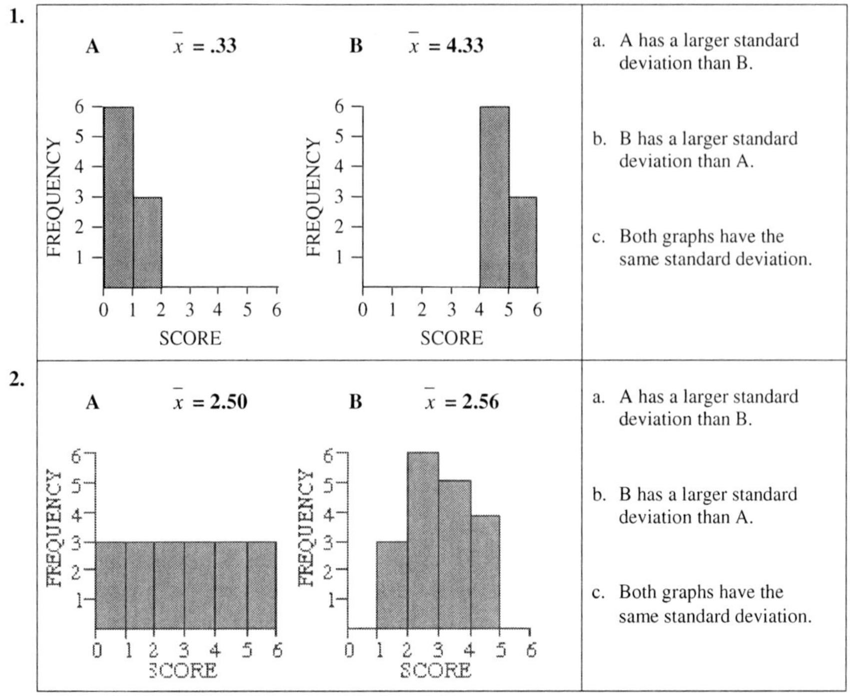

Fig. 10.1 Comparing standard deviations of pairs of histograms from the *What Makes the Standard Deviation Larger or Smaller* activity

Rationale for This Activity

The standard deviation is an important and much-used measure of variability, but one that is almost impossible for beginning statistics students to understand. While students learn what the standard deviation is and how it is calculated, they rarely have an understanding of what this measure is and how to interpret it. The activity described above is a culminating activity in the unit on variability, because it helps students recognize and integrate several sub ideas (e.g., a graphical representation of distribution, the mean of a distribution, spread, and deviation). The idea of standard deviation as an average distance from the center is developed by first having students estimate the mean of each graph, taking into account both value (on the number line) and density (frequency of each bar). Students are guided to reason about deviations from the mean and think about how values close to and far from the mean affect those deviations and squared deviations.

Entering the data from each histogram themselves (to find the actual standard deviations for each pair of graphs), helps students remember that each bar in the histogram represents one or more pieces of data of the same value, distinguishing these graphs from case value graphs (see more on this issue in Chapter 8). Seeing the squared deviations illustrates how deviations far from the mean have a greater impact on the size of the standard deviation.

The Importance of Understanding Variability

> Variability is ... the essence of statistics as a discipline and it is not best understood by lecture. It must be experienced.
>
> (Cobb, 1992)

A major goal of most introductory statistics courses is to help students understand and be aware of the omnipresence of variability and the quantification and explanation of variability (Cobb, 1992). These two topics are also highlighted in the GAISE report (2005a, 2005b):

> *The omnipresence of variability*: Recognizing that variability is ubiquitous. It is the essence of statistics as a discipline and it is not best understood by lecture. It must be experienced.
> *The quantification and explanation of variability*: Recognizing that variability can be measured and explained, taking into consideration the following: (a) Randomness and distributions; (b) patterns and deviations (fit and residual); (c) mathematical models for patterns; (d) model-data dialogue (diagnostics).

Understanding the ideas of spread or variability of data is a key component of understanding the concept of distribution, and is essential for making statistical inferences. While students develop informal ideas of spread in the earlier unit on graphing and describing distributions, they later encounter these ideas more formally as they learn about different measures of variability (e.g., range, standard deviation, and interquartile range), what they mean, how to interpret them, how they compare to each other as statistical summaries of data, and what information they provide and do not provide, and how we use them in analyzing data.

There has been increasing attention paid to the importance of students developing an understanding of and appreciation for variability as a core component of statistical thinking (Cobb, 1992; Moore, 1998). However, it is impossible to consider variability without also considering center, as both ideas are needed to find meaning in analyzing data.

The Place of Variability in the Curriculum

The idea of spread, or variability should permeate the entire curriculum. We advocate introducing ideas of spread first informally, and later formally. Ideas of variability can be introduced the first day of class (see lessons from the data unit) and revisited in the unit on distribution, where students describe the spread or clustering of values in a graph of a distribution. When center is introduced, the idea of deviation from the mean is used to help understand the meaning of the mean, and this idea of deviation from the mean is then revisited when studying standard deviation. While range is a fairly easy concept for students to understand, standard deviation is much more difficult. Interquartile range is also a difficult concept, and best introduced in the context of comparing groups with boxplots, when it is illustrated visually by the width of the box.

It is hard to imagine a situation where one would summarize a data set using only a measure of center or using only a measure of spread. When comparing groups or making inferences, we need to look at center and spread together: the signal, and the noise around the signal. Therefore, ideas of center and spread are most often seen and used together, whether informally describing distributions, looking at theoretical models such as the normal distribution and sampling distributions, or in making inferences.

Review of the Literature Related to Reasoning About Variability

Variation vs. Variability

Before we begin to summarize current research on reasoning about variability, we want to address the question of terminology. An inconsistent use of terminology is noticeable in research studies about variability. While some use "variability" and "variation" interchangeably, others distinguish between the meanings of these two words. Reading and Shaughnessy (2004) suggest the following distinctions: variation is a noun describing the act of varying, while variability is a noun form of the adjective "variable," meaning that something is likely to vary or change. Since this distinction has not yet been agreed upon in the statistics education research community, we note this argument but have chosen to use the term *variability* as the general, omnibus term for these ideas in this chapter.

The Emergent Research About Variability

Recent discussions in the statistics education community have drawn attention to the fact that statistics text books, instruction, public discourse, as well as research have been overemphasizing measures of center at the expense of variability (e.g., Shaughnessy, 1997). Instead, there is a growing consensus to emphasize general distributional features such as shape, center, and spread and the connections among them in students' early experiences with data. It is also suggested to focus students' attention on the nature and sources of variability of data in different contexts, such as variability in a particular data set, outcomes of random experiments, and sampling (Shaughnessy, Watson, Moritz, & Reading, 1999; Gould, 2004). These views are supported by a review of several studies by Konold and Pollatsek (2002) that has shown that "the notion of an average understood as a central tendency is inseparable from the notion of spread" (p. 263). Their well-known metaphor for data as *signal and noise* implies that students should come to see statistics as "the study of noisy processes – processes that have a signature, or signal" (p. 260).

Difficulties in Understanding Variability

Despite the widespread belief in the importance of this concept, current research demonstrates that it is extremely difficult for students to reason about variability and that we are just beginning to learn how reasoning about variability develops (Garfield & Ben-Zvi, 2005). Understanding variability has both informal and formal aspects, moving from understanding that data vary (e.g., differences in data values) to understanding and interpreting formal measures of variability (e.g., range, interquartile range, and standard deviation). While students can learn how to compute formal measures of variability, they rarely understand what these summary statistics represent, either numerically or graphically, and do not understand their importance and connection to other statistical concepts. What makes the understanding of the concept even more complex is that variability may sometimes be desired and of interest, and sometimes be considered error or noise (Gould, 2004; Konold & Pollatsek, 2002), as well as the interconnectedness of variability to concepts of distribution, center, sampling, and inference (Cobb et al., 2003b).

These difficulties are evident, for example, in a series of interview studies with undergraduate students who had earned a grade of A in their college statistics course, Mathew and Clark (2003) found that students could not remember much at all about the standard deviation. In another interview study of introductory statistics students' conceptual understanding of the standard deviation, delMas and Liu (2005) designed a computer environment to promote students' ability to coordinate characteristics of variation of values about the mean with the size of the standard deviation as a measure of that variation. delMas and Liu found that students moved from simple, one-dimensional understandings of the standard deviation that did not consider variation about the mean to more mean-centered conceptualizations that coordinated the effects of frequency (density) and deviation from the mean.

In a study investigating students' statistical reasoning, using the Statistical Reasoning Assessment (SRA), Garfield (2003) found that even students in introductory classes that were using reform textbooks, good activities, and high-quality technology had significant difficulty reasoning about different aspects of variability, such as representing variability in graphs, comparing groups, and comparing the degree of variability across groups.

Developing Students Reasoning About Variability

A variety of contexts have been used in statistics education to study students' reasoning about variability at all age levels. For example, in a study of elementary students, Lehrer and Schauble (2007) contrast students' reasoning about variability in two contrasting contexts: (a) measurement and (b) "natural" (biological). When fourth-graders were engaged in measuring the heights of a variety of objects, distribution emerged as a coordination of their activity. They were able to invent statistics as indicators of stability (e.g., center corresponded to "real" length) and variation of measure (e.g., spread corresponded to sources of error such as tool, person, trial-to-trial). In the context of natural variation activity (growth of plants), these same students (now fifth-graders) had difficulties handling sources of natural variation and related statistics. Activities that promoted investigations of sampling (e.g., "what would be likely to happen to the distribution of plant heights if we grew them again") and comparing distributions (e.g., "how one might know whether two different distributions of height measurements could be considered 'really' different") were found useful in developing students' understanding of variability.

In a design research conducted with students in grades 7 and 8 (Bakker, 2004b), instructional activities that could support coherent reasoning about key concepts such as variability, sampling, data, and distribution were developed. Two instructional activities were found to enable a conceptual growth: A "growing a sample" activity that had students think about what happens to the graph when bigger samples are taken, and an activity requiring reasoning about shape of data.

The advantage in discussing ideas of variability in connection with ideas of center was described by Garfield et al. (2007). In this study with undergraduate students, results indicated that students could develop ideas of "a lot" or "a little" variability when asked to make and test conjectures about a series of variables measuring minutes per day spent on various activities (e.g., time spent studying, talking on the phone, eating, etc.). They also found that by having students reason abut the distributions of these variables informally, they could move them toward comparisons of formal measures of variability (e.g., standard deviation, range, and interquartile range).

Other contexts examined include variability in data (Ben-Zvi, 2004a; Groth, 2005; Konold & Pollatsek, 2002; Petrosino, Lehrer, & Schauble, 2003), bivariate relationships (Cobb et al., 2003b; Hammerman & Rubin, 2003), comparing groups (Ben-Zvi, 2004b; Biehler, 2001; Lehrer & Schauble, 2002; Makar & Confrey, 2005), probability contexts (Reading & Shaughnessy, 2004), measures of spread such as the standard deviation (delMas & Liu, 2005), and sampling (Chance et al., 2004; Watson, 2004). These studies are mostly exploratory and qualitative, and their research goal is often to explore what and how students come to understand ideas of variability in the different contexts. The kinds of questions and activities used in these studies suggest ways we can help students develop reasoning about variability across an entire course as well as assess informal and formal aspects of students' understanding of variability.

Levels of Reasoning About Variability

Based on results from a large sample of students on a survey of variability tasks, Watson et al. (2003) propose a model of students' reasoning about variability. Using the SOLO model (Biggs & Collis, 1982), they qualitatively describe four hierarchical levels of progressively sophisticated understanding of reasoning about variability: Prestructural, Unistructural, Multistructural, and Relational. They suggested that this scale might be useful in tracking student improvement over time and in relation to particular sequences of learning activities. The descriptions of these levels, as well as the work on informal and formal ideas of variability (Garfield et al., 2007) suggest the need for carefully designed activities to lead students to develop higher or more formal levels of reasoning.

Implications of the Research: Teaching Students to Reason About Variability

Noticeably lacking in the current research literature are studies of how to best impact the learning of college students who are typically introduced to variability in a unit of descriptive statistics, following the units of graphing univariate data and measures of center. Measures of variability (or spread) are then introduced, and students learn to calculate and briefly interpret them. Typically, only the formal notion of variability as measured by three different statistics (i.e., the range, interquartile range, and standard deviation) is taught. Students often do not hear the term "variability" stressed again until a unit on sampling, where they are to recognize that the variability of sample means decreases as sample size increases. When students are introduced to statistical inference, variability is then treated as a nuisance parameter because estimating the mean becomes the problem of importance (Gould, 2004).

Given this typical introduction in textbooks and class discussion, it is not surprising that few students actually develop an understanding of this important concept. Good activities and software tools designed to promote an understanding of variability do exist. However, they are typically added to a lesson or given as an assignment instead of being integrated into a plan of teaching, and their impact on student understanding has not been subjected to systematic study. So, while there have been positive changes in introductory statistics classes, they still fall short of giving students the experiences they need to develop statistical thinking and a deep understanding of key statistical concepts.

We would like our students to follow the way statisticians think about variability. When statisticians look at one or more data sets, they often appraise and compare the variability informally and then formally, looking at appropriate graphs and descriptive measures. They look at both the center of a distribution as well as the spread from the center, often referring to more than one representation of the data to lead to better interpretations. Statisticians are also likely to consider sources of variability, including the statistical and measurement processes by which the data were collected.

Konold and Pollatsek (2002) offer the following suggestions about how we might help students and future teachers develop ideas of the signal-noise perspective of various statistical measures:

1. Using processes involving repeated measures;
2. Explorations of stability such as drawing multiple samples from a known population and evaluating particular features, such as the mean, across these replications;
3. Comparing the relative accuracy of different measurement methods;
4. Growing samples – students observe a distribution as the sample gets larger;
5. Simulating processes – students investigate why many noisy processes tend to produce mound-shaped distributions;
6. Comparing groups; or
7. Conducting experiments.

Garfield and Ben-Zvi (2005) outline a list of increasingly sophisticated ideas for constructing "deep understanding" of variability. This list offers a sequence through which students may be guided to develop a deep understanding of this concept, as shown below:

1. Developing intuitive ideas of variability
2. Describing and representing variability with numerical measures
3. Using variability to make comparisons
4. Recognizing variability in special types of distributions
5. Identifying patterns of variability in fitting models
6. Using variability to predict random samples or outcomes
7. Considering variability as part of statistical thinking

Progression of Ideas: Connecting Research to Teaching

Introduction to the Sequence of Activities to Develop Reasoning About Variability

Table 10.1 shows a series of steps that can be used to help students first build informal and then formal ideas of variability. These ideas are first introduced in earlier units on data, distribution, and center. Then the formal idea of standard deviation is introduced and is used to examine and reason about data. The concept of interquartile range is introduced in the later unit on comparing groups, a unit that helps connect ideas of center and spread visually and for the purpose of comparing sets of data to answer a research question. The idea of variability is visited again in the unit on models, when the normal distribution is introduced and the unique characteristics of the mean and standard deviation are shown as part of the Empirical Rule. The interconnections of center and spread are also demonstrated in the sampling, statistical inference, and covariation units. Each time the basic idea of variability is explicitly revisited in that particular context, emphasized, and discussed.

Introduction to the Lessons

While students have been informally introduced to the idea of spread and range earlier, this set of lessons looks more closely at variability and the standard deviation. First, students collect measurements to help them recognize two ways of looking at variability, as noise and as diversity. They informally think about a measure of spread from the center. The second lesson helps develop the idea of standard deviation and encourages students to reason about how this statistics is used to measure and represent variability and factors that affect the standard deviation, making it larger or smaller.

Table 10.1 Sequence of activities to develop reasoning about variability[1]

Milestones: ideas and concepts	Suggested activities
Informal ideas prior to formal study of variability	
• Data vary. Values of a variable illustrate variability	• Meet and Greet Activity (Lesson 1, Data Unit, Chapter 6)
• Variability in Results from a random experiment	• Activities in Lessons 1 and 2, Statistical Models Unit (Chapter 7)
• Informal idea of spread of data by examining a graph or comparing graphs	• Distinguishing Distributions Activity (Lesson 1, Distributions Unit, Chapter 8)
• Range as a simple measure of spread	❖ An activity where students describe distribution, and note range as a measure of spread. (The symbol ❖ indicates that this activity is not included in these lessons.)
Formal ideas of variability	
• Two ideas of variability: diversity or measurement error	• How Big is Your Head Activity (Lesson 1: "Variation")
• Sources of variability, a lot and a little variability	• How Big is Your Head Activity (Lesson 1)
• Averaging deviations from the mean as a measure of spread	• Comparing Hand Spans Activity (Lesson 2: "Reasoning about the Standard Deviation")
• Standard deviation as a measure of average distance from the mean	• Comparing Hand Spans Activity (Lesson 2)
• Understanding factors that cause the standard deviation to be larger or smaller	• What Makes the Standard Deviation Larger or Smaller Activity (Lesson 2)
• How center and spread are represented in graphs?	❖ An activity where students match a set of graphs to the corresponding set of statistics
Building on formal ideas of variability in subsequent topics	
• Range and IQR in a boxplot	• How Many Raisins in a Box Activity (Lesson 1, Comparing Groups Unit, Chapter 11)
• Variability within a group and variability between groups	• Gummy Bears Activity (Lesson 2, Comparing Groups Unit, Chapter 11)
• What makes the range and IQR larger and smaller?	• How do Students Spend their Time Activity (Lesson 4, Comparing Groups Unit, Chapter 11)
• Understanding how and why center and spread are used to compare groups	• How do Students Spend their Time Activity (Lesson 4, Comparing Groups Unit, Chapter 11)
• Role of mean and standard deviation in describing location of values in a normal distribution	• Activities in Lesson 3, Statistical Models Unit (Chapter 7)
• Understanding why and how variability decreases as sample size increases in sampling distributions	• The Central Limit Theorem Activity (Lesson 3, Samples and Sampling Unit, Chapter 12)
• Understanding ideas of variability between and within groups when comparing samples of data	• Gummy Bears Revisited Activity (Lesson 4, Statistical Inference Unit, Chapter 13)
• Variability of data in a bivariate plot	• Interpreting Scatterplots Activity (lesson 1, Co-variation Unit, Chapter 14)

[1] See page 391 for credit and reference to authors of activities on which these activities are based.

Lesson 1: Variation

This lesson is designed to help students reason informally about variability. Students compare measurements for two sets of repeated measurements, to discover two kind of variability: (1) variability as an error of measurement (repeated measures of the same head circumference); and (2) variability as an indicator of diversity (measurements of different people's head circumferences). Students are then introduced to the concept of signal and noise, and discuss the stability of the mean as more data are collected. Student learning goals for the lesson include:

1. Understand different types (sources) of variability (when it's desired and when it's noise).
2. Understand the ideas of mean as signal and variability as noise, from repeated measurements in an experiment.
3. Understand that it is desirable to reduce variability in measurement (by using experimental protocols).

Description of the Lesson

Students are asked to think about variability in the class, and in particular, of head sizes. In the *How Big is Your Head* activity, they plan a method to measure the circumference of each of their heads, keeping track of the decisions they make about measuring. A class discussion about this results in a common protocol to use. Students are given a measuring tape to use, and they measure each of their heads using the protocol established and they record the data on the board (or on a computer spreadsheet).

Next, as a class, the students choose one person who will have their head measured by every student in the class. These measurements are also recorded for the class. Students then work with a partner to obtain a set of additional body measurements (all in centimeters) listed in Fig. 10.2. These data will be used in other activities in the course. These data are later entered in *Fathom*.

Body Data Collection

Height (with shoes on): __________

Arm Span (from fingertip to fingertip with arms out-stretched): __________

Kneeling Height: __________

Hand Length (from the wrist to the tip of the middle finger): __________

Hand Span (from the tip of the thumb to the tip of the pinkie while hand is stretched): __________

Fig. 10.2 Record sheet for the body measurements survey

Fathom is used to create a plot of students' head sizes so that they may be examined and summarized. Unusual values are examined and discussed to see if they are legitimate or the result of a faulty measurement process.

Students then select two numbers that seem reasonable for completing the following sentence. (Note: There is more than one reasonable set of choices.)

> *The typical head circumference for students in this class is about -------- cm give or take about -------- cm.*

These answers as discussed and lead to a discussion of possible reasons for the variability in the measurements of students' head circumferences. Students then consider whether the observed variability could be reduced and if so, how that might happen. They offer suggestions for ways to make the measurements more standard.

Next, the class examines data for the repeated measurements of one student's head circumference. These data are entered into *Fathom* and are graphed and summarized in terms of shape, center, and spread. This graph is then compared to the first graph of all students' head circumferences and reasons for the differences are discussed. This time, the students suggest that the variability is solely due to the measurement process and talk about ways to reduce that variability.

A class discussion of the difference between these two sets of measurements of head circumference includes the different types and sources of variability, and when we might expect (and accept) variability in measurements and when we want to keep it as small as possible. The concepts of *signal and noise* are revisited, and the idea of variability as noise in the case of the repeated measurements of one head is discussed.

A wrap-up discussion includes suggestions for different sources of variability in data; the two kinds of variability are: "diversity" (spice of life) and "error or noise". Students are asked which type of variability we would like to have large and which we would like to have small, and why. They finally come up with some other examples of signal and noise, and to consider what is important in examining and interpreting signal and noise when we explore data.

Lesson 2: Reasoning About the Standard Deviation

This lesson encourages students to reason about the standard deviation. Students begin by visualizing and estimating average distances in an activity involving hand spans. The next part of the lesson is designed to help students improve their reasoning about and understanding of variability by thinking about what a standard deviation is and applying that knowledge to determine which of two graphs has a higher standard deviation. Student learning goals for the lesson include:

1. Understand and informally estimate deviations from the mean and "typical" deviation from the mean.
2. Understand standard deviation as a measure of spread.

3. Understand what makes standard deviations larger or smaller, what types of graphs reveal different amounts of variation.
4. Reason about connections between measure of center and spread, and how they are revealed in graphical representations of data.

Description of the Lesson

In the first activity, *Comparing Hand Spans*, students examine and compare their hands and think about variability in hand spans. Students find the hand span for every person in their group. They use a dot plot to examine how these values vary. They suggest two sources of variability in these measurements, i.e., two reasons why the measurements are not all the same.

Next, students record initials above the dots to identify each case. They find the mean and mark it with a wedge (▲) below the correct place on the number line. They estimate how far each of their hand span measurements is from the mean of their group. They make a second dot plot, this time of the differences (deviations) from the mean for each student in their group, and find the mean of these differences (deviations). Using the idea of deviations from the mean, students suggest a "typical" distance (deviation) from the mean.

Fathom is used to re-create the dot plot and to check their calculations and to compute the standard deviation of the group's hand span data. Students compare the actual standard deviation to the "typical" distance (deviation) the group found earlier and to speculate about the difference in these values.

Students then access the entire set of Hand Spans for the class that were gathered in the pervious activity and find the standard deviation of these measurements. This statistic is compared to the standard deviation of hand spans for the small group of four originally produced, and differences are discussed. Finally, students discuss the idea of a "typical" deviation and the standard deviation.

The second activity *What Makes the Standard Deviation Larger or Smaller* continues the discussion of a typical, or standard, deviation from the mean. First, students examine the following dot plot (Fig. 10.3), which has the mean marked by a vertical line. Students consider how large the deviations would be for each data point (dot).

Next students draw in the plot each deviation from the mean as shown below (Fig. 10.4).

Next, students reason about the average size (length) of all of those deviations, and use this to estimate the standard deviation. They draw the estimated length of the standard deviation. This process is repeated with a second dot plot as shown below (Fig. 10.5).

Then students are given a histogram (Fig. 10.6) and they use the same process, thinking about dots "hidden" by the bars, to draw and estimate the length of the standard deviation. The mean of the data set is given (2.57). Students are encouraged to draw in the appropriate number of dots in each bar of the histogram to make sure they have the appropriate number of deviations.

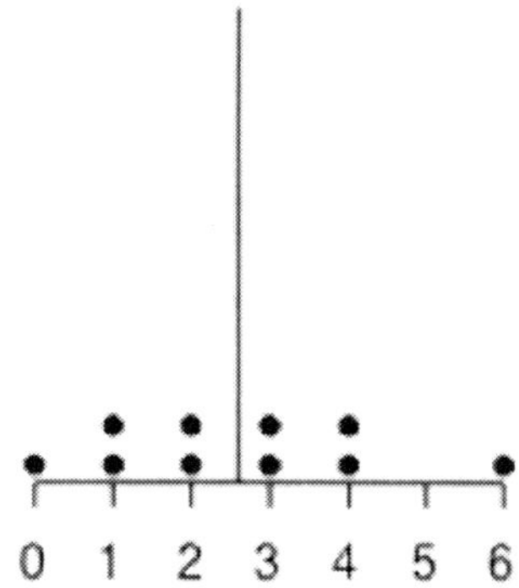

Fig. 10.3 Dot plot from the *What Makes the Standard Deviation Larger or Smaller* activity

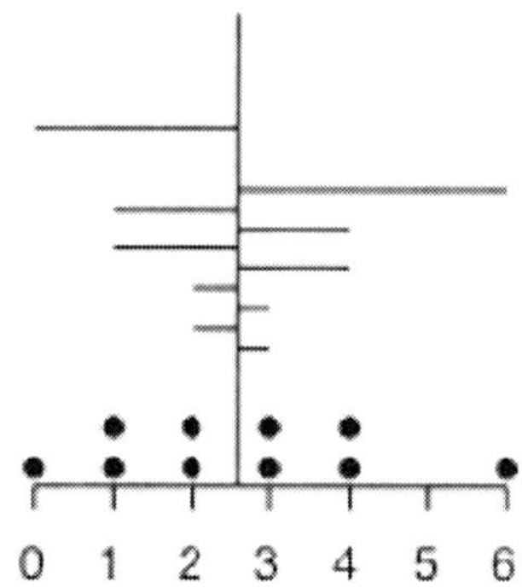

Fig. 10.4 Drawing deviations from the mean in a dot plot from the *What Makes the Standard Deviation Larger or Smaller* activity

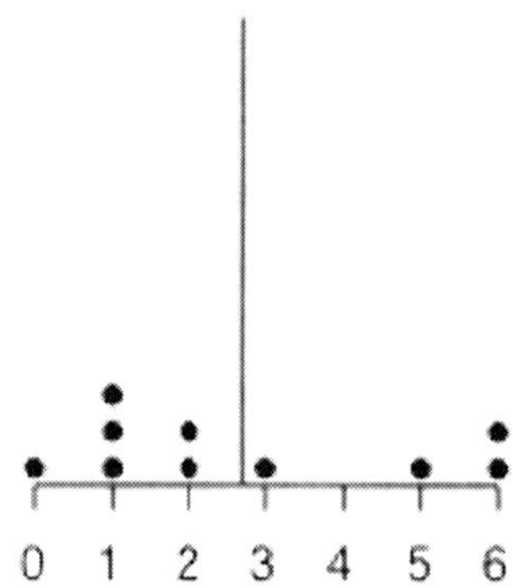

Fig. 10.5 Second dot plot from the *What Makes the Standard Deviation Larger or Smaller* activity

Students are then given six pairs of histograms, for which they are to try to determine which graph in the pair would have a larger standard deviation or would they be the same, and why. The mean for each graph is given just above each histogram. In doing so, students try to identify the characteristics of the graphs that make the standard deviation larger or smaller. Two such pairs of histograms are shown above in Fig. 10.1.

After students complete the set of comparisons, their answers can be discussed and compared as a class and correct answers provided (e.g., the actual size of

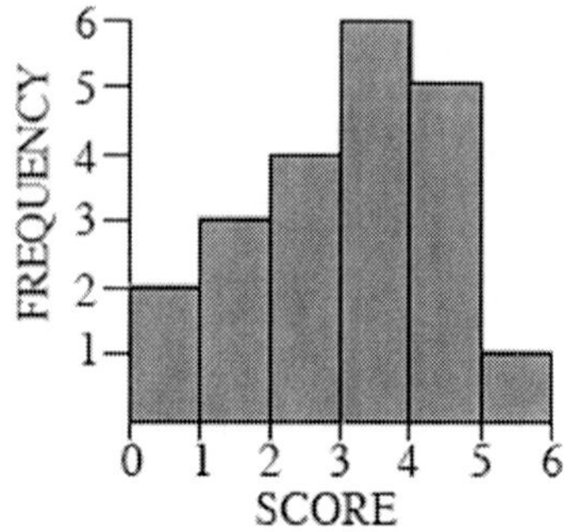

Fig. 10.6 A histogram from the *What Makes the Standard Deviation Larger or Smaller* activity

the standard deviation for each graph in the activity). Students elaborate on which graphs were harder to compare, which were easier and why.

In a wrap-up discussion, students comment on why we need measures of variability in addition to measures of center, and why variability is so important in data analysis. They speculate on why variability is the basis of statistical analysis and how we represent and summarize variability.

Summary

The two lessons in this unit focus mainly on the ideas of types of variability and the meaning of the standard deviation. If students can develop an understanding of this important measure of spread, it will help them learn and reason about the related concept of sampling error in the unit on sampling (Chapter 12) and margin of error in the unit on inference (Chapter 13). The next chapter (Chapter 11) introduces the range and interquartile range as measures of spread in the context of using boxplots to compare groups.

Chapter 11
Learning to Reason About Comparing Groups

> *As statistics moves to the forefront in education, much interest is developing around the process of comparing two groups ... (which) previews an important concept later developed in introductory college statistics courses: statistical inference.*
> *(Makar & Confrey, 2002, p. 2)*

Snapshot of a Research-Based Activity on Comparing Groups

Students are shown a bag of gummy bears (a rubbery-textured confectionery, roughly two cm long, shaped in the form of little bears) and two stacks of books: one is short (one book) and one is high (four stacked books). They are shown a launcher made with tongue depressors and rubber bands (see Fig. 11.1), and are asked to make a conjecture about how the height of a launching pad will result in different distances when gummy bears are launched. The students discuss different rationales for launches traveling farther from either of the height conditions. They are then randomly assigned to small groups to set up and gather data in one of the two conditions, each small group launching gummy bears 10 times to collect data for their assigned height (short or high stack of books).

Once the data are recorded, they are analyzed using boxplots to compare the results for the two conditions. The boxplots are used to determine that the higher launch resulted in further distances.

Students had previously completed an activity that showed them how dot plots can be transformed into boxplots, and are reminded again of the dots (individual data values) hidden within or represented by the boxplot. Their attention is drawn to two types of variability, the variability between the two sets of data (resulting from the two conditions) and the variability in the data: within each group (in each boxplot). Students recall earlier discussions in the variability unit on error variability (noise) and signals in comparing these groups, and they realize the need for an experimental protocol that will help to keep the noise small and reveal clearer signals, so that true difference can be revealed. This experiment is revisited in a later activity when they are able to use a protocol to gather data with less variability and analyze the difference using a t-test (in the Inference unit, see Chapter 13).

J.B. Garfield, D. Ben-Zvi, *Developing Students' Statistical Reasoning*,

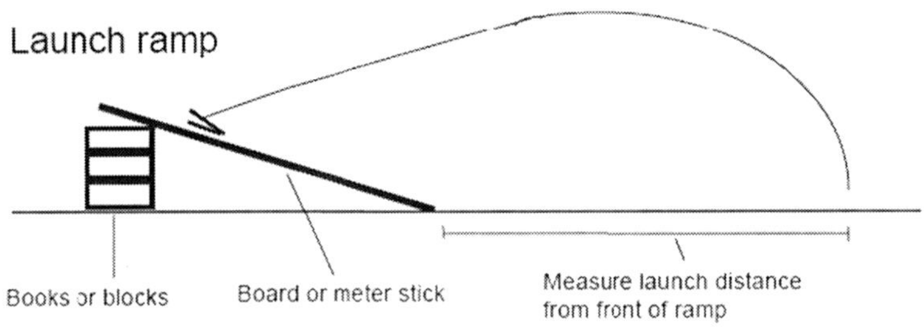

Fig. 11.1 Example of the launching set-up for gummy bears (from Cobb & Miao, 1998)

Rationale for This Activity

While this activity is often used to illustrate the principles of experimental design, we use it in this unit on comparing groups with boxplots for many reasons. First, it provides an interesting and motivating context for comparing two groups of data: to answer a question using an experiment. Second, this activity helps students deepen their understanding of variability, what causes it, how it affects an experiment, how it is revealed in graphs, and its role in comparing two groups of data. Finally, we believe that it is important to revisit principles of experimental design and methods of collecting data so that students can deepen their understanding of these concepts in different contexts and connect these principles to the new topics being studied.

The Importance of Reasoning About Comparing Groups

Comparing two groups of data is an intuitive and interesting task frequently used to engage students in reasoning about data. Many research studies compare two or more groups, either on an experimental variable (e.g., use of a new drug) or on an observational variable (e.g., gender, or age group). While many statistics courses first teach students to graph, summarize, and interpret data for a single group, often activities involving comparisons of more than one group are more interesting and provide the context for meaningful learning. While dotplots can be useful for comparing small data sets, we believe that boxplots are a very useful graphical representation for comparing larger data sets. Although boxplots are often very difficult for students to understand, we think that this graph is extremely useful because it facilitates the comparison of two or more groups, allowing for easy comparisons of center (median), variability, (range and interquartile range) and other measures of location (upper and lower quartile) as well as identifying outliers that may not be revealed in a histogram. While our lessons are designed to help students construct an understanding of boxplots as they may be used to compare data sets, we believe that it is always helpful for students to use different graphical representations in exploring and analyzing data.

The reasons for including "Comparing Groups" as a separate topic of instruction include:

1. Comparing two or more groups can be structured as an informal and early version of statistical inference, and can help prepare students for formal methods of statistical inference.
2. Problems that involve group comparisons are often more interesting than ones that involve a single group.
3. Research shows that students at all ages do not have good intuitive strategies for comparing groups and may have some common misconceptions regarding group comparisons, that need to be explicitly addressed in instruction (e.g., Konold et al., 1997).
4. Comparing groups motivates the need for and use of advanced data representations such as *boxplots*, a graphical display that is best employed in group comparison situations, but which is not easily understood or interpreted by students.

Comparing Groups with Boxplots

Boxplots are part of various graphical tools developed by Tukey (1977) for the purpose of analyzing data. In their review of the literature, Bakker, Biehler, and Konold (2004) suggested why educators began to introduce even young students to boxplots.

> First, the boxplot incorporates the median as the measure of center, and some early research had suggested that the median is easier for students to understand as a measure of center than is the mean (Mokros & Russell, 1995). Boxplots also provide, in the Interquartile Range (IQR), a measure of the degree of spread and an alternative to the computationally more challenging standard deviation (SD). (Besides, a clear geometrical interpretation of the SD can only be developed in the context of normal distributions.) Furthermore, boxplots depict both the measure of spread and center pictorially, which is largely why boxplots are such a powerful way to quickly compare several groups at once. Therefore the boxplot and the interquartile range promised to provide better tools for developing an initial feeling for spread than other graphs and measures of spread. (p. 164–5)

Bakker et al. (2004) describe boxplots as "conceptually rich" tools. To understand them, interpreters need at least to know what minimum, first quartile, median, third quartile, and maximum are. In many situations, they need to understand that the median is used as a measure of the center of a distribution; that the length of the box (not its width) is a measure of the spread of the data; and that the range is another measure of spread" (p. 166).

TinkerPlots, software for precollege-level students (Konold & Miller, 2005; http://www.keypress.com/tinkerplots), includes a simple graphic display called the "hatplot" that can be used to guide students to the more sophisticated idea of a boxplot. Each hat is composed of two parts: a central "crown" and two "brims" on each side of the crown. The "crown" is a rectangle that, in the case of *percentile hatplot*, shows the location of the middle 50% of the data – the Interquartile Range (IQR). The brims are lines that extend out to the minimum and maximum values of the data set. There are four different options for how the crown of a hatplot is formed: based on percentiles (the default, see example in Fig. 11.2), the range,

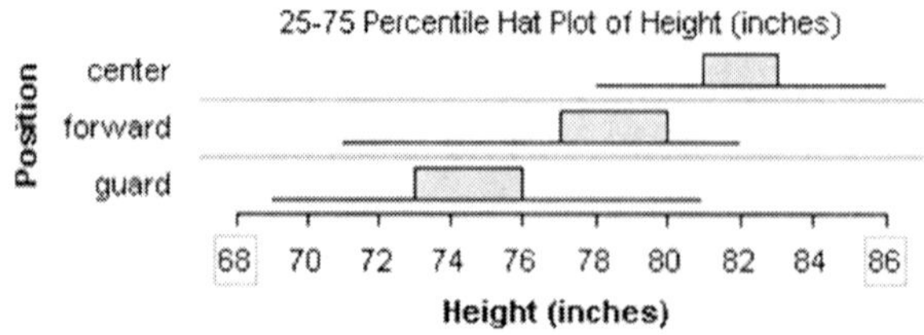

Fig. 11.2 Three parallel percentile hatplots of basketball player's height by their position in *TinkerPlots*

the average deviation, and the standard deviation. Thus, hatplots can be seen as a generalization of a boxplot, and may provide means for allowing students to build on intuitive ideas they have about distributions (Konold, 2002a).

The Place of Comparing Groups in the Curriculum

While this topic can be introduced early in a class, the formal study of comparing groups and boxplots usually takes place after students have studied measures of center and spread, as well as the topic of graphical representations of distributions. However, the ideas of quartile and interquartile range may be introduced at the same time students are learning about boxplots.

After students study this topic, it is helpful to combine all the topics of data analysis and examine them together before moving on to topics leading to statistical inference. We, therefore, offer suggestions in this chapter for activities that integrate ideas of distribution shape, center, and spread, along with comparison of different methods of graphically representing data. Informal inferences are made when comparing groups in this part of a course, laying the foundations for more formal study of statistical inference (see Chapter 13). When groups are compared later on, such as in two sample tests of significance, boxplots are used again to help examine variability between and within groups. Boxplots are revisited again in the unit on covariation (Chapter 14) when multiple boxplots are graphed over time, and the medians help students spot a linear trend.

Review of the Literature Related to Reasoning About Comparing Groups

Studies on comparing groups have focused on how learners approach this topic, what their typical strategies and difficulties are, and how to help them develop their reasoning about comparing groups. Early works indicated and demonstrated that the group comparison problem is one that students do not initially know how to approach and encounter many difficulties with negotiating comparison strategies. Various strategies to improve students reasoning about comparing groups were studied, as well as the role of graphical representations (emphasis on boxplots) in supporting making sense of comparing data sets situations.

Difficulties in Reasoning About Comparing Groups

Primary school students' use of intuitions and statistical strategies to compare simple data sets using line plots was explored by Gal et al. (1989, 1990). Although some students in their study used statistical strategies for comparison, many others focused only on some features of the data, but did not offer a complete synthesis. Others have used incorrect strategies, such as finding totals when they were inappropriate due to different sample sizes, or inventing qualitative explanations such as being better because data are more spread out.

In a follow up study, Watson and Moritz (1999) further identified and detailed categories of school students' reasoning in comparing two data sets. In their study, 88 students in grades three to nine initially compared data sets of equal sizes, but were not able to attend to the issue of unequal sample size. Only in higher reasoning levels, the issue of unequal sample size was resolved with some proportional strategy employed for handling different sizes. The researchers recommend the use of a combination of visual and numerical strategies in comparisons of data sets, "hopefully avoiding the tendency to "apply a formula" without first obtaining an intuitive feeling for the data sets involved" (p. 166).

This recommendation is supported by additional studies showing that students who appear to use averages to describe a single group or know how to compute means did not use them to compare two groups. Gal et al. (1990) found that sixth and ninth grade students did not resort to proportional reasoning or visual comparison of graphs to reach appropriate comparing groups conclusions. Difficulties were found in a case study of two pairs of high school students who were interviewed after a year-long course in which they had used a number of statistics including means, medians, and percents to make group comparisons (Konold et al., 1997). In this study, students did not use any of these comparison techniques during the interview. The researchers claimed that the students' failure to use averages when comparing two groups "was due in part to their having not made the transition from thinking about and comparing properties of individual cases, or properties of collections of homogeneous cases, to thinking about and comparing group propensities" (p. 165). It seems, therefore, that one challenge in instruction of this topic is to make students comfortable summarizing a difference by comparing two groups using some representative measure of center (see Chapter 9 on reasoning about center), a prerequisite to understand the rationale of statistical inference in their advanced studies.

Konold and Higgins (2003) suggested that students' difficulties in comparing groups stemmed from their initial inability to apply "aggregate-based reasoning" – understanding a distribution as a whole, an entity that has many features such as center, spread, and shape (See Chapter 8 on the concept of distribution). Bright and Friel (1998), for example, found that eighth grade students using a stem-and-leaf plot to compare groups, could identify a "middle clump" (where the majority of values are) in a single distribution, but could not use this information to make comparisons. Several students compared just selected individuals from each group. Ben-Zvi (2004b) similarly describes how seventh grade students attend to local details of the comparisons, such as comparing the difference between two cells in frequency table, the difference in heights of two adjacent bars in a double bar chart, or comparing disjoint edge values in the distributions, but find it hard to spot and describe the difference between the two distributions as a whole.

Instructional Approaches to Develop Reasoning About Comparing Groups

Researchers have offered different methods, instructional materials and sequences, and technological tools to overcome these difficulties. For example, Cobb (1999) suggests that the idea of "middle clumps" ("hills") helps students gradually develop their reasoning about comparing groups. Students in seven and eight grades, who used the *Minitools* software (Cobb et al., 1997), began to make decisions about group difference by comparing the numbers of cases in each group within narrow intervals of the range, and gradually moved to referring to global features of the distributions such as shape, center, and spread.

The introduction of new technological tools to support students' reasoning about comparing distributions has created new opportunities in the pedagogy and research of this topic. Hammerman and Rubin (2004) describe how teachers used a new dynamic data visualization tool (*TinkerPlots*) to divide distributions into slices and consequently compare frequencies and percentages within these slices to make inferences (see grey-shaded "slices" in Fig. 11.3). The type of thinking observed was "slice-wise comparison across groups," which tended to ignore the distribution as a whole. The researchers suggest that using this new tool engendered and made visible thinking that had previously lain dormant or invisible. These teachers' slice-wise comparison reasoning seemed to be an extension of the "pair-wise comparison" type of reasoning, which involves comparisons of two individual cases or data values, that other researchers have documented (e.g., Ben-Zvi, 2004b; Moritz, 2004).

In a follow up study, Rubin et al. (2005) found that teachers characterized data using both traditional aggregate measures such as the mean and median as well as novel methods for looking at data such as numbers or percentages around cut points, modal clumps, and overall shape. Teachers using *TinkerPlots* and *Fathom* (Key Curriculum Press, 2006; http://www.keypress.com/fathom) increased their confidence in what these measures were telling them when the stories each measure or characterization told pointed in the same direction. Similarly, when multiple samples from the same population gave some consistency in measures, their confidence in the measure was increased (such as in Fig. 11.3). By contrast, when measures pointed in different directions, teachers were found to spend time further exploring the data so that they can better understand what story was really being told, and

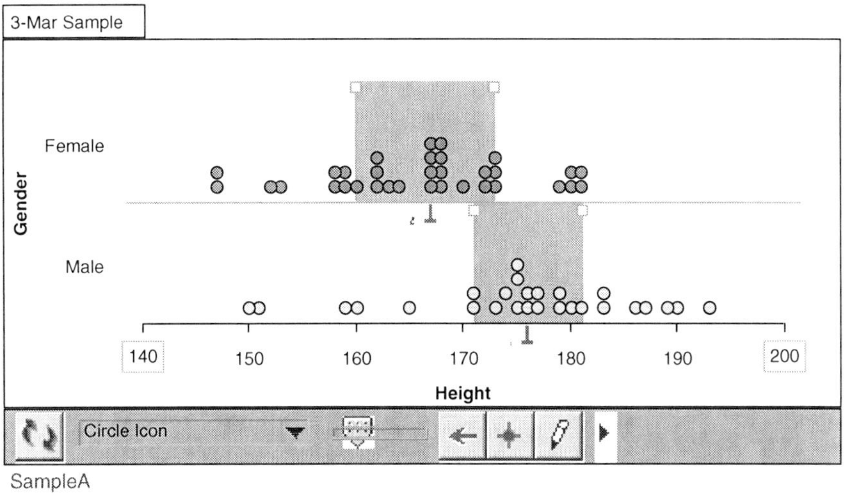

Fig. 11.3 Height data by gender, with mean, median, and IQR marked (From Rubin, Hammerman, Campbell, and Puttick, 2005)

exploring the meaning of the measures to understand what each was telling them. Such explorations become necessary in part when the shape of data is not bell-shaped and symmetric (e.g., skewed distributions), suggesting the importance of learners working with data sets of various shapes in order to more robustly understand the meaning behind various data analytic tools.

Two different kinds of measures of data distributions – *rule-driven* measures and *value-driven* measures were identified by Rubin et al. (2005) as they studied the development of teachers' reasoning about comparing groups. While both of these can describe data in an aggregate way, the researchers believe that value-driven measures are easier to use at first, perhaps until the meaning and implications of the rule that produce a rule-driven measure are clear. They also described how some people use rule-driven measures to create a value around which to make a value-driven comparison, and speculated about the relative power of using such a value rather than one chosen at random, although context driven values might be more powerful still.

Research on Learning to Understand Boxplots in Comparing Groups Situations

As described earlier, the boxplot is a valuable tool for data analysis. The use of boxplots allows students to compare groups of data by examining both center and spread, and to contrast from within group variability to between groups variability. However, several research studies have identified problems students having understanding and reasoning about boxplots. For example, Bakker et al. (2004) claim that several features of boxplots make them particularly difficult for young students to use in authentic contexts. For example, boxplots obscure information on individual cases, the median (shown by the line in the box) does not appear to be as intuitive to students as a measure of center, and the use of quartiles (to construct the box and show the upper and lower boundaries of the box) are difficult for students to fully understand. Bakker et al. (2004) suggest an explanation for these difficulties.

> Quartiles are particularly tricky. Not all integers can be divided by 4, and there is the additional complexity of how to deal with cases that have the same value. There are different ways of doing this, and thus different definitions of quartiles. Computer programs use different definitions, and these definitions are not always well-documented (Freund & Perles, 1987) … Quartiles do not match well the way students tend to conceive of distributions. In several recent studies, researchers noted that students tend to think of a distribution as comprising three parts, rather than four. They think about (a) the majority in the middle (which usually includes more than 50% of the cases); (b) low values; and (c) high values (Bakker & Gravemeijer, 2004; Konold, Robinson, Khalil, Pollatsek, Well, Wing, & Mayr, 2002). Students also referred to the center majorities as "clumps," which was why Konold and colleagues (2002) propose calling them "modal clumps." (p. 167–8)

In light of these major hurdles, Bakker et al. (2004) recommend that educators consider the various features of boxplots and carefully determine whether, how, and when to introduce boxplots to students at a particular grade level.

Boxplots are difficult even for teachers to fully understand. In a study of secondary teachers at the end of a professional development sequence, Makar and Confrey (2004) used interviews to study how teachers reasoned with boxplots with *Fathom* to address the research question, "How do you decide whether two groups are different?" The researchers found that the teachers were generally comfortable working with and examining traditional descriptive statistical measures as a means of informal comparison. However, they had major difficulties in regard to variability, in particular how to (1) interpret variability within a group; (2) interpret variability between groups; and (3) distinguish between these two types of variability.

Implications of the Research: Teaching Students to Reason About Comparing Groups

The research studies highlight that students have many difficulties understanding comparing groups, boxplots and the related ideas of quartiles, median, and interquartile range. It is not intuitive for students to look at data as an aggregate when comparing groups, so they need to be guided in this process. There are many times in an introductory statistics course when it is appropriate to compare two or more sets of data, and over time the guidance can be decreased. It may help to begin with more informal intuitive comparing methods first and then eventually move to more formal methods.

Group comparisons require students to revisit and integrate previously learned ideas about distribution: shape, center, and spread. Difficulties students have in roughness of quartiles and various methods for finding them, can be helped if students find quartiles by dividing data "roughly" into four groups, and not worrying about more precise computational details.

In order to understand the ideas of center and variability represented in the boxplots, students should have many opportunities to look at multiple graphs and multiple statistics for the same variables, so that they may see how these ideas are reflected using these different types of summaries. This also helps them to see that we do not just compare means or medians when comparing groups, but we also need to examine variability. Statistical thinking should be modeled for students as comparison of groups involves discussions of variability between and variability within groups, and how that affects inferences, even if they are informal. For example, even though one group has a high mean than the other, there is so much scatter and spread in the groups that it is hard to tell what the "trend" or "signal in the noise" is.

Since students often do not see the data values hidden in a boxplot, they tend to equate length of whiskers or width of the box with amount of data. Therefore, students need opportunities to see the data behind the box, using physical and computer examples. Students may confuse the height of a horizontal boxplot with frequency of data, so it is important to have students notice and play with this dimension so they realize that it does not indicate anything about the variable or its frequency. Finally, counterintuitive examples may help students improve their understanding and reasoning, such as presenting students with two groups of data where one has a higher interquartile range, but the other has a higher standard deviation, and why these are different.

The Role of Technology in Helping Students to Reason About Comparing Groups

Research suggests that students should be scaffolded to reason with boxplots through keeping the data in dotplot form, under the boxplots (Bakker et al., 2004). We find *TinkerPlots* useful for helping students learn to understand and reason about boxplots. This tool allows students to see how a dotplot can be transformed to a boxplot, first showing where the dots are in boxplot before they are hidden.

While different Web applets exist for boxplots, they typically show the five number summary for a boxplot of data, allow one group to be expanded into several boxplots based on a categorical variable (e.g., http://www.shodor.org/interactivate/activities/boxplot/?version=1.4.2 & browser=MSIE & vendor=Sun_Microsystems_Inc.0), or rotate back and forth between a boxplot and histogram of data (e.g., http://nlvm.usu.edu/en/nav/frames_asid_200_g_4_t_5.html?open=instructions). These applets can be useful to help students interpret boxplots and learn how different features of data sets are presented differently in a histogram or a boxplot.

Progression of Ideas: Connecting Research to Teaching

Introduction to the Sequence of Activities to Develop Reasoning About Comparing Groups

Students begin comparing groups informally when they examine graphs for different variables and data sets in the earlier unit on distribution, and as they examine graphs of data in units on center and variability. Boxplots are introduced as a more formal method of comparing two groups of data, once students have already studied basic ideas of center and spread. Discussions can be focused on what we compare, when we compare two dotplots, and which is center and spread. Students discuss where the middle half of the data is in the groups being compared, which can be graphically illustrated using a hatplot (from *TinkerPlots*). This allows students to see the data set as an entity rather than as points or slices of data (see Chapter 8), and to compare the middles of the two data sets. Medians can then be added to the hatplots, which transforms them into boxplots. Once boxplots have been introduced, students should be encouraged to make the connections back to dotplots, seeing how the dots in a dotplot map to a boxplot, an idea that is often hidden and confusing to students. Advantages of using boxplots to compare groups can be examined, as students see that it is easy to compare both center and spread simultaneously when comparing boxplots for data sets.

In order to further develop students' reasoning about boxplots, students can be given sets of boxplots and histograms and match the two graphs that are for the same set of data, allowing then to think about how features of a histogram would show up in a boxplot (e.g., symmetry, skewness, outliers) and vice versa. Students can then be given different sets of boxplots to compare as they answer research questions about how these plots reveal group differences. This can lead to informal inferences with boxplots as students consider differences in means relative to variability. Table 11.1 shows a series of steps that can be used to help students first build informal ideas and then formal ideas of comparing groups with boxplots.

Introduction to the Lessons

There are four lessons that lead students to compare groups and develop the idea of boxplot as a graphical representation of data that reveals both center and spread and facilitates comparisons of two or more samples of data. The lessons begin with a comparison of two brands of raisins to show that boxplots help in making comparisons and informal inferences. Then students are guided to examine more carefully the characteristics of a boxplot, moving from a dotplot to a hatplot to boxplot, to show how the dots are hidden by the plot, and what the parts of the box represent. The second lesson has students make informal inferences using boxplots to compare distances for Gummy Bears launched using two different heights for launching pads and focuses on comparing groups of data using boxplots. The third lesson develops students' understanding and use of boxplots by having them interpret boxplots in

Table 11.1 Sequence of activities to develop reasoning about comparing groups with boxplots[1]

Milestones: ideas and concepts	Suggested activities
Informal ideas of comparing groups	
• Informal comparisons of dot plots and histograms	• Activities in Lessons 1 and 2 of the Distribution Unit (Chapter 8)
• Comparison of graphs to determine which has a higher and lower standard deviation	• What Makes the Standard Deviation Larger or Smaller Activity? (Lesson 2, Variability Unit, Chapter 10)
Formal ideas of comparing groups with boxplots	
• Data as an aggregate rather than points and slices when comparing groups	• How Many Raisins in a Box Activity (Lesson 1: "Understanding Boxplots")
• How a boxplot represents a data set, how points are "hidden" in a boxplot	• How Many Raisins in a Box Activity (Lesson 1)
• Coordination of comparisons of center and spread in comparing groups	• Gummy Bears Activity (Lesson 2: "Comparing Groups with Boxplots")
• How variability between groups and variability within groups are used in comparing groups	• Gummy Bears Activity (Lesson 2)
• Advantages of using boxplots to compare groups	• Comparing Boxplots Activity (Lesson 2)
• How to make informal inferences from comparisons of samples of data using boxplots	• Interpreting Boxplots Activity (Lesson 3: "Reasoning about Boxplots")
• Understanding how features of data are revealed in different graphs of the same data	• Matching Histograms to Boxplots Activity (Lesson 3)
• Integrating reasoning about shape, center, and spread in different graphical representations	• How do Students Spend Their Time Activity (Lesson 4: "Comparing Groups with Histograms, Boxplots, and Statistics")
Revisiting the idea of comparing groups in subsequent units	
• Variability between groups and variability within groups when making formal inferences involving two samples of data	• Gummy Bears Revisited Activity (Lesson 4, Statistical Inference Unit, Chapter 13)

answering different research questions, and then match boxplots to histograms. The final lesson, integrates all the main ideas in data analysis as students use boxplots (and other graphs and statistics) to analyze a multivariate data set, exploring which variables have larger and smaller amounts of variability.

[1] See page 391 for credit and reference to authors of activities on which these activities are based.

Lesson 1: Understanding Boxplots

This lesson introduces the boxplot as a way to graphically compare two or more groups of data. It has students progress from comparing groups with dotplots, to using hat plots (a feature of *TinkerPlots*) and finally moving to boxplots. By using *TinkerPlots*, students are able to see the data values "hidden" in a boxplot. Students then examine and compare two groups of data in a series of questions using boxplots. Student learning goals for this lesson include:

1. Understand that a boxplot shows where certain percentages of data lie.
2. Understand that a boxplot offers a good way to compare groups of data.
3. Begin to reason about comparing groups using boxplots.
4. Learn how to read and interpret boxplots.
5. Become more fluent in comparing groups of data by comparing shapes, centers, and spreads of two data sets given in boxplots.

Description of the Lesson

The lesson begins with a question about how different brands of the same food product vary, and whether all similar products (of the same size) give the same amount (e.g., number of M&M candies in a small bag, or "does the same size bag of potato chips from two competing companies, give the same amount of chips in each box?"). Students are asked how they can make an informed decision about which product to purchase, and this leads to the need to collect and examine some data.

In the *How Many Raisins in a Box* activity, students are given small boxes of raisins and data are collected on the number of raisins in each box for two competing brands. The data are first collected as two dotplots, but then the class discusses a better way to graphically compare the two data sets. *TinkerPlots* is used to help develop an understanding of a boxplot. First students talk about ways to compare the two data sets; one option is to compare where most of the data are, and then where the middle halves of the data are. The hatplots graphs in *TinkerPlots* are used, where the "hat" is the middle fifty percent of the data, and the outer brims are the remaining quarters of the data set, as shown in Fig. 11.4.

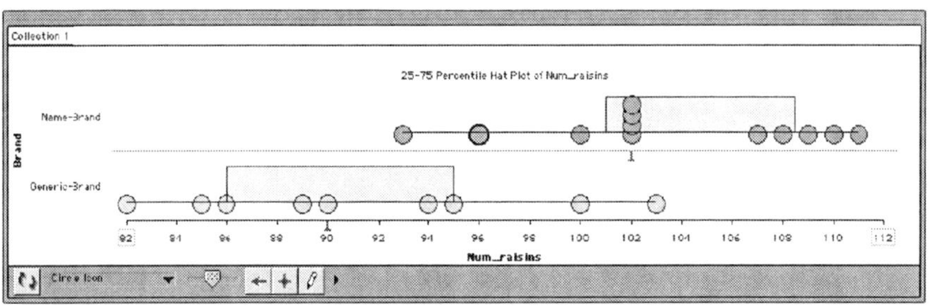

Fig. 11.4 Comparing two data sets (brand of raisins) using the hatplot graph in *TinkerPlots*

Next, the medians are added to the plots by clicking on the *Median* icon (shown in Fig. 11.4 as red "⊥"). Students can count the data values above and below the median and the values in each part of the hatplot.

Finally, the hatplots are converted to boxplots and the individual data values can be hidden (Fig. 11.5). By going back and forth between the hatplot that shows the data values and the boxplot that hides the data values, students are led to see that the two plots include the same data points (number of raisins in a box), that the box of the boxplot is the hat of the hatplot, and that the "whiskers" of each boxplot also includes the same data points as the "brims" of the hat in the hatplot. They can see that the median is now included inside the box of the boxplot as well, and that this is the only important difference between the two plots other than hiding the data values. Students see the individual data points disappear as they go from hatplots to boxplots, illustrating how the boxplot represents the same number of data points (boxes of raisins), the median is still in the same place, and that there are equal numbers of boxes of raisins on either side of the median and in each whisker.

A discussion follows on how boxplots help compare the two brands of raisins showing differences in the center and spread of the numbers of raisins per box. They discuss why this difference exists as well as why there is variability from box to box, and come up with reasons for the two types of variability, within and between brands of raisins. They also make inferences about what they believe to be true for the larger population of boxes of raisins for each brand, based on these samples of data, making informal inferences.

Students then try to reason about and draw two boxplots, with 20 data values each, so that one has a long tail and one has a short tail, but both have five data values in the tail. Next, they reason about and draw a boxplot that would have the mean equal to a quartile, and then two different boxplots that both have ten data

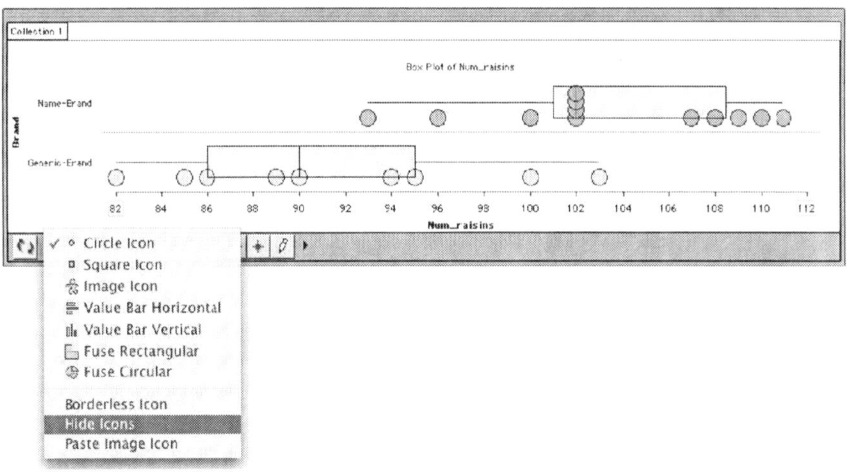

Fig. 11.5 Converting hatplots to boxplots and the option of hiding the individual data values in *TinkerPlots*

values lower than the median. How outliers are determined and represented may also be discussed along with how *Fathom* and other statistical software packages represent outliers on boxplots.

In a wrap-up discussion, students consider pairs of dotplots, histograms and boxplots, and discuss which type of graphical display makes it easier to identify shape, center, and spread, and which type makes it easier to compare groups of data.

Lesson 2: Comparing Groups with Boxplots

This lesson continues to use boxplots to compare groups, but this time the focus is on an experiment. Students make and test conjectures about how the height of a launching pad will result in distances when gummy bears are launched. Two types of variability are examined: the variability within each group and the variability between groups. This also help students distinguish between error variability (noise) and signals (trends), in comparing groups, and then realize the need for little noise and clearer signals, revisiting these ideas from the center and variability units (see Chapters 9 and 10). Student learning goals for this lesson include:

1. Use boxplots as a way to compare results of an experiment.
2. Deepen understanding of boxplots as a graphical representation of data.
3. Use boxplots to visually represent different types (sources) of variability (when it is desired and when it is noise).
4. Revisit the ideas of mean as signal and variability as noise, from repeated measurements in an experiment.
5. Recognize stability of measures of center as sample size increases. When sample grows, see how measure of center predict center of larger population, and how it stabilizes (varies less) as sample grows.
6. Distinguish between variability within treatments and variability between treatments.
7. Understand that it is desirable to reduce variability within treatments (by using experimental protocols).
8. Revisit idea that the only way to show cause and effect is with a randomized experiment.

Description of the Lesson

Students are shown a gummy bear and a launching system made from a tongue depressor and rubber bands. They make conjectures about the following question:

> *Will gummy bears travel a farther distance if they are launched from a steeper height or a lower height? (A stack of four books, or one book)*

They are then given supplies and told how to launch gummy bears and how to measure the distances that they travel. They are asked how students should be assigned to conditions, so that the results may be used to infer cause and effect relationships. Then, randomization is used to assign them to a group that will gather data for one of the two conditions. Students working in groups gather data for their condition: height of one book or height of four books. Data are gathered for 10 launches, and recorded in a table. Data are collected from each group and entered into *Fathom*. Students are asked how they think the data should be summarized and graphed so that they can compare the difference in distances for the two conditions. Various summaries can be generated and various graphs can be examined. Boxplots of a set of data are shown in Fig. 11.6.

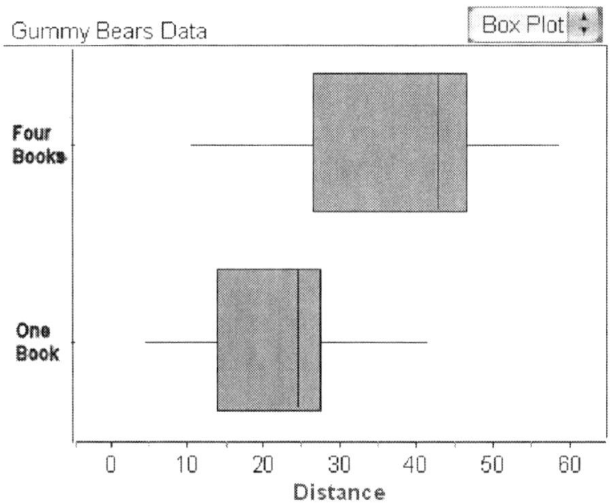

Fig. 11.6 Boxplots comparing the difference in gummy bear distances for the two conditions in *Fathom*

The following questions are used to guide the discussion of results:

- Is there variability in the measurements for each condition? How do we show that variability?
- Is there variability between the two groups (conditions)? How do we look at and describe that variability?
- Why did we get different results for each group within a condition?
- What represents the signal and what represents the noise for each condition?
- How could we get the signal clearer? What would we have to do? (e.g., add more teams to each condition? Have each team launch more bears?)
- If we made a plot of the sample means from each group, how much variability would you expect to see in the distribution? Why?
- Based on our experiment, are we willing to say that a higher launch ramp caused the gummy bears to go farther? What are important parts of an experiment that are needed in order to show causation?

- What are some different sources of variability? There are two kinds of variability: "diversity" and "error, or noise." Which do we like to have large? Which do we like to have small? Why?

The second activity, *Comparing Boxplots,* focuses student's attention on the different kinds of information in a boxplot (e.g., quartiles) and how these can be used in comparing groups. In a wrap-up discussion, students summarize and explain how boxplots help make the comparison of results more visual and apparent, and how they help us examine signal and noise in this experiment.

Lesson 3: Reasoning About Boxplots

This lesson consists of activities that can be used to help students develop their reasoning about boxplots and to deepen their understanding of the concepts of distribution, center, and spread, and how they are interrelated. There are two activities. One has students practice comparing boxplots and second has students try to compare and match histograms to boxplots for the same variables. Student learning goals for this lesson include:

1. Gain experience in using boxplots to compare data sets and draw informal inferences about the populations represented.
2. Move from scaffolded questions to guide their interpretation and comparison of boxplots to situations where the scaffolding is removed and having to analyze the boxplot comparison without guidance.
3. Deepen their reasoning about different representations of data by having to match different graphs of the same data.

Description of the Lesson

In the first activity, *Interpreting Boxplots*, students compare and interpret boxplots. They are given different research questions along with two side by side boxplots. They are asked questions that guide them to make comparison based on the boxplots. The early questions direct their attention to percentages of data in different parts of the boxplot as shown in Fig. 11.7.

The following graph shows the distribution of ages for 72 recent Academy Award winners split up by gender (36 females and 36 males). Use the graph to help answer the following questions.

a) Estimate the percentage of female Oscar winners that were younger than 40.
b) The oldest 50% of male Oscar winners are between which two ages?
c) What would you expect the shape of the distribution to be for male Oscar winners? Explain.

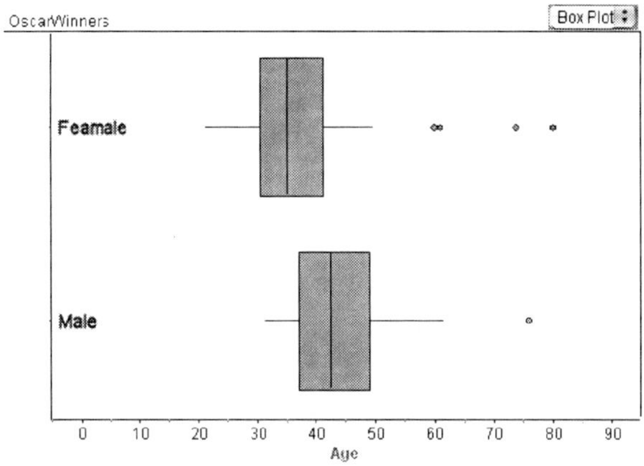

Fig. 11.7 The distribution of ages for 72 recent Academy Award winners split up by gender

d) Explain how to find the Interquartile Range (IQR) for the female Oscar winners.
e) Now, find the IQR for the female Oscar winners.
f) What information does the IQR of the female Oscar winners offer us? Why would a statistician be more interested in the IQR than in the range?
g) Compare the medians for male and female Oscar winners. What do you conclude about the ages of male and female Oscar winners? Explain.
h) Compare the IQR for the male and female Oscar winners. What do you conclude about the ages of male and female Oscar winners now? Explain.

Then other graphs are given with more open ended questions and students work in pairs to discuss and answer these questions. A class discussion allows comparison of answers and explanations of student reasoning.

In the second activity, *Matching Histograms to Boxplots*, students are given a set of five histograms and a set of five boxplots as shown in Fig. 11.8.

Students match each histogram to a boxplot of the same data. This activity requires them to think about how shape of a histogram might be represented in a boxplot, how the median shown in a boxplot might be located in a histogram, and how spread from the center is represented in both types of graphs (e.g., a histogram that is more bell shaped has more clustering to the center and therefore would show a smaller IQR as represented by the width of a boxplot).

A group discussion follows where students are asked which graphs were the easiest to match and why, and which were the most difficult to match and why. They identify how they made the matches, making their reasoning explicit. In a final wrap-up discussion, students are asked what different information is given by histograms and boxplots, and what similar information each provides. They comment on when it is better to use a histogram or a boxplot for a data set and they come to realize the importance of looking at more than one graph when analyzing data.

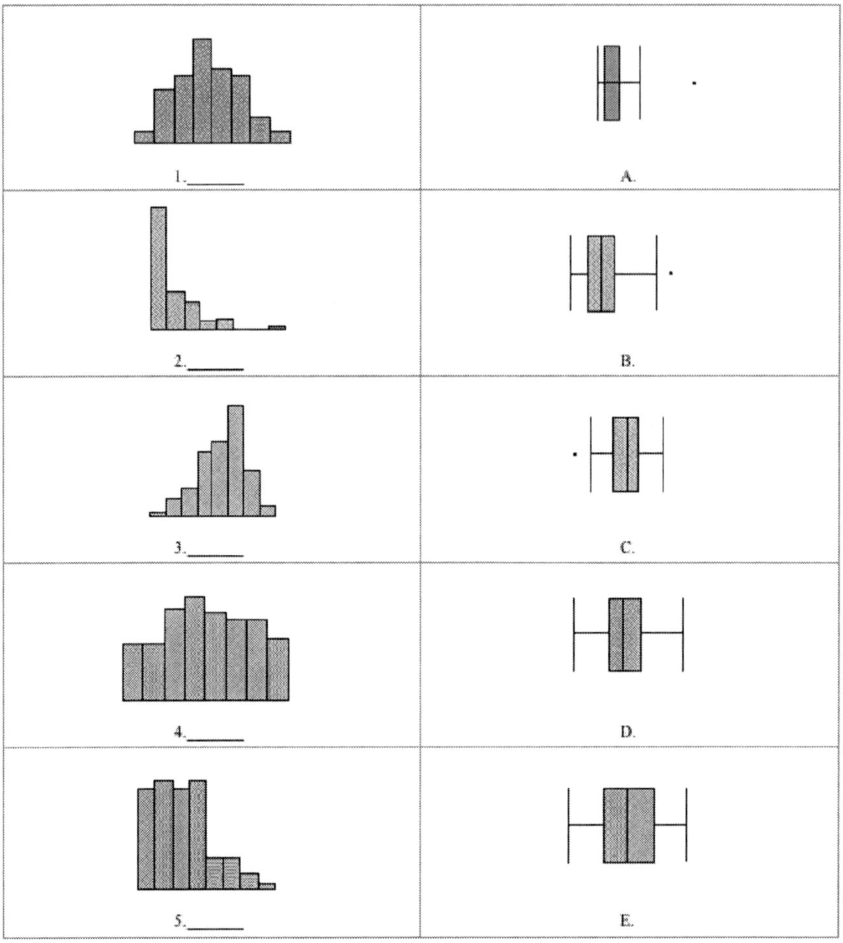

Fig. 11.8 Matching histograms to boxplots activity

Lesson 4: Comparing Groups with Histograms, Boxplots, and Statistics

This lesson builds upon and integrates the ideas of distribution: shape, center, and spread as they analyze a multivariate data set. Students make and test conjectures about variability expected for different variables, and then use graphs and statistics to test their conjectures. The lesson shows that in analyzing real data, we draw on a variety of methods and the answers we give depend on the methods we use. The analysis of multivariate data challenges students to see what they can learn from these data about how students spend their time. Student learning goals for this lesson include:

1. Review the concepts of distribution, center, and spread.
2. Understand how the concepts of distribution, center, and spread are related.
3. Know when to use each type of measure of center and variability.
4. Use boxplots to compare groups.
5. Realize that more than one graph is necessary to understand and analyze data, and that while boxplots are useful to compare groups, histograms (or dotplots) are also needed to better see the shape of the data.
6. Informally analyze a multivariate data set to find answers to open-ended questions that have different possible solutions.

Description of the Lesson

In the activity, *How do Students Spend their Time,* students consider and discuss how similar or different students are in their class in how they spend their time each day. Students are then divided into groups of three or four and predict the average number of minutes per day that students in this class spend on various activities. They record their predictions in a table (Table 11.2).

Students consider the variables and discuss with their group how much variability they would expect to see for each one as well as the shape of the distribution. Then they identify one of the variables that they think will have *little* variability and why they would expect this variable to have a *little* variability, discuss the shape of this distribution and then draw an outline of what they expect this graph to look like, labeling the horizontal axis with values and the variable name and where they expect the mean or median to be. After sharing their results in a whole class discussion, they repeat this activity for a variable that think would have *a lot* of variability.

Data gathered on the First Day of Class Survey (described in Chapter 6 on data, and converted from hours per week to minutes per day) are examined, using software, so that students can compare their predictions to the actual results, discussing any differences they found. In the last part of the activity, students compare side by side boxplots, histograms, and summary statistics for the entire data set of daily times. Students consider and discuss what information is shown in each graph, about

Table 11.2 Students' prediction table in the *How do students spend their time?* activity

Variable	Activity	Prediction of Average Time Spent (minutes per day)
Travel	Traveling to school	
Exercise	Exercise	
Parents	Communications with parents by email, phone, or in person	
Eating	Meals and snacks	
Internet	Time on the Internet	
Study	Study time	
Cell phone	Talk on cell phone	

the variability of the data, and what each summary statistics tells them as well as which graphs and statistics are most appropriate for summarizing each variable. Using all of these data, students then discuss and determine which variable has the smallest and which has the largest amount of variability and justify their answers.

In a wrap-up class discussion, results are compared and the issue emerges that you can answer this question in different ways, depending on the choice of graphs or statistics used. For example, interquartile range may be larger for one variable when you show boxplots, but the standard deviation may be larger for another variable because of outliers in the data set. Students then revisit what each measure of variability tells and how these relate to measures of center and shape of distribution. Students come to explain that there is no simple answer, and the shape, center, and spread are all interconnected. For example, for a skewed distribution with outliers, it is not helpful to use the standard deviation as a measure of variability. Also, it is not helpful to only consider variability; these measures need to be examined along with measures of center in order to meaningfully describe and analyze data. Students also may comment that side by side boxplots were much easier for comparing all the variables than individual dotplots or histograms.

Summary

The activities in this unit provide an important bridge from concepts of distribution, center, and variability (the elements of data analysis) to the ideas of statistical inference. At the same time, the topic of comparing groups helps students integrate and build on ideas of shape, center, and spread, learned in the previous units. Because research has suggested that students often fail to understand or correctly interpret boxplots, we have described a full sequence of activities that are designed to help students better understand and reason about boxplots as a method of graphically representing data as well as an efficient way to compare groups. Without such a careful progression of ideas along with software to help students see the points hidden by the graph, we do not believe most students will understand and correctly use and interpret these graphs.

Chapter 12
Learning to Reason About Samples and Sampling Distributions

Why is the role of sample size in sampling distributions so hard to grasp? One consideration is that ... the rule that the variability of a sampling distribution decreases with increasing sample size seems to have only few applications in ordinary life. In general, taking repeated samples and looking at the distribution of their means is rare in the everyday and only recent in scientific practice.

(Sedlmeier & Gigerenzer, 1997, p. 46)

Snapshot of a Research-Based Activity on Sampling

The activity begins with the teacher holding up a bag of Reese's Pieces candies, which includes a mixture of three different colors (orange, yellow, and brown) and asking students to make conjectures about what they might expect the number of orange candies to be if they take repeated samples of 25 candies from the bag. Students propose different numbers such as 10 or 15. They are also asked if they would be surprised (and complain) if they had only 5 orange candies in their sample of 25 candies. Students consider the variation they would expect to see from sample to sample in the proportion of orange candies, acknowledging that they would not expect every sample of 25 to have exactly the same number of orange candies. Then they count proportions of orange Reese's Pieces in samples of 25 candies presented to them in small paper cups, reporting their counts to be numbers like 11, 12, 13, and 14.

The students plot their individual sample proportions of orange candies in a dot plot on the board and use the data to discuss the variability between samples, as well as to estimate the proportion of candies in the large bag from which the candies were sampled. Then students use a Web applet that simulates sampling Reese's Pieces from a hypothetical population (Reese's Pieces Sampling applet from the *Rossman-Chance* Website, http://rossmanchance.com, Fig. 12.3). Students then compare their individual proportions to distributions of sample proportions produced by a claim about the population proportion of orange candies. They predict and test the effect of taking larger or smaller samples on the closeness of each sample proportion to the population parameter, using simulated data from the Web applet. By the end of this activity, students have determined if a result of 5 orange candies in a random sample of 25 Reese's Pieces is surprising and would be cause for complaint.

J.B. Garfield, D. Ben-Zvi, *Developing Students' Statistical Reasoning*,

Rationale for This Activity

This *Reese's Pieces Activity* helps students focus on how they may expect sample statistics to vary from sample to sample when taking repeated samples from a population, an idea that develops slowly with repeated experience with random samples. Students' intuitions about variability between samples are often misleading: many students think samples should be more similar to each other and to the population parameter or more variable than what we can expect due to chance. In this lesson, students also develop an informal sense of statistical inference when they determine if a particular sample result is surprising or unusual, by comparing this result to a simulated distribution of sample statistics.

The approach used in this lesson is to develop conceptual thinking before students are introduced to the formal ideas of sampling distribution and statistical inference. By preceding the computer simulation with a parallel concrete activity, students are more likely to understand and believe the results of the simulation tool, and are able to develop reasonable ideas of sampling variability. Finally, by having students make conjectures that they test with data, they become more engaged in reasoning about the statistical concepts than if the concepts were presented to them.

The Importance of Understanding Samples and Sampling Distributions

Taking samples of data and using samples to make inferences about unknown populations are at the core of statistical investigations. Although much of data analysis involves analyzing a single sample and making inferences based on this sample, an understanding of how samples vary is important in order to make reasoned estimates and decisions. Most introductory statistics classes include distinctions between samples and populations, and develop notions of sampling variability by examining similarities and differences between multiple samples drawn from the same population. In high school and college classes, the study of sampling variability is extended to examining distributions of sample statistics. Looking at distributions of sample means for many samples drawn from a single population allows us to see how one sample compares to the rest of the samples, leading us to determine if a sample is surprising (unlikely) or not surprising. This is an informal precursor to the more formal notion of P-value that comes with studying statistical inference. We note that simulations based on randomization activities, such as those described in Chapter 7 on models, can also be used to examine the place of an observed value in a distribution of sample statistics to judge whether this result may be explained by chance or due to a particular treatment or condition.

Comparing means of samples drawn from the same population also helps build the idea of sampling variability, which leads to the notion of sampling error, a fundamental component of statistical inference whether constructing confidence intervals or testing hypotheses. Sampling error indicates how much a sample statistic may be expected to differ from the population parameter it is estimating. The sampling error is used in computing margins of error (for confidence intervals) and is used in computing test statistics (e.g., the *t* statistic when testing hypotheses)

The idea of a sample is likely to be a familiar one to students. They have all taken samples before (for example, tasting a food sample at a grocery store) and have an idea of a sample as something that is drawn from or represents something bigger. Students seem to have an intuitive sense that each sample may differ from the other samples drawn from the same larger entity. It may, therefore, seem surprising that students have such difficulty understanding the behavior of samples when they study statistics, how they relate to a population, and what happens when many samples are drawn and their statistics accumulated in a sampling distribution.

The two central ideas of sampling: sampling representativeness and sampling variability have to be understood and carefully balanced in order to understand statistical inference. Rubin, Bruce, and Tenney (1991) cautioned that over reliance on sampling representativeness leads students to think that a sample tells us everything about a population, while over reliance on sampling variability leads students to think that a sample tells us nothing useful about a population. In fact, the ideas of sample and sampling distribution build on many core concepts of statistics, and if these concepts are not well understood, students may never fully understand the important ideas of sampling. For example, the fundamental ideas of distribution and variability underlie an understanding of sampling variability (how individual samples vary) and sampling distribution (the theoretical distribution of a sample statistic computed from all possible samples of the same size drawn from a population). The idea of center is also involved (understanding the mean of the sampling distribution) as is the idea of model (the Normal Distribution as a model that fits sampling distributions under certain conditions). We also interpret empirical sampling distributions of simulated or collected data in similar ways, sometimes referring to these as, for example, a distribution of 500 sample means, rather than referring to it as a (theoretical) sampling distribution. Finally, samples and sampling variability also build on basic ideas of randomness and chance, or the study of probability.

The Place of Samples and Sampling Distributions in the Curriculum

The teaching of sampling variability typically comes later in introductory statistics courses, after the study of the foundational concepts of distribution, center, and variability. It is often introduced after the formal study of the Normal Distribution and its characteristics, and after studying probability. However, in recent years, some textbooks introduce ideas of sampling early in a course, along with informal ideas of inference that are revisited throughout the course until formalizing them later in chapters on significance tests and confidence intervals.

Sampling and sampling distributions are prerequisite topics that precede the formal methods of making statistical inferences (constructing confidence intervals and finding P-values used to compare a sample of data to a distribution based on a null hypothesis). Often the culminating lesson on sampling distributions is one that introduces and illustrates the Central Limit Theorem (CLT).

The implications of this important theorem are that for large sample sizes, distributions of sample statistics (typically, means and proportions) will be normal or approximately normal. This fact allows us to use the normal probability distribution to estimate probabilities for different intervals of values of sample means (given a particular population or hypothesis) and allows us to use the z or t distribution when making formal inferences. Although this theorem is usually included in every introductory statistics course as a prerequisite for learning statistical inference, most students never appear to understand this theorem during a course, although some are able to repeat or identify correct definitions.

Before technology tools were readily available, most text books showed some pictures of different shaped curves, and then presented what sampling distributions would look like for a few different sample sizes. The purpose was to show that curves of these sampling distributions became more normal and narrower as the sample size increased. Then students were told how to calculate the standard error (*sigma* over the square root of n, $\frac{\sigma}{\sqrt{n}}$, for the sample mean) and to use this to find probabilities (e.g., that a sample mean is larger than 3, under a particular null hypothesis). Students often confuse the standard error with the sample standard deviation (s) and the population value (σ). Some typical textbook questions ask students to calculate a probability for a single value (convert x to z using σ) and then repeat the problem for the same value but of a sample mean (convert $\bar{x}$ to z using $\frac{\sigma}{\sqrt{n}}$). Students often fail to notice or understand the difference between these two procedures.

Review of the Literature Related to Reasoning About Samples and Sampling Distributions

Because sampling and sampling distributions are so confusing to students, there has been a considerable amount of research on this topic, particularly with college students. However, even at the elementary and secondary levels, studies have examined how students understand and misunderstand ideas of samples and sampling.

Studies of Students in Precollege Level Settings

A study of students' conceptions of sampling in upper elementary school by Jacobs (1999) suggested that students understood the idea that a sample is a small part of a whole and that even a small part can give you an idea of the whole. Watson and Moritz (2000a, 2000b) also studied children's intuitive ideas of samples, and identified six categories of children's thinking about this concept. They point out that while students have a fairly good "out of school" understanding of the concept of sample, they have difficulty making the transition to the formal, statistical meaning of this term and the related connotations. For example, one can make good generalizations from a small sample of food or blood to the larger entity from which it was drawn, but these intuitive ideas do not generalize to the notion of sampling variation and the need for large, representative samples in making statistical estimates. Watson and Moritz (2000a) suggest making explicit these differences (e.g., between taking a small cube of cheese which represents a homogeneous entity, with taking a sample from the population of fifth grade students to estimate a characteristic such as height, which is a population that has much variability). Watson (2004), in a summary of research on reasoning about sampling, describes how students often concentrate on fairness, and prefer biased sampling methods such as voluntary samples because they do not trust random sampling as a process producing fair samples. Saldanha and Thompson (2002) found both of these types of conceptions on sampling in high school students in a teaching experiment conducted in a statistics class. Not surprisingly, they found that only the concept of sampling as part of a repeated process with variability from sample to sample supported the notion of distribution needed to understand sampling distributions.

In a teaching experiment with eighth grade students, Bakker (2004b) was able to help students understand that larger samples are more stable (less variable) and better represent the population, using a sequence of "growing samples" activities. In a growing a sample activity (see Chapter 8), students predict and explain what happens to a graph when bigger samples are taken (Konold & Pollatsek, 2002). The goal of the growing samples activity was to use imagined and computer-simulated sets of data to build students' reasoning about sampling in the context of variability and distribution. Activities were designed to begin with students' own ideas and guide them toward more conventional notions and representations. Bakker (2004b) suggests that asking students to make conjectures about possible samples of data push them to use conceptual tools to predict the distributions, which helps them develop reasoning about samples.

Studies Involving College Students

Confusion about sampling has been found in college students and professionals. In their seminal paper, "Belief in the Law of Small Numbers," psychologists Tversky and Kahneman (1971) wrote:

> The research suggests that people have strong intuitions about random sampling; that these intuitions are wrong in fundamental aspects; that these intuitions are shared by naïve subjects and by trained scientists, and that they are applied with unfortunate consequences in the course of scientific inquiry ... People view a sample randomly drawn from a population as highly representative, that is, similar to the population in all essential characteristics. Consequently, they expect any two samples drawn from a particular population to be more similar to one another and to the population than sampling theory predicts, at least for small samples. (p. 24).

Since the publication of this article, many researchers have examined and described the difficulties students have understanding samples, sampling variability, and inevitably, sampling distributions and the Central Limit Theorem. In a summary of articles by psychologists on this topic, Well, Pollatsek, and Boyce (1990) noted that people sometimes reason correctly about sample size (e.g., that larger samples better represent populations) and sometimes do not (e.g., thinking that both large and small samples equally represent a population). To reveal the reasons for this discrepancy, they conducted a series of experiments that gave college students questions involving reasoning about samples and sampling variability. For example, one problem described a post office that recorded the heights of local males when they turned 18. The average height of 18 year old males is known to be 5 ft 9 in. (1.75 m). Information is given on two post offices: A which registered 25 males and B which registered 100 males. Students are asked which set of heights would have a mean closer to 5 ft 9 in. Most were able to correctly pick B, which had a larger set of males. Students were also asked to estimate the percentage of days for each post office that produced an average height greater than 6 ft (1.83 m). Fewer students were able to reason correctly about this problem.

The first part of this problem looked at the accuracy of small samples compared to large samples and the second part asked students to think about which sample mean would be more likely to fall in the tail of a distribution of sample means, far away from the population mean. The researchers found that students used sample size more wisely when asked the first type of question (which sample size is more accurate) than on the question that asked them to pick which sample would produce a value in the tail of the population distribution, indicating that they do not understand the variability of sample means. They also noted that students confused distributions for large and small samples with *distributions of averages* based on large and small samples. The authors concluded that students' statistical intuitions are not always incorrect, but may be crude and can be developed into correct conceptions through carefully designed instruction.

Summarizing the research in this area as well as their own experience as statistics teachers and classroom researchers, delMas, Garfield, and Chance (2004) list the following common misconceptions about sampling distributions:

- The sampling distribution should look like the population (for $n > 1$).
- Sampling distributions for small and large sample sizes have the same variability.
- Sampling distributions for large samples have more variability.
- A sampling distribution is not a distribution of sample statistics.
- One sample (of real data) is confused with all possible samples (in distribution) or potential samples.
- The Law of Large Numbers (larger samples better represent a population) is confused with the Central Limit Theorem (distributions of means from large samples tend to form a Normal Distribution).
- The mean of a positive skewed distribution will be greater than the mean of the sampling distribution for large samples taken from this population.

In addition, students have been found to believe that a sample is only good (e.g., representative) if the sample size represents a large percentage when compared to the population (e.g., Smith, 2004). To confront the common misconceptions that develop and to build sound reasoning about samples and sampling distributions, statistics educators and researchers have turned to technological tools to illustrate the abstract processes involved in repeated sampling from theoretical populations and help students develop statistical reasoning.

In a series of studies, Sedlmeier and Gigerenzer (1997) revealed that when subjects seem to have a good understanding of the effect of sample size, they are thinking of one frequency distribution (for a sample). When subjects show confusion about sample size, they are struggling with the more difficult notion of sampling distribution (statistics from many samples). Sedlmeier (1999) continued this research, and found that if he converted items that required subjects to consider sampling distributions, to ones that instead required frequency distributions, higher percentage of correct solutions were obtained.

Building on this work, Saldanha and Thompson (2002) studied high school students' reasoning about samples and sampling distributions in a teaching experiment. They identified a multiplicative concept of samples that relates the sample to the population as well as to a sampling distribution in a visual way. This interrelated set of images is believed to build a good foundation for statistical inference, which suggests that instructors clearly help students distinguish between three levels of data: the population distribution, the sample distribution, and the sampling distribution. Lane-Getaz (2006) provides such a visual model in her Simulation Process Model, which we have adapted and called the *Simulation of Samples (SOS)* Model. This model, shown in Fig. 12.1, distinguishes between the first level of data (population), many random samples from the population (level 2) along with sample statistics for each sample, and the distribution of sample statistics (Level 3). In the last level, a sample outcome can be compared to the distribution of sample statistics to determine if it is a surprising outcome, an informal approach to statistical inference.

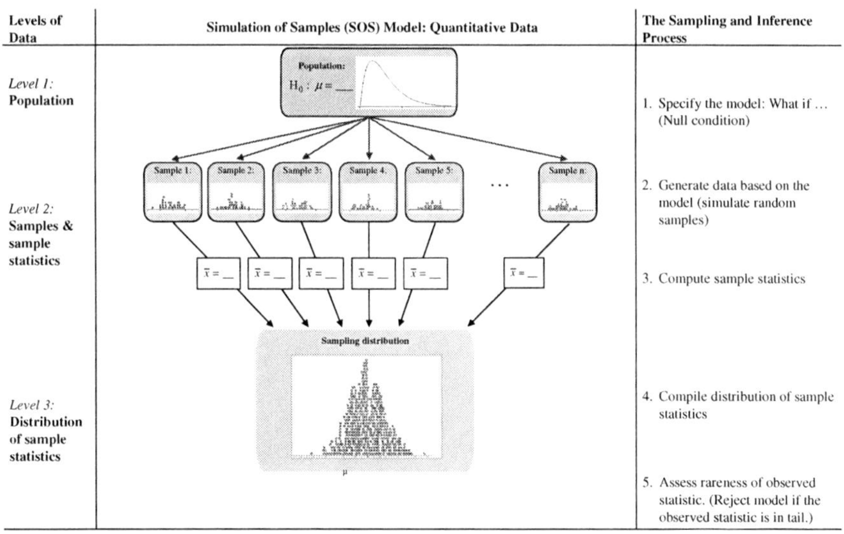

Fig. 12.1 The *Simulation of Samples* model (SOS) for quantitative data

Use of Simulations to Develop Reasoning About Sampling

There are several articles (see Mills, 2002 for a review of these articles) that discuss the potential advantage of simulations in providing examples of the process of taking repeated random samples and allowing students to experiment with variables that affect the outcomes (sample size, population parameters, etc.). In particular, technology allows students to be directly involved with the "building up" of the

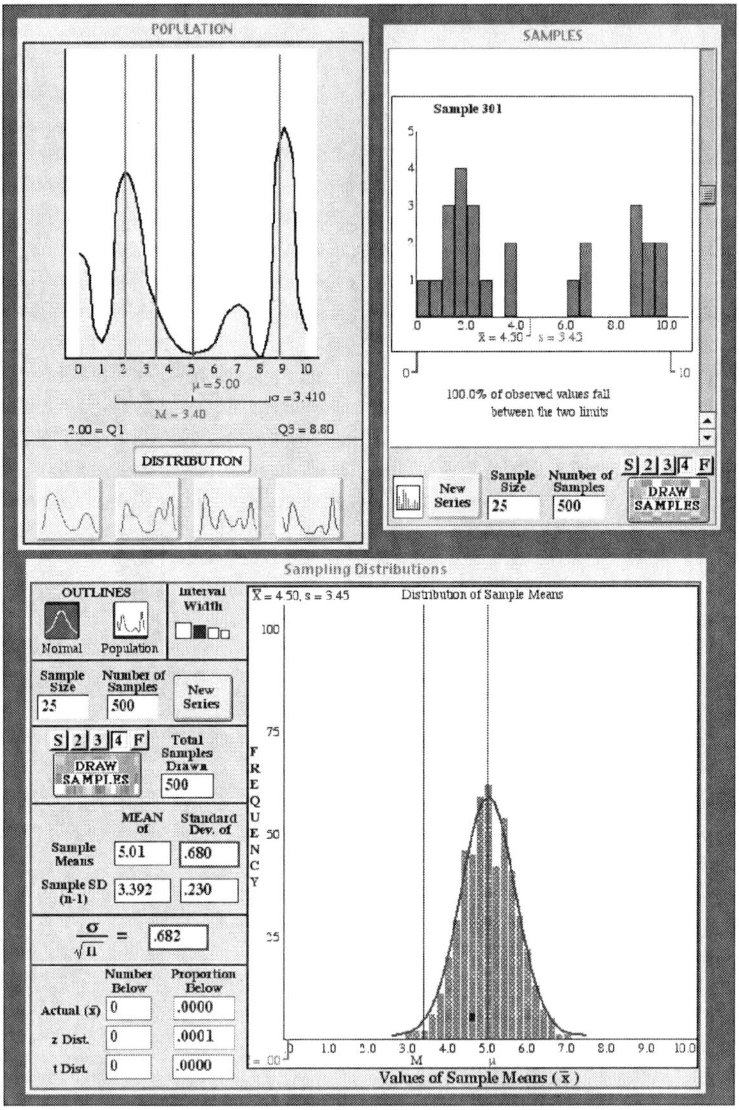

Fig. 12.2 Screen shots of the *Sampling SIM* software

sampling distribution, focusing on the process involved, instead of being presented only the end result. Recently, numerous instructional computer programs have been developed that focus on the use of simulations and dynamic visualizations to help students develop their understanding of sampling distributions and other statistical concepts (e.g., Aberson, Berger, Healy, Kyle, & Romero, 2000). However, despite the development of flexible and visual tools, research suggests that just showing students demonstrations of simulations using these tools will not necessarily lead to improved understanding or reasoning.

Chance, delMas, and Garfield (2004) report the results of a series of studies over a 10-year period that examined various ways of having students interact with the *Sampling SIM* software (delMas, 2001a). *Sampling SIM* software, described in more detail later in this chapter, is a program that allows students to specify different population parameters and generate random samples of simulated data along with many options for displaying and analyzing these samples (see Fig. 12.2). They found that it worked better to have students first make a prediction about a sampling distribution from a particular population (e.g., its shape, center, and spread), then to generate the distribution using software, and then to examine the difference between their prediction and the actual data. They then tried different ways to embed this process, having students work through a detailed activity, or be guided by an instructor. Despite students appearing to be engaged in the activity and realizing the predictable pattern of a normal looking distribution for large samples from a variety of populations, they nonetheless had difficulty applying this knowledge to questions asking them to use the Central Limit Theorem to solve problems. An approach to using the software combines a concrete activity (Sampling Reese's Pieces) with the use of some Web applets, before moving to the more abstract *Sampling SIM* Software. This sequence of activities will be described in the following section.

Implications of the Research: Teaching Students to Reason About Samples and Sampling Distributions

We believe that it is important to introduce ideas of sample and sampling to students early in a statistics course, preferably in a unit on data production and collection. By the time students are ready to study the formal ideas of sampling distributions, they should have a good understanding of the foundational concept of sample, variability, distribution, and center. They should also understand the model of the Normal Distribution and how that model may be used to determine (or estimate) percentages and probabilities (e.g., use the Empirical Rule).

As students learn methods of exploring and describing data, they should be encouraged to pay attention to ideas of samples and to consider sampling methods (e.g., where did the data come from, how was the sample obtained, how do different samples vary). By the time students begin the formal study of sampling variability, they should understand the nature of a random sample and the idea of a sample being representative of a population. They should understand how to choose a good sample and the importance of random sampling.

The study of sampling variability typically focuses on taking repeated samples from a population and comparing sample statistics, (such as the sample mean or the sample proportion). There is a lack of agreement among college statistics teachers and textbooks about whether to begin the study of sampling distributions with proportions or means. Both have their advantages and disadvantages. Chance and Rossman (2001) present both sides of the disagreement in the form of a debate. They conclude that it is more important to pay careful attention to ideas of data collection and sampling throughout the introductory course, than which statistic is introduced first. They recommend that much time should be spent on sampling distributions so that students will be able to use these ideas as a basis for understanding statistical inference.

DelMas et al. (2004) list desired learning outcomes for the situation where distributions of means are used. These learning outcomes include understanding that:

- A sampling distribution for means (based on quantitative data) is a distribution of sample means (statistics) of a given sample size, calculated from samples that are randomly selected from a population with mean μ and standard deviation σ. It is a probability distribution for the sample mean.
- The sampling distribution for means has the same mean as the population (parameter).
- As n gets larger, variability of the sample means gets smaller (as a statement, a visual recognition, and as a prediction of what will happen or how the next picture will differ).
- The standard error of the mean is a measure of variability of sample statistic values.
- The building block of a sampling distribution is a sample statistic. In other words, the units shown in the distribution are sample statistics, rather than individual data values (e.g., measurements).
- Some values of statistics are more or less likely than others to result from a sample drawn from a particular population.
- It is reasonable to use a normal approximation for a sampling distribution under certain conditions.
- Different sample sizes lead to different probabilities for the same value (know how sample size affects the probability of different outcomes for a statistic).
- Sampling distributions tend to look more normal than the population, even for small samples unless the population is Normal.
- As sample sizes get very large, all sampling distributions for the mean tend to have the same shape, regardless of the population from which they are drawn.
- Averages are more normal and less variable than individual observations. (Again, unless the population is Normal.)
- A distribution of observations in one sample differs from a distribution of statistics (sample means) from many samples (n greater than 1) that have been randomly selected.

- A sampling distribution would look different for different populations and sample sizes (in terms of shape, center, and spread, and where the majority of values would be found).
- Some values of the sample mean are likely, and some are less likely for different sampling distributions. For sampling distributions for a small sample size, a particular sample mean that is farther from the population mean may not be unlikely, but this same value may be unlikely for a sample mean for a larger sample.
- That the size of the standard error of the mean is determined by the standard deviation of the population and the sample size, and that this affects the likelihood of obtaining different values of the sample mean for a given sampling distribution.

Moving from Concrete Samples to Abstract Theory

The research suggests that when students view simulations of data, they may not understand or believe the results, and instead watch the simulations without reasoning about what the simulation represents. Therefore, many instructors find it effective to first provide students with concrete materials (e.g., counting Reese's Pieces candies or pennies) before moving to an abstract simulation of that activity. One nice feature of the Reese's Pieces Samples applet described below is that it provides an animation to show the sampling process that can later be turned off, to provide data more quickly when students understand where the data values are coming from.

One of the challenges in teaching about sampling distributions is that students are already familiar with analyzing samples of data, a concrete activity. When they are asked to imagine many, many samples of a given sample size, they are forced to grapple with the theory that allows them to later make inferences. Many students become confused and think the point is that they should always take many samples. It is difficult for students to understand that in order to later make inferences from one single sample from an unknown population; they must first observe the behavior of samples from a known population. This is far from a trivial task, and using technology seems to help. It is also helpful to show a model of the sampling process that distinguishes between three levels: the population, the samples, and the sample statistics calculated from those samples (see the *SOS Model* in Fig. 12.1 above).

Technological Tools to Visualize Sampling and Sampling Distributions

Perhaps the first technology tool designed to illustrate sampling and the Central Limit Theorem (CLT) was the *Sampling Laboratory*, which ran only on the Macintosh platform, described by Rubin and Bruce (1991). This program visually illustrated the process of taking samples from a population and accumulating distributions of particular characteristics (e.g., mean or median). Individual samples could also be displayed. Since then, many Web applets and simulation programs have been developed and are often used to visually illustrate the CLT. *Sampling SIM* software (delMas, 2001a) as mentioned earlier is a free-standing program that was developed to develop students' reasoning about sampling distributions. Students may select different types of populations (e.g., right-skewed, bimodal, normal) as well as different means and standard deviations, and then explore the impact of changing these parameters on the resulting simulated data. Figure 12.2 shows three windows that may be displayed simultaneously in the *Sampling SIM* program (delMas, 2001a). The Population window shows a population that may be continuous (as shown), discrete (bars), or binomial. Users may select particular populations of interest (e.g., normal, skewed, bimodal), choose one of the irregular shapes shown at the bottom of the Population window, or create their own distributions by raising and lowering the curve or bars using the mouse. The Samples window shows a histogram of each random sample that is drawn along with the sample statistics. The Sampling Distribution window shows the distributions of sample statistics as they accumulate, during the sampling process. Lines show the placement of the population mean and median, and a curve representing the Normal Distribution can be superimposed on the distribution of sample statistics, as well as the outline of the original population.

Among the best of the many Web applets that simulate sampling and sampling distributions are applets from the *RossmanChance* Website (http:// rossmanchance.com/) that help students make the bridge between sampling objects to

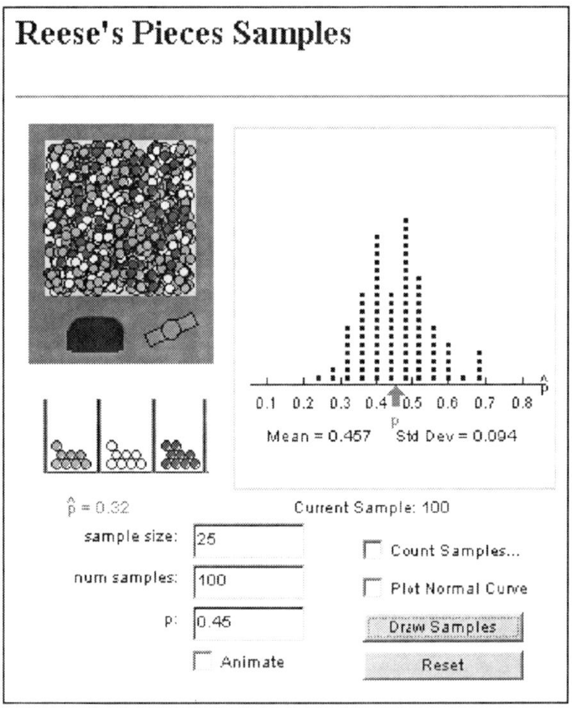

Fig. 12.3 Reese's pieces sampling applet from the *RossmanChance* Web site

sampling abstract data values. For example, the Reese's Pieces Samples applet (see Fig. 12.3) samples colored candies from a population that has a specified percent of orange candies. Different sample sizes and numbers of samples can be drawn. Animation shows the candies being sampled, but this can be turned out after a while to expedite the sampling. There are other *RossmanChance* Web applets that illustrate sampling coins to record their ages, and sampling word lengths from the Gettysburg Address (see Fig. 12.4, used in the *Gettysburg Address Activity*, Lesson 3, Unit on Data, Chapter 6). An advantage in using these applets is that they may be preceded by actual physical activities in class (taking samples of candies, coins, and words), which provide a real context for drawing samples.

Programs such as *Fathom* (Key Curriculum Press, 2006) can also be used to illustrate the sampling process, allowing for parameters to be varied such as sample size, number of samples, and population shape. Although not as visually effective as a computer screen, samples may also be taken and accumulated using graphing calculators. Despite the numerous software tools that currently exist to make this difficult concept more concrete, there is still little research on the most effective ways to use these tools. For example, if a teacher shows a demonstration of the software to the students, is that effective? Is it better for students to interact with the

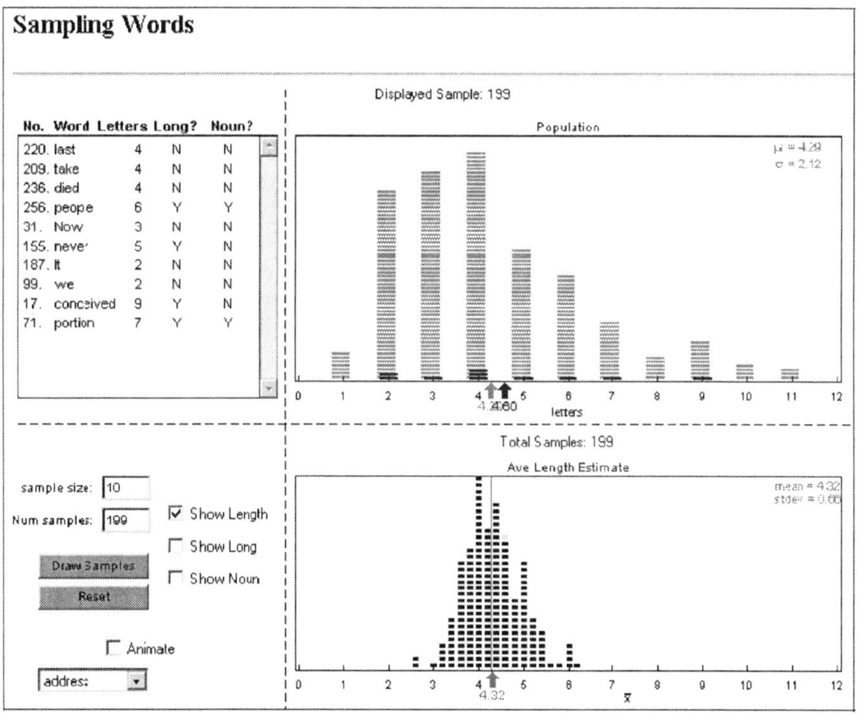

Fig. 12.4 Sampling words applet from the *RossmanChance* Website

software directly, taking their own samples? How much guidance should be given to students when using the software? Are some tools more effective than others? All of these research questions are waiting to be investigated.

Progression of Ideas: Connecting Research to Teaching

Introduction to the Sequence of Activities to Develop Reasoning About Samples and Sampling Distributions

Back in 1991, Rubin and Bruce proposed a sequence of ideas to lead high school students to understand sampling as they participated in activities using *Sampling Lab* software. Their list included:

1. A sample statistic is not necessarily the same as the corresponding population parameter, but it can provide good information about that parameter.
2. Random samples vary, especially for small samples. Therefore, the sample statistics will vary for each sample as well.
3. The variation from sample to sample is not due to error, but is a consequence of the sampling process. It occurs even with unbiased sampling methods and carefully chosen samples.
4. Although sample statistics vary from population parameter, they vary in a predictable way. Most sample statistics are close to the population parameter, and fewer are extremely larger or smaller than the population value.
5. Despite sampling variation, a large enough random sample can be used to make a reasonably good prediction for a population parameter.
6. The goodness of a particular estimate is directly dependent on the size of the sample. Samples that are larger produce statistics that vary less from the population value.

The research and related literature reviewed suggest a progression of activities that can be used to help students develop the ideas of sampling variability and sampling distributions described above. Table 12.1 contains a progression of ideas that build on those suggested by Rubin and Bruce (1991) along with types of activities that may be used to develop these ideas.

One important implication from the research is that it takes time to help students develop the ideas related to sampling distribution, longer than just one class session which is the amount of time typically allotted. In addition, prior to a formal unit on sampling distribution, students need experience in taking samples and learning how samples do and do not represent the population. This may be part of an early unit on collecting data through surveys and experiments, where they learn characteristics of good samples and reasons for bad samples (e.g., bias). Another implication is that a visual model (e.g., the *SOS model*, Fig. 12.1 above) may help students develop a deeper understanding of sampling distribution and inference, if it is repeatedly used when dealing with repeated samples and simulations.

Table 12.1 Sequence of activities to develop reasoning about samples and sampling distributions[1]

Milestones: ideas and concepts	Suggested activities
Informal ideas prior to formal study of samples and sampling distributions	
• Population parameter is fixed, but sample statistics vary from sample to sample	• The Gettysburg Address Activity (Lesson 3, Data Unit, Chapter 6)
• The idea of a random sample	• The Gettysburg Address Activity (Lesson 3, Data Unit, Chapter 6)
• As a sample grows, or as more data are collected, at some point the sample provides a stable estimate of the population parameter	• Growing a Distribution Activity (Lesson 1, Distribution Unit, Chapter 8)
• Larger random samples are more likely to be representative of the population than small ones	❖ An Activity where samples are taken from a specified population and the size of the sample is increased to determine at what point the estimates of the population are stable. (The symbol ❖ indicates that this activity is not included in these lessons.)
• The size of a representative sample is not related to a particular percentage of the population. A large well-chosen sample can be a good one even if it is a small percent of the population	❖ An activity where different sample sizes are examined in light of how well they represent the population in terms of shape, center, and spread
Formal ideas of samples and sampling distributions	
• Sample variability: Samples vary for a given sample size, for a random sample from the same population	• Reese's Pieces Activity (Lesson 1: "Sampling from a Population")
• Variability of sample statistics from sample to sample	• Reese's Pieces Activity (Lesson 1)
• There are three levels of data involved in taking random samples: the population, the individual samples, and the distribution of sample statistics	• Reese's Pieces Activity (Lesson 1)
• How and why statistics from small samples vary more than statistics from large samples	• Reese's Pieces Activity (Lesson 1)
• Sample statistics can be graphed and summarized in a distribution, just as raw data may be graphed and summarized	• Reese's Pieces Activity (Lesson 1)

[1] See page 391 for credit and reference to authors of activities on which these activities are based.

Table 12.1 (continued)

Milestones: ideas and concepts	Suggested activities
• Understanding that a simulation of a large number (e.g., 500) sample statistics is a good approximation of a sampling distribution	❖ An activity using a simulation computer tool that draws students' attention to these ideas
• Understanding that for a large number of trials (simulations) what is important to focus on is the change in sample size, not the change in number of simulations	❖ An activity using a simulation computer tool that draws students' attention to these ideas
• Although sample statistics vary from population parameter, they vary in a predictable way	• Body Temperatures, Sampling Words, and Sampling Pennies Activities (Lesson 2: "Generating Sampling Distributions")
• When and why a distribution of sample statistics (for large enough samples) looks bell shaped	• Central Limit Theorem Activity (Lesson 3: "Describing the Predictable Pattern: The Central Limit Theorem")
• Distributions of sample statistics tend to have the same predictable pattern for large random samples	• Central Limit Theorem Activity (Lesson 3)
• Understanding how the Central Limit Theorem describes the shape, center, and spread of sampling distributions of sample statistics	• Central Limit Theorem Activity (Lesson 3)
Building on formal ideas of samples and sampling distributions in subsequent topics	
• Understand the role of sample variability in making statistical inferences	• Activities (Lessons 1, 2, 3, and 4, Statistical Inference Unit, Chapter 13)

Introduction to the Lessons

Building on the basic idea of a sample, these lessons provide students with experience across different contexts with how samples vary and the factors that affect this variability. This leads to the idea of accumulating and graphing multiple samples from the same population (of a given sample size), which leads to the more abstract idea of a sampling distribution. Different empirical sampling distributions are generated and observed, to see the predictable pattern that is a consequence of the Central Limit Theorem (CLT). Finally, students use the CLT to solve problems involving the likelihood of different values of sample means. Believing that it is more intuitively accessible to students, we begin the study of sampling with proportions and then move to sample means.

Lesson 1: Sampling from a Population

In this lesson, students make and test conjectures about sample proportions of orange-colored candies. They take physical samples from a population of colored candies (Reese's Pieces) and construct distributions of sample proportions. Students then use a Web applet to generate a larger number of samples of candies, allowing them to examine the distribution of sample proportions for different sample sizes. Students map the simulation of sample proportions to the *Simulation of Samples (SOS)* Model (Fig. 12.1), a visual scheme that distinguishes between the population, the samples, and the distribution of sample statistics (see also Chapter 6 where this model is first introduced). Student learning goals for this lesson include:

1. Understand variability between samples (how samples vary).
2. Build and describe distributions of sample statistics (in this case, proportions).
3. Understand the effect of sample size on: how well a sample resembles a population, and the variability of the distribution of sample statistics.
4. Understand what changes (samples and sample statistics) and what stays the same (population and parameters).
5. Understand and distinguish between the population, the samples, and the distribution of sample statistics.

Description of the Lesson

The lesson begins with a discussion of questions relating to what information a small sample can provide about a population. Students discuss their opinions about whether a small sample of Reese's Pieces can provide a good estimate of the proportion of orange Reese's Pieces candies produced by Hershey Company, and whether or not they would be surprised if they found only 5 Orange Reese's Pieces in a cup of 25 candies. These questions lead to a general discussion of sampling that reviews previous material covered in the introductory course, such as: What is a sample? Why sample? What do we do with samples? How should we sample? What is a good sample?

Students are then guided through the *Reese's Pieces* activity. They are first asked to guess the proportion of each color candy in a bag of Reese's Pieces and predict the number of orange candies that they would expect in 10 samples of 25 candies. Next, each student is given a small paper cup of Reese's Pieces candies, and is instructed to count out 25 without paying attention to the color. Then they count the number of orange candies and find the proportion of orange candies for their sample (of 25).

These proportions are collected and graphed and the class is asked to describe the graph in terms of shape, center, and spread. Note: this is not an actual sampling distribution because it does not consist of all possible samples from the population, but it is a *distribution of sample means* (and can serve as an approximation to the sampling distribution). This is an important distinction to make when generating or simulating sample data.

Students are asked to consider what they know and do not know: they do know the sample statistics but do not know the population parameter. They are asked to consider what is fixed and what changes (i.e., the sample statistics change from sample to sample; the population proportion stays the same regardless of the sample). Attention is drawn to the variability of the sample statistics, and students refer back to their predicted sample statistics for 10 samples earlier in the activity. Finally, students are asked to produce an estimate of the proportion of orange candies in the population of all Reese's Pieces candies.

Next, students use a Web applet from *RossmanChance* Website that has a picture of a large container of Reese's Pieces and allows them to draw multiple samples of any size. They use the proportion of orange candies given at the Website (.45), and then draw samples size 25 and watch as distributions of samples proportions are visually created (see Fig. 12.3 above). They are asked to consider what kinds of sample statistics they might expect if there were 10 candies in each sample instead of 25, or 100 candies in each sample. It is easy to change the sample size in the applet and simulate data to see how it affects the variability of the sample proportions for different sample sizes. Students are asked to complete a blank copy of the *SOS Model* for the simulation of Reese's Pieces they completed.

After completing the activity using the applet, a wrap-up discussion is used to draw students' attention to the important aspects of the concept they have just seen demonstrated. Students are asked to distinguish between how samples vary from each other, and variability of data *within* one sample of data. The Law of Large Numbers is revisited as it relates to the fact that larger samples better represent the population from which they were sampled. There can be a discussion of how big a sample needs to be to represent a population. (Students mistakenly think a sample has to be a certain percentage of the population, so a random sample of 1000 is not big enough to represent a population of one million.) Finally, students discuss the use of a model to simulate data, and the value of simulation in allowing us to determine if a sample value is surprising (e.g., 5 orange candies in a cup of 25 candies).

Lesson 2: Generating Sampling Distributions

In this lesson, students first speculate about the distribution of normal body temperatures and then contrast typical and potentially unusual temperatures for an individual person with a typical and potentially unusual values of means for samples of people. Students contrast the variability of individual values with the variability of sample means, and discover the impact on variability for different sample sizes. Web applets are used to simulate samples and distributions of sample means from two additional populations, revealing a predictable pattern as they generate sample means for increasingly large sample sizes, despite the differing population shapes. At the end of the lesson, students discuss how to determine if one value of a sample statistic is surprising or unusual, a precursor to formal statistical inference. Student learning goals for this lesson include:

1. Be able to generate sampling distributions for the sample mean, for different populations shapes.
2. Observe a predictable pattern (more normal, narrower, centered on the population mean) as the sample size increases.
3. Be able to distinguish between the population distribution, sample distribution, and the distribution of sample means.
4. Use the *Simulation of Samples (SOS)* Model to explain the process of creating sampling distributions.

Description of the Lesson

Class discussion begins with discussion about what is meant by a "normal" body temperature (e.g., healthy), what might be an unusual "normal" body temperature for someone who is not sick, and at what point they would consider someone sick as their temperature is too far beyond the expected range of natural variability. Students are also asked to consider and discuss at what point they would consider the mean temperature for a group of students to be unusual or suspicious. They are asked to consider any one person and how likely a person would be to have a body temperature of 98.6°F (37°C). This leads to the first activity (*Body Temperatures*), where students make conjectures about what they would expect to see in a distribution of normal body temperatures for the population of healthy adults in terms of shape, center, and spread. They draw graphs of what they would expect this distribution to look like.

Then, students are asked to consider where some particular values would be located on their graphs: 99.0, 98.0, and 97.2°F (37.2, 36.7, and 36.2°C). They are asked to think about whether or not any of these values would be considered unusual or surprising. They are asked to consider a temperature of 96°F (35.6°C). Most students will most likely say that this is a surprising body temperature for a normal adult. Then they are asked to think about how to determine what a surprising or unusual value is.

This question leads to a discussion about how much statistical work has to do with looking at samples from populations and determining if a particular result is surprising or not, given particular hypotheses. If a result is surprising, students are told, we often call that a *statistically significant* result. Students are told that they will be looking at what leads to a sample or research results that are "statistically significant."

Students are guided to begin by looking at z scores, which are computed using population values for μ and σ. They are given these values and are asked to find z scores for the body temperatures considered earlier: 99.0, 98.0, and 97.2°F. Students are next asked to think about z scores that would be more or less surprising, if they represent an individual's body temperature. The next part of the activity is designed to help move students from considering individual values to considering averages.

As the discussion continues, students are asked to consider a random sample of 10 students who have their temperatures taken using the same method. They are asked whether they would predict if the *average temperature* for this sample would be exactly 98.6°F or close to this value, and to give their reasons. They are then asked what about another sample of 10 different students and if that sample would have the same mean as the first sample of 10 or whether it would produce a different mean. Students are asked to think about what would cause differences in sample means.

Working in pairs, students are asked to write down plausible values for five students' body temperatures and then to write down plausible values of sample mean body temperatures for five random samples of ten college students ($n = 10$). Students are asked to think about and describe how they would expect these two sets of temperatures to compare and which set would have more variability and why. They are asked to think about what would happen if they took five random samples of 100 people, what the means of those samples of 100 people might be. *Fathom* is used to simulate a population of body temperatures with the given values of μ and σ, and samples sizes are varied to allow students to compare their estimates with simulated data (see Fig. 12.5 for a graph of the population).

In the next activity (*Sampling Words*), students return to the Web applet at *RossmanChance* Website that visually illustrates sampling words from the Gettysburg Address, as shown in Fig. 12.4 (Note, Lesson 3 in Chapter 6 uses this applet to take random samples). Students use the applet to sample five words and list the word lengths, then discuss how they vary. They find the mean and standard deviation for their set of five word lengths and consider what would be an unusual value (word length).

Students are asked to compare the mean and standard deviation for their sample of five word lengths to the given values of μ and σ for the population. They next begin taking samples of five words at a time, using the software, and examine the

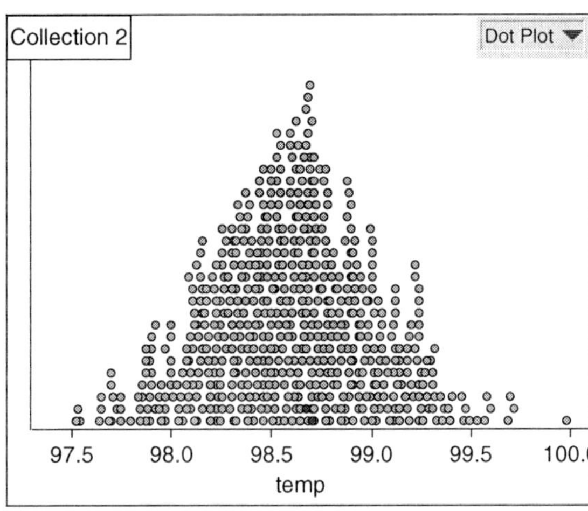

Fig. 12.5 *Fathom* simulation of human body temperatures

sample means. They are asked to consider what would be unusual values of means for samples of this size and to consider criteria to use in determining what a surprising value would be. Next, they generate a distribution of 500 sample means for samples of size 5 and use it to determine where their hypothetical "unusual" value would be. They are instructed to change the sample size to 10 and determine what differences will result for a new distribution of 500 sample means. They then repeat this for sample size of 20 and 50.

A whole class discussion follows where students are asked to compare these distributions in terms of shape, center, and spread, to discuss what is different about them, and how their unusual value fits on each. They are then guided to compute z scores for this value and compare them.

In the third activity (*Sampling Pennies*), students take simulated samples from a third population (ages of pennies) to see if they notice the same predictable pattern. They use the Sampling Pennies applet at RossmanChance.com, which shows a skewed distribution of dates on pennies from which they sample and use to generate distributions of sample means. Students vary parameters on the applet starting with small sample sizes and increasing to large sample sizes. Each time, they take 500 samples and look at the distribution of sample statistics (means or proportions) to describe a predictable pattern.

In a wrap-up discussion, students refer back to the beginning activity that looked at the variability of individual body temperatures from a normal (distribution) population and note that:

- Some temperatures are more likely than others.
- To see if a value is more or less likely (or surprising) we needed to look at their relative position in the distribution.
- Number of standard deviations above and below the mean (z scores) can tell us if something is unlikely or unusual. This can also be done for sample means after we learn the appropriate way to find z scores for a sample mean.
- Sample means vary too, but they tend to vary less than individual values.
- Means from smaller samples vary more than from large ones.
- There was a predictable pattern when we took larger sample sizes and plotted their means. The predictable pattern was: Symmetric, bell shaped (even when the populations were skewed), centered on μ, and smaller spread (less variability).
- To determine if a sample mean is unusual or surprising, we need to compare it to many other sample means, of the same size, from the same population. This is a distribution of sample means.
- If a distribution of sample means is normal, we can use z scores to help us see if values are unlikely or surprising.

Students also discuss how each of the simulations is modeled by the *SOS model*. The next lesson helps us determine when it is appropriate to assume that a distribution of sample means is normal so that we may use z scores to see if values are unlikely or surprising.

Lesson 3: Describing the Predictable Pattern – The Central Limit Theorem

This lesson moves students from noticing a predictable pattern when they generate distributions of sample statistics to describing that pattern using mathematical theory (i.e., the Central Limit Theorem, CLT). Students investigate the impact of sample size and population shape on the shape of the sampling distribution, and distinguish between sample size and number of samples. Students then apply (when appropriate) the Empirical Rule (68, 95, 99.7% within 1, 2, and 3 standard deviations from the mean) to estimate the probability of sample means occurring in a specific interval. Student learning goals for this lesson include:

1. Discover the Central Limit Theorem by examining the characteristics of sampling distributions.
2. See that the Central Limit Theorem describes the predictable pattern that students have seen when generating empirical distributions of sample means.
3. Describe this pattern in terms of shape, center, and spread; contrasting these characteristics of the population to the distribution of sample means.
4. See how this pattern allows us to estimate percentages or probabilities for a particular sample statistic, using the Normal Distribution as a model.
5. Understand how the *SOS Model* represents the Central Limit Theorem.
6. Understand how we determine if a result is surprising.

Description of the Lesson

The lesson begins with a review discussion of the previous lessons on samples and distributions of sample statistics. This includes revisiting ideas of variability in data values and in sample means, how to determine if a particular value or sample statistics is unlikely (surprising) using z scores, that statistics from small samples have more variability than those from large samples, and there is a predictable pattern when plotting the means of many large random samples of a given sample size from a population.

Students are asked to compare z scores for individual words to z-scores for means of word lengths (from the Gettysburg Address) and use z-scores as a yardstick for distance from a mean. They are asked to consider which is important: the number of samples or sample size, and what these terms mean. This is because students often confuse number of samples (an arbitrary number when performing simulations) with sample size. The point is made that we often used 500 samples because the resulting distribution of sample means is very close to what we expect the sampling distribution to look like in terms of shape, center, and spread.

The focus of the main activity of this lesson (*Central Limit Theorem* activity) is to examine and describe in detail the predictable pattern revealed with different populations, parameters, and sample sizes. The students are asked to make predictions, generate simulations to test them, and then evaluate their predictions. Their predictions are about what distributions of sample means will look like as they change the population shape and sample size. Each time, they take 500 samples, which provides a picture that is very close to what they would get if they took all possible samples.

This activity can be done using different simulation tools, but we prefer to use *Sampling SIM* software (delMas, 2001a) and a set of stickers that can be printed using a template available at the Website. These stickers show three columns of distributions, each headed by a specified population (see Fig. 12.6). Students are

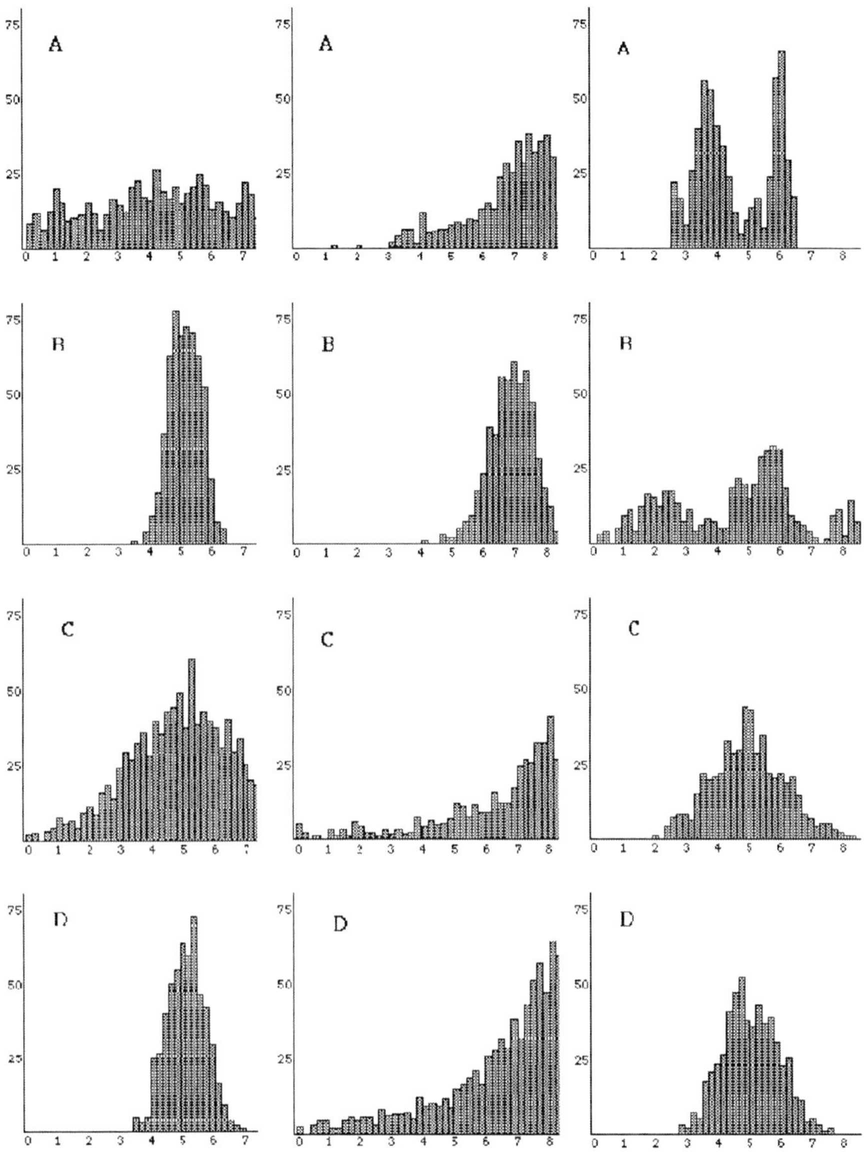

Fig. 12.6 Stickers for the *Central Limit Theorem Activity*

asked to consider three populations, one at a time, and to predict which of five distributions shown on the stickers will correspond to a distribution of 500 sample means for a particular sample size. They test their conjecture by running the simulation, then affix the sticker that matches that simulation result in a "Scrapbook." When the activity is finished, students have striking visual record that allows them to see and describe what happens when different sized samples are taken from a

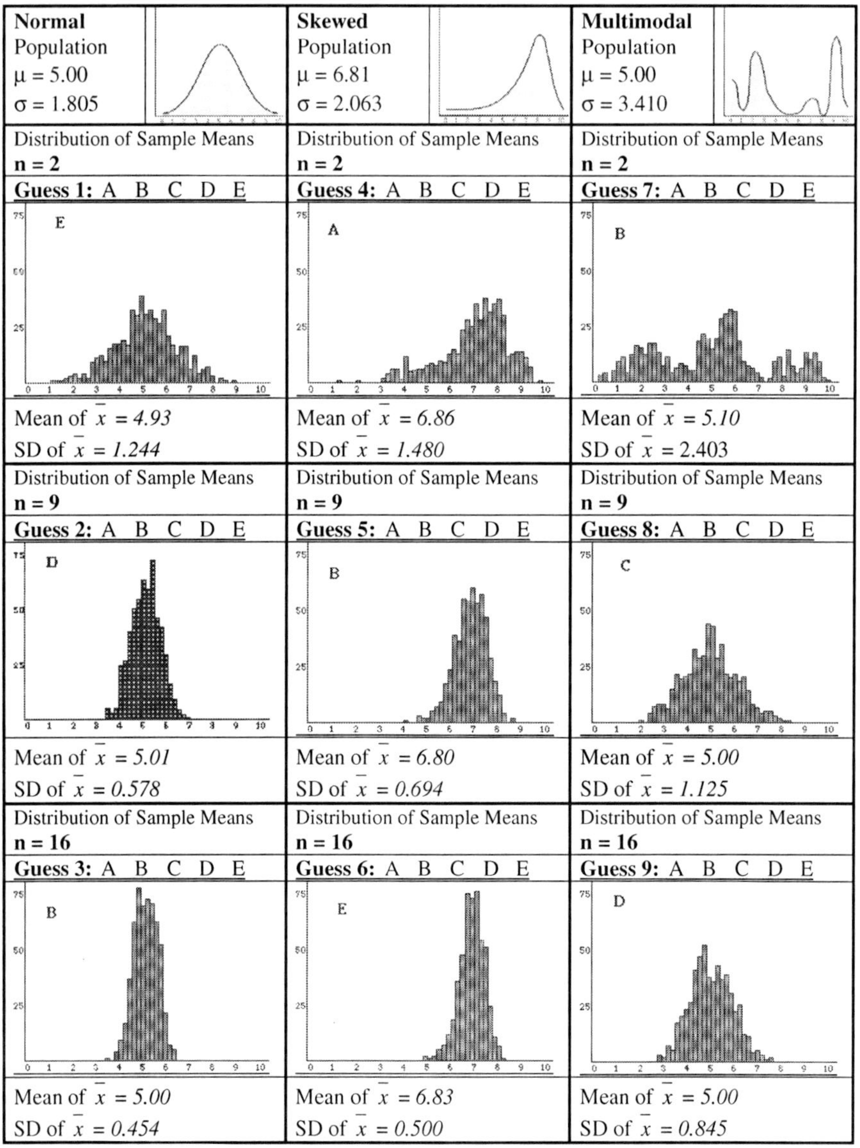

Fig. 12.7 Sample of a sampling scrapbook at the end of the *Central Limit Theorem Activity*

Normal, a skewed, and a trimodal distribution (see Fig. 12.7). If it is not possible to use stickers, copies of the pages shown in Figs 12.6 and 12.7 can be used instead.

As students work through this activity, they see that the predictable pattern is true regardless of population shape, and that (surprisingly) as the sample size increases, the distributions look more normal and less like the population (for non Normal populations). They are encouraged to develop a theory of how to evaluate whether a sample mean is surprising or not. Students are then guided to use the Central Limit Theorem to apply the Empirical Rule (68, 95, and 99.7% of data within 1, 2, and 3 standard deviations in a Normal Distribution) to sampling distributions that appear to be normal.

In a wrap-up discussion, students are asked to contrast the Law of Large Numbers to the Central Limit Theorem and also to discuss how they are connected. For example, when a large sample size is generated, individual samples better represent the population (Law of Large Numbers) and their sample statistics are closer to the population parameters. When graphing 500 of these sample means, they will cluster closer to μ resulting in less variability (smaller standard error) and a symmetric bell shape. Students are asked to distinguish between: populations, samples, and sampling distributions. They are asked to discuss what is similar and what is different and why. Finally, they work in small groups to respond to questions such as: How can we describe the sampling distribution of sample means without running simulations? For random samples size 100 from a population of human body temperatures, what would be the shape, center, and spread of the distribution of sample means? And which would be more likely to be closest to 98.6°F: A mean temperature based on 9 people or a mean temperature based on 25 people? Finally, students are asked to use the *SOS Model* to represent the Central Limit Theorem.

Summary

We believe that ideas of samples and sampling distributions should be introduced early in a statistics course and that by the time students reach the formal study of sampling distributions, they have already generated and examined different distributions of sample statistics while making informal inferences. Even given that background, we believe that the full set of activities described in this chapter are needed in order for students to understand and appreciate the Central Limit Theorem. Then, when the formal study of inference is introduced we encourage the revisiting of ideas of sampling distribution and the reference to the *Simulation of Samples (SOS)* Model (Fig. 12.1) so that students can see the role of a sampling distribution in making statistical inferences.

Chapter 13
Learning to Reason About Statistical Inference

> *Despite all the criticisms that we could offer of the traditional introductory statistics course, it at least has a clear objective: to teach ideas central to statistical inference.*
> *(Konold & Pollatsek, 2002, p. 260)*

Snapshot of a Research-Based Activity on Statistical Inference

Students revisit an activity conducted earlier in the semester in the unit on comparing groups with boxplots (*Gummy Bears Activity* in Lesson 2, Chapter 11). Once again, they are going to design an experiment to compare the distances of gummy bears launched from two different heights. The experiment is discussed, the students form groups, and the conditions are randomly assigned to the groups of students. This time a detailed protocol is developed and used that specifies exactly how students are to launch the gummy bears and measure the results. The data gathered this time seem to have less variability than the earlier activity, which is good. The students enter the data into *Fathom* (Key Curriculum Press, 2006), which is used to generate graphs that are compared to the earlier results, showing less within group variability this time due to the more detailed protocol.

There is a discussion of the between versus within variability, and what the graphs suggest about true differences in distances. *Fathom* is then used to run a two sample *t test* and the results show a significant difference, indicated by a small P-value. Next, students have *Fathom* calculate a 95% confidence interval to estimate the true difference in mean distances. In discussing this experiment, the students revisit important concepts relating to designing experiments, how they are able to draw casual conclusions from this experiment, and the role of variability between and within groups. Connections are drawn between earlier topics and the topic of inference, as well as between tests of significance and confidence intervals in the context of a concrete experiment.

The metaphor of making an argument is revisited from earlier uses in the course, this time in connection with the hypothesis test procedure. Links are shown between the claim (that higher stacks of books will launch bears for farther distances), the evidence used to support the claim (the data gathered in the experiment), the quality and justification of the evidence (the experimental design, randomization, sample size), limitations in the evidence (small number of launches) and finally, an indicator of how convincing the argument is (the P-value). By discussing the idea of the

J.B. Garfield, D. Ben-Zvi, *Developing Students' Statistical Reasoning*,

P-value as a measure of how convincing our data are in refuting a contradictory claim (that the lower height resulted in farther distances), students see that the farther they are from this contradictory claim, the more likely we are to win our argument. As they have seen in earlier uses of informal inference throughout the course, the farther in the tails, the smaller the probability of observing what was seen in the sample if the contradictory claim is true and the smaller the P-values. So they link small P-values with convincing evidence and a more convincing argument.

Rationale for This Activity

Unlike many of the topics in previous chapters of this book, there is little empirical research on teaching concepts of inference to support the lessons described in this chapter. However, there are many studies that document the difficulties students have reasoning and understanding inferential ideas and procedures. Therefore, we are much more speculative in this chapter, basing our lessons and activities more on writing by influential statistics educators as well as general research-based pedagogical theories. Later in this chapter, we address the many questions we have about appropriate ways to help students develop good reasoning about statistical inference and some promising new directions that are just beginning to be explored.

This particular activity is introduced near the end of a course that is designed to lead students to understand inferences about one and two means. We use it at a time where the material often becomes very abstract and challenging for students, a time where it is often hard to find a motivating activity for students to engage in. Now that students have already conducted this experiment, they are more aware of the need to use good, consistent protocols for launching gummy bears, to decrease the variability within each condition, and to provide a convincing argument supporting their claim and refuting the alternative claim. Also, now that students are acquainted with formal methods of making statistical inferences, they can do a statistical comparison of the difference in distances using a two-sample test of significance. The use of the argument metaphor helps students connect the confusing terminology used regarding hypothesis tests to something they can understand and relate to, and builds upon earlier uses of this metaphor and associated terms throughout the course.

The Importance of Understanding Statistical Inference

> Drawing inferences from data is now part of everyday life but it is a mystery as to why and how this type of reasoning arose less than 350 years ago.
>
> (Pfannkuch, 2005b, p. 267)

Drawing inferences from data is part of everyday life and critically reviewing results of statistical inferences from research studies is an important capability for all adults. Methods of statistical inference are used to draw a conclusion about a particular population using data-based evidence provided by a sample.

Statistical inference is formally defined as "the theory, methods, and practice of forming judgments about the parameters of a population, usually on the basis of random sampling" (Collins, 2003). Statistical inference "moves beyond the data in hand to draw conclusions about some wider universe, taking into account that variation is everywhere and the conclusions are uncertain" (Moore, 2004, p. 117). There are two important themes in statistical inference: parameter estimation and hypothesis testing and two kinds of inference questions: generalizations (from surveys) and comparison and determination of cause (from randomized comparative experiments). In general terms, the first is concerned with generalizing from a small sample to a larger population, while the second has to do with determining if a pattern in the data can be attributed to a real effect.

Reasoning about *data analysis* and reasoning about *statistical inference* are both essential to effectively work with data and to gain understanding from data. While the purpose of exploratory data analysis is exploration of the data and searching for interesting patterns, the purpose of statistical inference is to answer specific questions, posed before the data are produced. Conclusions in EDA are informal, inferred based on what we see in the data, and apply only to the individuals and circumstances for which we have data in hand. In contrast, conclusions in statistical inference are formal, backed by a statement of our confidence in them, and apply to a larger group of individuals or a broader class of circumstances. In practice, successful statistical inference requires good data production, data analysis to ensure that the data are regular enough, and the language of probability to state conclusions (Moore, 2004, p. 172).

The Place of Statistical Inference in the Curriculum

The classical approach to teaching statistical inference was a probability theory-based explanation couched in formal language. This topic was usually introduced as a separate topic, after studying data analysis, probability, and sampling. However, most students had difficulty understanding the ideas of statistical inference and instructors realized something was wrong about its place and portion of the curriculum. For example, an important part of Moore's (1997) plea for substantial change in statistics instruction, which is built on strong synergies between content, pedagogy, and technology, was the case to depart from the traditional emphasis of probability and inference. While there has been discussion on whether to start with means or proportions first in introducing inference (see Chance & Rossman, 2001), there has been some mention about ways to bring ideas of inference earlier in a course. The text book *Statistics in Action* (Watkins et al., 2004) does a nice job of introducing the idea of inference at the beginning of the course, asking the fundamental question - 'is a result due to chance or due to design', and using simulation to try to address this question.

We believe that ideas of inference should be introduced informally at the beginning of the course, such as having students become familiar with seeing where a sample corresponds to a distribution of sample statistics, based on a theory or hypothesis. Thus, the informal idea of P-value can be introduced. These types of informal inferences can be part of units on data and on distribution (does this sample represent a population? would it generalize to a population?), comparing groups (do the observed differences lead us to believe there is a real difference in the groups these samples represent?), sampling (is a particular sample value surprising?), and then inference (significance tests and confidence intervals). By integrating and building the ideas and foundations of statistical inference throughout the course, we believe that students should be less confused by the formal ideas, procedures, and language when they finally reach the formal study of this topic; however, there is not yet empirical research to support this conjecture. We also recommend revisiting the topic of inference in a subsequent unit on covariation, where students build on applying their inference knowledge to test hypotheses about correlation coefficients and regression slopes.

Review of the Literature Related to Reasoning About Statistical Inference[1]

> Historically, there were huge conceptual hurdles to overcome in using probability models to draw inferences from data; therefore, the difficulty of teaching inferential reasoning should not be underestimated.
>
> (Pfannkuch, 2005b, p. 268)

Difficulties in Inferential Reasoning

Research on students' informal and formal inferential reasoning suggests that students have many difficulties in understanding and using statistical inference. These results have been obtained across many populations such as school and college students, teachers, professionals, and even researchers. Many types of misunderstandings, errors, and difficulties in reasoning about inference have been studied and described (e.g., Carver, 1978; Falk & Greenbaum, 1995; Haller and Krauss, 2002; Mittag & Thompson, 2000; Oakes, 1986; Vallecillos and Holmes, 1994; Wilkerson and Olson, 1997; Williams, 1999; Liu, 2005; Kaplan, 2006). In addition to studies documenting difficulties in understanding statistical inference, the literature contains studies designed to help explain why statistical inference is such a difficult topic for people to understand and use correctly, exhortations for changes in the way inference is used and taught, and studies exploring ways to develop students reasoning about statistical inference.

---

[1] We gratefully acknowledge the contributions of Sharon Lane-Getaz as part of her dissertation literature review with Joan Garfield.

Survey Studies on Assessments of Students' Understanding Statistical Inference

In a study of introductory students' understandings about "proving" the truth or falsity of statistical hypotheses, Vallecillos and Holmes (1994) surveyed more than 400 students from different fields who responded to a 20-item survey. One of the interesting results in this study was that nearly one-third of the answers reflected a faulty belief that hypothesis tests logically prove hypotheses. Additional misunderstandings were found among introductory statistics students at the end of a one-semester introductory statistics course by Williams (1997, 1999). Williams interviewed eighteen respondents and found that statistical ideas of P-values and significance were poorly understood. In an earlier study, Williams (1997) identified several sources of students' misunderstanding of P-values such as inadequate or vague connections made between concepts and terms used, and confusion between P-value and significance level. Williams (1999) also found that many introductory students believed that the P-value is always low.

To assess graduate students' understanding of the relationships between treatment effect, sample size, and errors of statistical inference, Wilkerson and Olson (1997) surveyed 52 students. They found many difficulties students had, such as misunderstanding the role of sample size in determining a significant P-value. Similar results were documented in a study by Haller and Krauss (2002), who surveyed instructors, scientists, and students in psychology departments at six German universities. The results showed that 80% of the instructors who taught courses in quantitative methods, almost 90% of instructors who were not teaching such courses, and 100% of the psychology students identified as correct at least one false meaning of P-value (Haller and Krauss, 2002).

Additional difficulties in reasoning about inference were identified such as confusion about the language of significance testing (Batanero et al., 2000) and confusion between samples and populations, between α and Type I error rate with P-value (Mittag & Thompson, 2000). In sum, survey studies have identified persistent misuses, misinterpretations, and common difficulties people have in understanding of inference, statistical estimation, significance tests, and P-values.

Students' responses to inference items were described as part of an examination of data from a national class test of the Comprehensive Assessment of Outcomes in a first Statistics course (CAOS – delMas et al., 2006). A total of 817 introductory statistics students, taught by 28 instructors from 25 higher education institutions from 18 states across the United States, were included in this study. While the researchers found a significant increase in percentage of correct scores from pretest to posttest on items that assessed understanding that low P-values are desirable in research studies, ability to detect one misinterpretation of a confidence level (95% refers to the percent of population data values between confidence limits), and ability to correctly identify the standard interpretation of confidence interval, there were also items that showed no significant gain from pretest to posttest. For these items, less than half the students gave correct responses, indicating that students did not appear to learn these concepts in their courses. These items included ability to detect two misinterpretations of a confidence level (the 95% is the percent of sample data between confidence limits, and 95% is the percent of all possible sample means between confidence limits), and understanding of how sampling error is used to make an informal inference about a sample mean. There was also a significant increase in students selecting an incorrect response (26% on pretest and 35% on posttest), indicating that they believed that rejecting the null hypothesis means that the null hypothesis is definitely false. In addition, although there was statistically significant gain in correct answers to an item that assessed understanding of the logic of a significance test when the null hypothesis is rejected (37% correct on the pretest to 47% correct on the posttest), there were still more than half the students who answered this item incorrectly on the posttest.

Why Is Statistical Inference so Difficult to Learn and Use?

Reasoning from a sample of data to make inferences about a population is a hard notion to most students (Scheaffer, Watkins & Landwehr, 1998). Thompson, Saldanha and Liu (2004) examined this difficulty, noting that literature on statistical inference "smudges" two aspects of using a sample.

> The first aspect regards attending to a single sample and issues pertaining to ensuring that an individual sample represents the population from which it is drawn. The second aspect regards the matter of variability amongst values of a statistic calculated from individual samples. The two aspects get "smudged" in this way: (1) we (researchers in general) hope that people develop an image of sampling that supports the understanding that increased sample size and unbiased selection procedures tend to assure that a sample will look like the population from which it is drawn, which would therefore assure that the calculated statistic is near the population parameter; (2) we hope that people develop an image of variability amongst calculated values of a statistic that supports the understanding that as sample size increases, the values of a statistic cluster more tightly around the value of the population parameter.
>
> (Thompson et al., 2004, p. 9)

Thompson et al. (2004) state that they see ample evidence from research on understanding samples and sampling that suggests that students tend to focus on individual samples and statistical summaries of them instead of on how collections of samples are distributed. There is also evidence that students tend to base predictions about a sample's outcome on causal analyses instead of statistical patterns in a collection of sample outcomes. They view these orientations as problematic for learning statistical inference because they appear to "disable students from considering the relative unusualness of a sampling process' outcome" (Thompson et al., 2004, p. 10). These authors report on a study that explored students developing reasoning about inference in two teaching experiments in high school mathematics classes that involve activities and simulations to build ideas of sampling needed to understand inference. They found that those students who seemed to understand the idea and use a margin of error for a sample statistics had developed what. Saldanha and Thompson (2002) called a "multiplicative conception of sample" – a conception of sample that entails recognition of the variability among samples, a hierarchical image of collections of samples that simultaneously retain their individual composition, and the idea that each sample has an associated statistic that varies as samples varied. This study suggested that if students could be guided to develop this reasoning, they would be better able to understand statistical inference. Indeed, Lane-Getaz (2006) developed a visual diagram to help students develop this type of reasoning that has been adapted and used in the lessons in this book (*Simulation of Samples Model*, see Chapters 6 and 12).

Other studies designed to reveal why students have difficulty learning statistical inference have examined how this reasoning develops and offer suggested ways to help students move toward formal inference (e.g., Biehler, 2001; Konold, 1994b; Liu, 2005; Pfannkuch, 2006a).

Using Simulation to Illustrate Connections Between Sampling and Inference

Recent research suggests that improving the instruction of sampling will help students better understand statistical inference (e.g., Watson, 2004). This can be done by using good simulation tools and activities for teaching sampling distribution and the Central Limit Theorem (e.g., delMas et al., 1999; Chance et al., 2004).

However, using these simulation tools is not enough; they need to be linked to ideas of statistical inference. Lipson (2002) used computer simulations of the sampling process and concept maps to see how college students connected sampling concepts to statistical inference. She found that while the simulations appeared to help students understand some aspects of sampling distributions, students did not appear to be linking these ideas to hypothesis testing and estimation. In a subsequent study, Lipson, Kokonis, and Francis (2003) devised a computer simulation session to support the development of students' conceptual understanding of the role of the sampling distribution in hypothesis testing. The researchers identified four developmental stages through which students progress while using the visual simulation software: (a) *recognition* of the software representations, (b) *integration* of the three concepts of population, sample, and sampling distribution; (c) *contradiction* that the sample may not be typical of the hypothesized population, and (d) *explanation* of results from a statistical perspective. A stumbling block for the students appeared to be that that they looked for a contextual explanation rather than a statistical explanation, even when they acknowledged the low probability of the sample coming from hypothesized population. The researchers concluded that current software supported the recognition stage only, and suggested that students need to have a substantial experience in thinking about samples and sampling.

Some statistics educators (e.g., Biehler, 2001; Gnanadesikan et al., 1987; Jones, Lipson & Phillips, 1994; Konold, 1994b; Scheaffer, 1992) advocate that inference should be dealt with entirely from an empirical perspective through simulation methods to help students understand how statistical decisions are made. One such approach is the *resampling* method. Konold (1994b) used his *DataScope* Software (Konold & Miller, 1994) tool to introduce resampling methods to help students develop a more intuitive idea of a *P-value*. Mills (2002) summarizes papers that give examples of how simulation can be used to illustrate the abstract ideas involved in confidence intervals; however, it is difficult to locate research studies that document the impact of these methods on students' reasoning.

Informal Reasoning About Statistical Inference

A topic of current interest to many researchers as well as teachers of statistics is informal inferential reasoning rather than formal methods of estimation and tests of significance (e.g., Pfannkuch, 2005a). As new courses and curricula are developed, a greater role for informal types of statistical inference is anticipated, introduced early, revisited often, and developed through use of simulation and technological tools.

Informal Inferential Reasoning is the cognitive activities involved in informally drawing conclusions or making predictions about "some wider universe" from data patterns, data representations, statistical measures and models, while attending to the strength and limitations of the drawn conclusions (Ben-Zvi et al., 2007). Informal inferential reasoning is interconnected to reasoning about distribution, measures of centre, variability, and sampling within an empirical enquiry cycle (Pfannkuch, 2006a; Wild & Pfannkuch, 1999).

Rubin et al. (2006) conceptualize informal inferential reasoning as statistical reasoning that involves consideration of multiple dimensions: properties of data aggregates, the idea of signal and noise, various forms of variability, ideas about sample size and the sampling procedure, representativeness, controlling for bias, and tendency. Bakker, Derry, and Konold (2006) suggest a theoretical framework of inference that broadens the meaning of statistical inference to allow more informal ways of reasoning and to include human judgment based on contextual knowledge.

Using the Logic of an Argument to Illustrate Hypotheses Testing

Ben-Zvi (2006) points out that informal inference is closely related also to argumentation. Deriving logical conclusions from data – whether formally or informally – is accompanied by the need to provide persuasive explanations and arguments based on data analysis. Argumentation refers to discourse for persuasion, logical proof, and evidence-based belief, and more generally, discussion in which disagreements and reasoning are presented (Kirschner, Buckingham-Shum, & Carr, 2003). Integration and cultivation of informal inference and informal argumentation seem to be essential in constructing students' statistical knowledge and reasoning in rich learning contexts. This view is supported by Abelson (1995), who proposes two essential dimensions to informal argumentation: The act or process of deriving conclusions from data (inference), and providing persuasive arguments based on the data analysis (rhetoric and narrative).

Part of making a statistical argument is to know how to examine and portray the evidence. In statistical inference, this means understanding how a sample result relates to a distribution of all possible samples under a particular null hypothesis. Therefore, one type of informal inference involves comparing samples to sampling distributions to get a sense of how surprising the results seem to be. This type of informal reasoning is based on first having an understanding of sampling and sampling distributions (see Chapter 12).

Students' Dispositions Regarding Statistical Inference

Another important research topic is students' dispositions and their relation to statistical proficiency. Kaplan (2006) studied the extent to which differences in psychological dispositions can explain differences in the development of students' understanding of hypothesis testing. Kaplan investigated undergraduate students who have taken an algebra-based statistics course. She used large samples to find relationships between statistics learning and dispositions and smaller samples to uncover themes and common conceptions and misconceptions held by undergraduate statistics students. No relationships were found between the statistics learning and the dispositions that were studied: "Need for Cognition," and "Epistemological Understanding." The research did identify three emergent themes in the student discussions of hypothesis testing: how students consider the experimental design factors of a hypothesis test situation, what types of evidence students find convincing, and what students understand about P-values.

Teachers' Understanding of Statistical Inference

Content and pedagogical-content knowledge of statistics teachers have a considerable influence on what and how they teach in the classroom. Liu (2005) explored and characterized teachers' understanding of probability and statistical inference, and developed a theoretical framework for describing teachers' understanding. To this end, she analyzed a seminar with eight high school teachers. Liu revealed that the teachers experienced difficulties in understanding almost every concept that is entailed in understanding and employing hypothesis testing. Beyond the complexity of hypothesis testing as a concept, Liu conjectured that teachers' difficulties were due to their lack of understanding of hypothesis testing as a tool, and of the characteristics of the types of questions for which this tool is designed. Although the teachers were able to root the interpretation of margin of error in a scheme of distribution of sample statistics, some of them were concerned with the additive difference between a population parameter and a sample's estimate of it. This study revealed a principle source of disequilibrium for these teachers: They were asked to develop understandings of probability, sample, population, distribution, and statistical inference that cut across their existing compartments.

Implications of the Research: Teaching Students to Reason About Statistical Inference

> Deepen the understanding of inferential procedures for both continuous and categorical variables, making use of randomization and resampling techniques.
>
> (Scheaffer, 2001)

The research suggests that understanding ideas of statistical inference is extremely difficult for students and consists of many different components. Many of these components themselves are difficult for students to understand (e.g., sampling distributions). Simulation and resampling methods are viewed as having the potential to offer a way to build informal inferences without focusing on the details of mathematics and formulas. In addition, using data sets and questions in early data analysis units to have students consider informal inferences (e.g., what does this sample suggest about the population, what do we believe about the difference in means for these two groups that these two samples come from) may help develop formal ideas of inference in later units.

In studying the difficulties students have reasoning about statistical inference, many different types of errors and misunderstanding have been identified, as well as a detailed description about what it means to reason about different aspects of statistical inference. Being aware of the complexities of the ideas as well as the common misunderstandings can help teachers be on the alert for student difficulties through formal and informal assessments that can be used for diagnostic purposes.

Some of the ideas related to correct (and incorrect) reasoning about two aspects of statistical inference: P-values and confidence intervals have been detailed by the *Tools for Teaching and Assessing Statistical Inference Project* (see http://www.tc.umn.edu/~delma001/stat_tools/). For example, some common misconceptions about P-values and confidence intervals are summarized as follows:

Misconceptions about P-values

- A P-value is the probability that the null hypothesis is true.
- A P-value is the probability that the null hypothesis is false.
- A small P-value means the results have significance (statistical and practical significance are not distinguished).
- A P-value indicates the size of an effect (e.g., strong evidence means big effect).
- A large P-value means the null hypothesis is true, or provides evidence to support the null hypothesis.
- If the P-value is small enough, the null hypothesis must be false.

Misconceptions about Confidence Intervals

- There is a 95% chance the confidence interval includes the sample mean.
- There is a 95% chance the population mean will be between the two values (upper and lower limits).
- 95% of the data are included in the confidence interval.
- A wider confidence interval means less confidence.
- A narrower confidence interval is always better (regardless of confidence level).

Suggestions for Teaching Statistical Inference

As mentioned at the beginning of this chapter, there is little empirical research on the effectiveness of different instructional strategies, sequences of activities, or technological tools in helping students develop correct reasoning about statistical inference. However, there are many strong and often conflicting beliefs among statistics educators about optimal methods of teaching these ideas. Arguments have been made for teaching inferences on proportions before means, teaching confidence intervals before tests of significance, not teaching students the method of pooling variances in comparisons of two-sample means, and abandoning t-tests altogether and instead using resampling and randomization methods. We describe below some of the suggestions that we believe to be aligned with the approaches described in our book and which we have used to build our suggested sequences of activities, acknowledging that they are not necessarily based on empirical research studies, and that their effectiveness is untested at this point.

Connecting Statistical Inference to Data Collection, Description, and Interpretation

Rossman and Chance (1999) offer "Top Ten" list of recommendations for teaching the reasoning of statistical inference. Their goal is to help students to focus on investigation and discovery of inferential reasoning, proper interpretation and cautious use of results, and effective communication of findings. The list includes the following recommendations:

1. Have students perform physical simulations to discover basic ideas of inference.
2. Encourage students to use technology to explore properties of inference procedures.
3. Present tests of significance in terms of P-values rather than rejection regions.
4. Accompany tests of significance with confidence intervals whenever possible.
5. Help students to recognize that insignificant results do not necessarily mean that no effect exists.
6. Stress the limited role that inference plays in statistical analysis.
7. Always consider issues of data collection.
8. Always examine visual displays of the data.
9. Help students to see the common elements of inference procedures.
10. Insist on complete presentation and interpretation of results in the context of the data.

Presenting Statistical Inference as Argumentation

A more recent approach to teaching statistical inference is to connect these ideas to the making of an argument, as described earlier by Ben-Zvi (2006). The logic of arguments can be used to explain the reasoning of a hypothesis test as follows:

- In statistics, we argue about claims (hypotheses) we believe to be true or false. While we cannot prove they are true or false, we can gather evidence to support our argument.
- A hypothesis test can be viewed as a method for supporting an argument.

- An argument (hypothesis test) may originate from two different perspectives: wanting to argue *against* a claim (i.e., the null hypothesis) or wanting to argue for a claim (i.e., the research (alternative) hypothesis).
- Just as in real life, even if we convince someone by our argument, we are only convincing them with evidence, we cannot really establish if our claim is actually true or not. In a hypothesis test, we only decide if the evidence is convincing enough to reject the null hypothesis, but not *prove* it is true or false.
- In order to make a good argument, we need four building blocks:

    1. A clear claim we are making (and a counterclaim that includes all other possibilities).
    2. Data to support our argument.
    3. Evidence that the data are accurate and reliable, not misleading.
    4. A good line of reasoning that connects our data to our argument.

- In real life when we make an argument, the resolution is that we win or lose the argument based on how convincing our argument is. This is based on the strength of our evidence, and how we use the evidence to support our case. In a hypothesis test, the result is to reject or fail to reject the null hypothesis, which is based on the size of the obtained P-value.
- We need to see how far away our data are from the claim we are arguing against. Therefore, we look for data that are far from what we would expect if the claim we are arguing against is true. A low P-value results from data that are far from the claim we are arguing against, and the lower (farther) they are, the stronger the evidence.

Introducing the idea of an argument would seem to be a useful way to help students understand the process of making and testing hypotheses, and may help students better understand this complex and often counterintuitive procedure.

Basing Inference on Simulation and Randomization

While many educators have advocated the use of simulations to help students understand the connections between sample, population, and sampling distribution in inference, to illustrate the abstract ideas of confidence interval (e.g., Mills, 2002) others have suggested that traditional approaches to inference be replaced entirely with resampling methods (e.g., Simon, Atkinson, & Shevokas 1976; Simon, 1994; Konold, 1994b). More recently, in light of flexible and accessible technological tools, educators such as Cobb (2007) and Kaplan (2007) have suggested radically different approaches to statistical inference in the introductory course. Their suggestions place inference as the focus of a course that teaches three R's: Randomize data production, Repeat by simulation to see what's typical, and Reject any model that puts your data in the tail of the distribution (see Cobb, 2007). We find these ideas very appealing but have not yet explored ways to build a sequence of lessons around them and experimented with them in our classes.

Progression of Ideas: Connecting Research to Teaching

Introduction to the Sequence of Activities to Develop Reasoning About Statistical Inference

The sequence of ideas and activities for inference represent one of many possible ways to guide students to develop good inferential reasoning, and we do not have a strong conviction that this sequence is an optimal one. Although we have used these lessons and find them to work well in engaging students, we believe that it might be better to adopt more of an informal and conceptual approach, rather than leading students to learn the formal aspects of testing hypotheses and constructing confidence intervals. However, we provide examples in this chapter of how to build lessons about inference on the previous big ideas and activities, and make connections between foundational concepts and the formal aspects of statistical inference.

We suggest that ideas of informal inference are introduced early in the course and are revisited in growing complexity throughout the course. Underlying the development of this inferential reasoning is a fundamental statistical thinking element, *consideration of variation* (Moore, 1990; Wild & Pfannkuch, 1999), and how variability of data and samples is a key part of making inferences. This means that students have opportunities to see and describe variability in samples throughout the course as they make informal inferences about how these samples relate to the population from which they were drawn, and whether these samples lead us to infer about what that population might be. When ideas of formal inference are eventually introduced, they are devoid of computations and formulas so that students can focus on what the ideas of null and alterative hypothesis mean, the idea of P-value, and types of errors. The computer is used to run tests and generate confidence intervals before students see the formulas. The culmination of this progression of ideas is giving students a set of research questions and associated data and having them use their statistical thinking to choose appropriate procedures, test conditions, arrive at conclusions, and provide evidence to support these conclusions.

In addition to the progression from informal to formal methods of statistical inference, we suggest the use of two important pedagogical methods. One is the modeling by the teaching of statistical reasoning and thinking in making statistical inference. This means, making their thinking visible as they go from claims to conclusions, checking conditions, considering assumptions, questioning the data, choosing procedures, etc. The second is the use of the argumentation metaphor for hypothesis testing as described earlier. This means using the language of arguing about a claim, whether we believe a claim is true, the role of evidence and using that evidence well, and what it takes to be convinced that the claim is true or false. Table 13.1 shows a suggested series of ideas and activities that can be used to guide the development of students' reasoning about statistical inference.

Table 13.1 Sequence of activities to develop reasoning about statistical inference[2]

Milestones: Ideas and concepts	Suggested activities
Informal ideas prior to formal study of statistical inference	
• Making inferences and generalizations from a sample of simulated data	• One Son Activity (Lesson 1, Statistical Models and Modeling Unit, Chapter 7)
• Statistical inference as an argument	❖ An informal discussion early in a course about the nature of statistical inference, and comparing this to making an argument and providing evidence to support your claim. (The symbol ❖ indicates that this activity is not included in these lessons.)
• Random sample and how it is representative of a population	• The Gettysburg Address Activity (Lesson 3, Data Unit, Chapter 6)
• Results being due to chance or due to design (some other factor)	• Taste Test Activity (Lesson 4, Data Unit, Chapter 6)
• As a sample grows, the characteristics become more stable, that with more data you can better generalize to a population	• Growing a Distribution Activity (Lesson 1, Distribution Unit, Chapter 6)
• Two samples of data may or may not represent true differences in the population	• Activities in Lessons 1–4, Comparing Groups Unit (Chapter 11)
• When comparing groups, you must take into account the variability between groups relative to the variability within each group	• Gummy Bears Activity (Lesson 2, Comparing Groups Unit, Chapter 11)
• If the normal distribution provides a good model for a data set we may make inferences based on the Empirical Rule	• Normal Distribution Applications Activity (Lesson 3, Statistical Models and Modeling Unit, Chapter 7)
• We can make inferences by comparing a sample statistic to a distribution of samples based on a particular hypothesis	• Activities in Lessons 1 and 2, Samples and Sampling Unit (Chapter 12)
Formal ideas of statistical inference	
• Hypothesis test as making an argument	• Modeling Coin Tosses Activity (Lesson 1: "Testing Statistical Hypotheses")
• Hypothesis test, null and alternative hypothesis	• Balancing Coins Activity (Lesson 1)
• The idea of a P-value	• P-values Activity (Lesson 2)
• Types of errors and correct decisions	• Types of Errors Activity (Lesson 2)
• What is needed to test a hypothesis?	• Types of Errors and P-values Activities (Lesson 2)

[2] See page 391 for credit and reference to authors of activities on which these activities are based.

Table 13.1 (continued)

• Confidence interval as an estimate of parameter, with margin of error	• Introduction to Confidence Intervals (Lesson 2)
• Understanding how confidence intervals may be presented in different ways	• Introduction to Confidence Intervals (Lesson 2)
• Understanding what 95% refers to in a confidence interval	• Estimating with Confidence, Estimating Word Lengths, and What Does the 95% Mean Activities (Lesson 3: "Reasoning about Confidence Intervals")
• A statistically significant difference between two groups where randomization of conditions has taken place	• Gummy Bears Revisited Activity (Lesson 4: "Using Inference in an Experiment")
Building on formal ideas of statistical inference in subsequent topics	
• Statistically significant correlation coefficient	• Activities in Lesson 3, Covariation Unit (Chapter 14)
• Statistically significant regression slope	• Activities in Lesson 3, Covariation Unit (Chapter 14)
• There are many types of statistical inferences, and software may be used by correctly choosing the commands	• Research Questions Involving Statistical Methods Activity (Lesson 5: "Applying Methods of Statistical Inference")
• Understanding that the interpretation of P-values and confidence depends on assumptions being met	• Research Questions Involving Statistical Methods Activity (Lesson 5)

Introduction to the Lessons

There are five lessons on statistical inference that begin with informal ideas and lead to running tests of significance and confidence intervals on the computer. The focus is on understanding the ideas and methods and interpreting the results, rather than on formulas and computing test statistics. The lessons proceed very slowly, building on informal ideas from previous lessons and also integrating ideas of argumentation. The final lesson provides students with an opportunity to think statistically and to integrate and apply their knowledge, as they are given only research questions and a data set and need to answer the questions using the data and software.

Lesson 1: Testing Statistical Hypotheses

This lesson uses the context of balancing a coin on its edge to introduce formal ideas of testing hypotheses. The proportion of heads obtained when a balanced coin falls is used to test a null distribution based on equally likely outcomes. The idea of the P-value is examined visually and conceptually, and then P-values are found using simulation software. The argumentation metaphor is used to explain the logic of testing hypothesis. Student learning goals for this lesson include:

1. Connect informal to formal ideas of statistical inference.
2. Introduce the process and language of significance tests.
3. Use *Sampling SIM* to conduct an informal test of significance.
4. Understand the use of P-value in a test of significance.

Description of the Lesson

In the *Modeling Coin Tosses* activity, the instructor holds up a penny and asks what students expect if the coin is tossed. It is agreed while the outcome of a toss is unpredictable, that they expect a fair penny to land with Heads up half the time and with Tails up half the time. Students make a conejcture about what would happen if they balance a coin on its edge and let it fall, and if this is done many times, would it also land Heads and Tails in fiarly equal numbers. They are asked how to determine if a balanced coin is just as likely to land Heads up as it is to land Heads down.

Students discuss in pairs and then write down possible numbers of Heads they might expect to get for 8 sets of 10 tosses of a fair penny (e.g., list the number of Heads out of 10 for eight repetitions of this experiment). They are asked whether they expect to get 5 Heads each time, or if they expected some variability between results of each set of 10 tosses, and how variable they expected each set of 10 to be in the number of Heads produced. Students also reason about what outcomes they would consider to be less likely if using a "fair" coin and why.

Next, students use *Sampling SIM* to model tossing a fair coin ten times. They sketch the resulting distribution of sample proportions and describe it in terms of shape, center, and spread. Students shade in areas of the distribution that include what they would consider to be surprising results, so that if they obtained one of those results, they might question the assumption that the coin is equally likely to land Heads up or down (probability of Heads is 0.5).

In the *Balancing Coins* activity, students are asked what they think will happen if they balance sets of 10 pennies on their edge and let them fall, and if they expect the same number of Heads and Tails when flippoing a coin ($p = 0.5$). They are introduced to the idea of testing a *statistical hypothesis*, as shown below:

Idea 1: Balancing a coin is a "fair" process: Heads and Tails are equally likely to result.
Idea 2: Balancing a coin is an "unfair" process: There will be a higher percent of Heads or Tails.

These ideas are then written as statistical hypotheses:

> Null hypothesis: The proportion of Heads when we balance a coin repeatedly is 0.5.
> Alternative hypothesis: The proportion of Heads when we balance a coin repeatedly is not 0.5. (In other words the proportion is more, or less, than 0.5.)

The *null hypothesis* is discussed as an *idea of no difference* from the norm or prior belief (e.g., getting the same results as tossing fair coins). The *alternative hypothesis* is discussed as a statement that there will *not* be an equal number of Heads and Tails, something contrary to the first idea.

Students are told that we gather evidence (data) and determine whether or not it supports the null hypothesis or whether it provides convincing support for an alternative hypothesis. To do this, students design an experiment to lead them to make a decision about which of the two hypotheses are supported by the data. They discuss what is needed to test a hypothesis or to make a good argument given this context:

1. *A hypothesis to test* (e.g., the proportion of Heads is 0.5) (The claim).
2. *A sample of data which gives us a sample statistic* (e.g., a sample proportion).
3. *A sampling distribution* for that statistic (based on the null-hypothesis) so we can see how unusual or surprising it is, by seeing if it is far off in one of the Tails (surprising) or in the middle (not surprising). This sampling distribution is based on the null hypothesis and the sample size for our sample data. If our sample statistic is in one of the Tails, that would lead us to reject H_0 (A method to test the claim).
4. *A decision rule*: how far is far off in the Tails? How far in one of the Tails does our sample statistic need to be for us to decide it is so unusual and surprising that we reject the idea stated in H_0, that the coin is equally likely to land Heads up or Heads down when we balance it? (How to evaluate the strength of the evidence.)

Students then get in groups and balance coins, counting the result when the coins fall. The numbers of Heads and Tails are tallied, proportions of Heads for each set of 10 balances are found and gathered for the class. The sample proportions typically range from 0.5 to 0.9.

The next discussion regards an appropriate sampling distribution to use to judge whether their results are due to chance or whether the chances of getting Heads when balancing a coin is greater than 0.5. They decide to refer to the simulation created earlier (in the *Modeling Coin Tosses Activity*, Lesson 1), which allows a comparison of their sample statistics to what they would expect if the coin is equally likely to turn up Heads or Tails when balanced. Students use their sketches made earlier in the activity to determine whether or not this result is in a tail. They mark the sample proportion for their group in the graph and discuss whether they think this result is surprising, and why or why not. This leads to an examination of what percent of the distribution has values more extreme than theirs. They use *Sampling SIM* to find this area.

This value is discussed as the chance of getting the result students got or a more extreme one, and is referred to as a *P-value*. The role of the P-value in making a decision is seen as helping determine which of the two hypotheses seems more likely. Students discuss how small a P-value must be to be judged surpassing and leading them to reject the null hypothesis. Again, the argument metaphor is used, and the P-value is described as an indicator of how convincing the evidence is against the claim (null hypothesis). The farther it is in the tails, the more we are convinced that the null hypothesis (claim) is false. So the smaller the P-value, the stronger is the evidence. Students evaluate their P-values and determine whether they reject the claim that the coin is equally likely to land Heads up or Heads down when balanced on its edge. The class then combines their data to get a better, more stable estimate of the proportion of Heads, and test this result using the *Sampling SIM* software and finding the P-value via simulation.

Students are asked what conclusion can be drawn about the original research question, and then apply the same procedure in determining whether or not they believe a Euro coin is equally likely to land Heads up or down when tossed, using data from a series of 100 tosses of a Euro coin.

A wrap-up discussion reviews the process of hypothesis testing (hypotheses, data-sample statistic, sampling distribution, and decision rule) and how this process maps to making a convincing argument. The *Simulation of Samples (SOS)* Model is revisited and used to map the different levels of data: population, sampling distribution, and sample value.

Lesson 2: *P*-values and Estimation

This lesson builds on the previous lesson, using the context of balancing coins to test hypothesis and learn the language of tests of significance. This lesson also introduces the idea of a confidence interval, helping students see the two parts of the interval (e.g., sample statistic and margin of error) and different ways or reporting confidence intervals. Students also begin to interpret a confidence interval. Student learning goals for this lesson include:

1. Review use of simulations for inference.
2. Review the process for hypothesis testing.
3. Learn about the two types of errors when conducting tests of significance.
4. Use *Fathom* to conduct a test of significance.
5. Understand the idea of a confidence interval as a way to estimate a parameter.

Description of the Lesson

After balancing pennies in the previous lesson, students are asked if they think that balancing the Euro coin will yield equally likely chances of getting Heads and Tails. In the *P-values* activity, they are given a sample result from a person who balanced a Euro 100 times and got 31 Heads. First, students repeat the process they used earlier, of finding this sample proportion and comparing it to the simulated sampling distribution for a null hypothesis of equally likely outcomes. Next they use *Fathom* software to find P-values without simulation. These two P-values are compared and students reason about why the P-value from the simulation is not exactly the same as the one produced by *Fathom* (*Sampling SIM* ran 500 simulations while *Fathom* is basing their result on the true sampling distribution of all possible samples).

In the *Types of Errors* activity, students review the steps of the previous lesson (Lesson 1 on *Testing Statistical Hypotheses*) discussing the components needed to test a hypothesis and how these compare to the argumentation process. They map the types of errors (Type 1 and Type 2) to this context of balancing the Euro coin. For example:

1) We select H_a but it is the wrong decision because H_0 is true (Type 1 error).
2) We select H_0 but it is the wrong decision because $H_0$is not true (Type 2 error).

Another context is provided and students try to reason about what the different types of errors would mean in that context and the importance of keeping the chance of making these errors small. The idea of alpha as the chance of making a Type 1 error is contrasted to the idea and role of the P-value, and what is meant by the term "statistically significant." This term is compared to winning an argument because the evidence is strong, and compelling. However, winning an argument by presenting strong evidence may also result in an error, if the claim being disputed is actually true. So this parallel is drawn to rejecting a hypothesis when it is actually true.

The next activity, *Introduction to Confidence Intervals*, examines what happens after a null hypothesis is rejected. In this case, balancing a Euro will result in an equal number of Heads and Tails. Students are referred back to the Euro data and make a conjecture about the proportion of Heads they would expect to find in a large number of repetitions of this experiment. When students give different answers or ranges of answers, it is suggested that because we are unsure about giving a single number as our estimate, due to variability of our sample data, we might feel more confident about offering a range of values instead. Students are asked what interval, or range of values, might give an accurate estimate of possible values for this "true" proportion of Heads when a Euro coin is balanced on its edge and falls down. To move to the formal idea of a confidence interval, students are given the following news clip to read:

> *A recent poll of people in the military stated: While 58% say the mission (being in Iraq) is clear, 42% say that the U.S. role is hazy. The survey included 944 military respondents interviewed at several undisclosed locations throughout Iraq. The margin of error for the survey, conducted from Jan. 18 through Feb. 14, 2006, is ± 3.3 percentage points.*

Students are guided to use the information stated above to obtain an interval estimate for the percentage of *all people in the military* who believe the mission is hazy. They construct a confidence interval using this information. They see that they need two pieces of information that are given in this article: This information is then related back to the problem of finding a confidence interval for the proportion of Heads when balancing a Euro coin. This includes:

- A *sample statistic* (e.g., the class proportion of Heads when balancing coins), and,
- A *margin of error* (an estimate of how much this statistic varies from sample to sample for a given sample size, calculated from the sample data and information from the sampling distribution for the sample statistic).

Students are shown two ways to present confidence intervals:

- The *sample average, plus or minus a margin of error* (e.g., estimating the average textbook price for statistics, \$80 ± \$15).
- The *two endpoints* (low and high values) of the interval. (e.g., \$65–\$95).

The relationship of the confidence level to the idea of error is examined, and students reason about what a confidence interval tells about estimating a parameter and possibly making an error about that estimate. Students see that a confidence interval provides two kinds of information: an *interval estimate* for the population parameter (rather than a single number estimate) and a *level of confidence* (how confident we are that our interval includes the population value we want to estimate).

A wrap-up discussion includes what the term "margin of error" means, and how this term is used when interpreting results from a poll. Students describe the sample and the population for the survey reported above and critique it, referring back to material from the unit on Data related to designing good surveys (Lessons 1 and 2 in the unit on Data, Chapter 6). Students also consider and discuss different interpretations of the poll results, such as: Can we use our interval to give a guess about the true percentage of all people in the military that believe the mission is hazy? How? How sure are we? Are there any problems with generalizing from our sample of 944 military respondents to all people in the military?

Lesson 3: Reasoning About Confidence Intervals

This lesson helps students develop their reasoning about confidence intervals by using simulation to make and test conjectures about factors that affect confidence intervals. They also have opportunities to discuss common misconceptions as they critique interpretations of confidence intervals. Student learning goals for this lesson include:

1. Develop reasoning about confidence interval.
2. Understand what 95% *confident* actually means.

3. Understand how sample size and confidence level affect the length of the confidence interval.
4. Become familiar finding a confidence interval using *Fathom* software.
5. Understand connections between confidence intervals and hypothesis tests.

Description of the Lesson

In the *Estimating with Confidence* activity, students return to the question from the previous lesson: "What is the true (expected) proportion of Heads when a Euro is balanced?" Now that they believe that the proportion of Heads when a Euro balanced is *not* equal to 0.5, then what is it? Students now know the idea of a confidence interval. *Fathom* is used to produce a confidence interval for the sample of data based on balancing a Euro coin. The class discusses how to interpret this result and are asked what type of estimate might be more informative about the location of the actual population parameter, a narrower or wider interval, and why.

Connections are then made between testing a hypothesis and estimating with a confidence interval, and students see how a confidence interval can be used to test a hypothesis. Students make a conjecture about how the confidence interval would be different if they had only 50 pieces of data rather than 100, and then if they had 1,000 data values and why. This conjecture will be examined later in a simulation activity. Students reflect on the previous unit on sampling and distinguish between the sample statistic and a population parameter for the Euro coin example, and how much they would expect a sample statistic to vary from a population parameter.

In the *Estimating Word Lengths* activity, students return to the *Gettysburg Address* activity from the unit on Data (Lesson 3 in Chapter 6) in which they sampled words from the Gettysburg Address. They use the Gettysburg Address as a population and take samples and construct confidence intervals to see how they behave and how to interpret them. They use the Gettysburg Address Web applet to take a random sample of 25 words and then use *Fathom* to find a 95% confidence interval to estimate the true mean word length for all of the words in the Gettysburg Address. Next, the students draw their confidence intervals on the board, one on top of another. These intervals are compared to the true population mean word length, and students examine how many of the intervals generated by the class overlap the true population mean. Students are asked what percentage of all the intervals in the class they would expect to *not* overlap the population mean and find it is close to what they have generated.

The next activity (*What Does the 95% Mean?*) leads students use *Sampling SIM* to make and test conjectures about confidence intervals. They sample data from different populations such as a normal curve as well as for a skewed distribution, which is shown in Fig. 13.1.

Students generate 200 95% confidence intervals for samples of size 25 and examine how many do not include the population mean (shown as red lines) and how close the proportion of intervals that include the mean is to 95% (see Fig. 13.2).

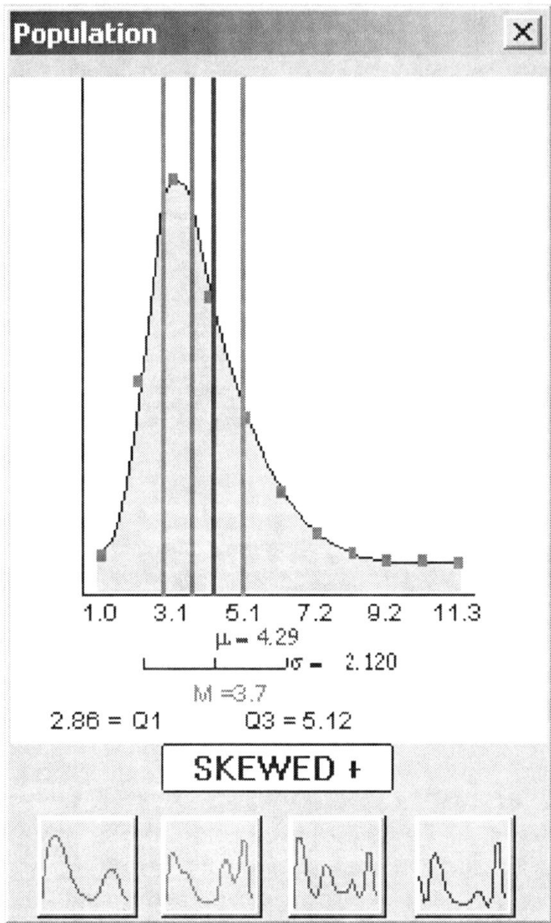

Fig. 13.1 A right-skewed population produced by *Sampling SIM*

They use the results from *Sampling SIM* to help answer the following questions that target common misconceptions about confidence intervals:

1. Does the level of confidence, 95%, refer to the percent of data values in the interval?
2. Does the level of confidence, 95%, refer to the location of the *sample mean* or locating the *population mean*? Explain.
3. Does the level of confidence, 95%, refer to a *single interval* (e.g., the one you found in *Fathom*) or to the *process or creating many intervals* (e.g., all possible intervals)? Explain.

Next, students use the *Sampling SIM* to make and test conjectures about what factors affect the width of the confidence interval. They then test these conjectures by increasing and decreasing the level of confidence, and changing the sample size,

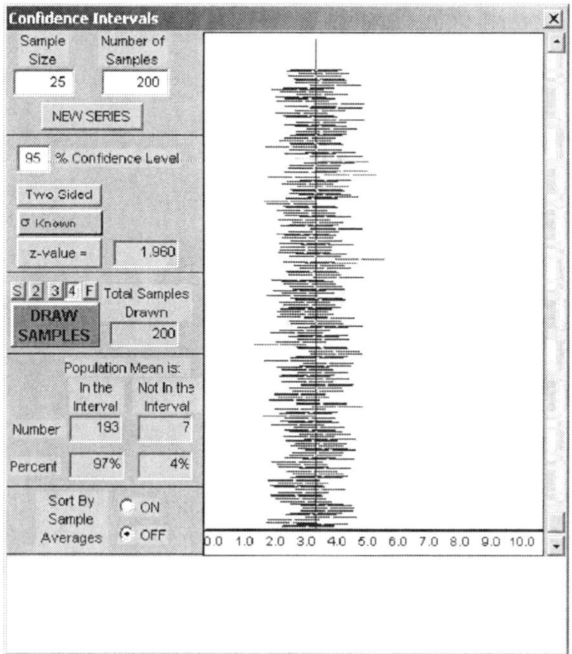

Fig. 13.2 Two hundred 95% confidence intervals (sample size 25) from a right-skewed population in *Sampling SIM*

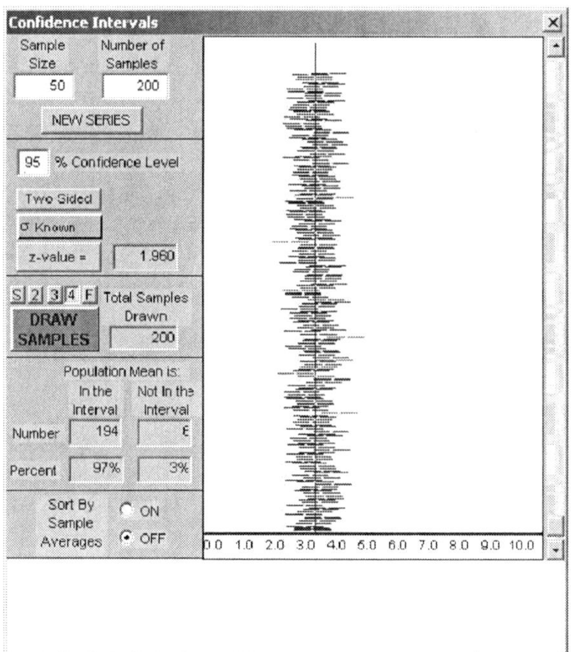

Fig. 13.3 Two hundred 95% confidence intervals (sample size 50) from a right-skewed population

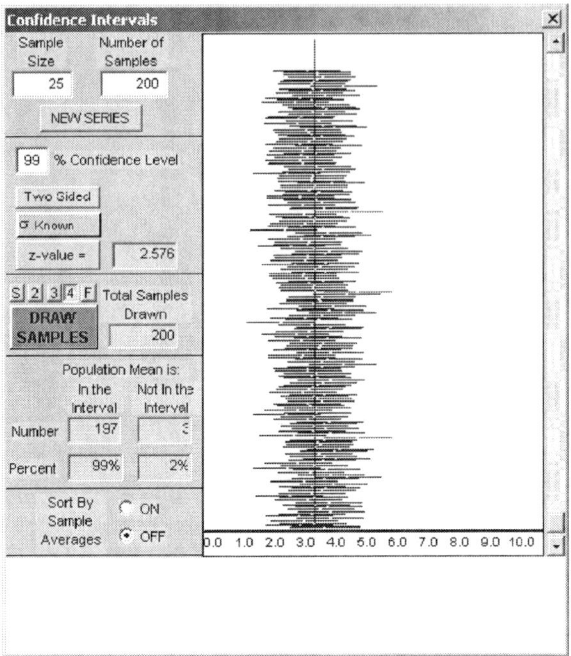

Fig. 13.4 Two hundred 99% confidence intervals (sample size 25) from a right-skewed population

generating new simulated intervals each time. See Fig. 13.3 for larger sample size and Fig. 13.4 for larger confidence level.

A discussion follows about what type of width (narrow or wide) gives the most precise estimate of the population parameter, and what level of confidence (lower or higher) most often includes the true population parameter being estimated.

A wrap-up discussion includes when and why we use a confidence interval in a statistical analysis and why we say "95% confident" instead of "95% probability." Students consider why and how confidence intervals and hypothesis tests are connected, and what is unique about each approach and the information it provides.

Lesson 4: Using Inference in an Experiment

This lesson described at the beginning of this chapter revisits an earlier experiment, giving students a chance to try to reduce within group variation and better detect a difference in the two conditions. Data are gathered and analyzed first graphically and then using *Fathom* to run a two sample t-test. The logic of hypothesis tests and comparison to making an argument are revisited for this context. Student learning goals for the lesson include:

1. Understand the idea of a two-sample hypothesis test.
2. Differentiate between a one-tailed and a two-tailed test.

3. Use *Fathom* to conduct a two-sample test.
4. Understand the idea of a two-sample confidence interval (difference in means).
5. Use *Fathom* to conduct a confidence interval to estimate a difference in means.
6. Revisit the ideas of designing an experiment and making cause and effect inferences.
7. Revisit ideas of within and between group variation and how they affect a two sample comparison.
8. Revisit ideas of how to reduce variation within a condition, and ideas of signal and noise in repeated measurements within each condition.

Description of the Lesson

In the *Gummy Bears Revisited Activity*, students reflect on the earlier *Gummy Bear* activity (Lesson 2 in the Comparing Groups Unit, Chapter 11) and discuss how to determine if there is a difference between two conditions in an experiment, in this case, if there are different average launching distances for the one book or four book launching pads. Students are asked, in light of recent discussions and activities on statistical inference, to suggest how, if a difference in sample means is observed, this is not just due to chance.

The students redo the experiment after first discussing a careful and systematic protocol to follow in launching the Gummy bears. Treatments are assigned to groups and each group produces data for 10 launches. Students use *Fathom* to produce side by side boxplots, discussing what the boxplots suggest about the differences in flight distances for the two conditions. Students are asked how to determine if the observed difference in group means is statistically significant and what this means. The null and alternative hypotheses are constructed and *Fathom* is used to run the test. Students contrast one and two tailed tests for this experiment, and run the test both ways using Fathom, contrasting the difference in results. Students explain what the results of the hypothesis test suggest about the difference between the two launching heights. Next, students use a confidence interval to estimate the mean difference in average launch. They discuss what it means if a difference of 0 is in the interval or is not in the interval. Since 0 was not in the interval, they concluded that this is a statistically significant difference in flight distances.

In a wrap-up discussion, students suggest reasons to use a one-tailed or two-tailed test of significance, and advantages and disadvantages of each method. They reason about how the type of test (one or two tailed) affects the P-values obtained and which method is more conservative. Finally, students give a full statistical conclusion about the comparison of flight distances for short vs. high launching pads.

Lesson 5: Solving Statistical Problems Involving Statistical Inference

This lesson comes at the end of a course, after the study of covariation (see Chapter 14) and helps students connect and integrate concepts and processes in statistical inference, developing their statistical thinking. Student learning goals for the lesson include:

1. Review the process of conducting and interpreting a test of significance.
2. Review the process for finding, reporting, and interpreting confidence intervals.
3. Review the conditions/assumptions that are necessary for our inferences to be valid.
4. Be able to research questions to appropriate inferential procedures.
5. Practice using *Fathom* to conduct tests of significance and to find confidence intervals.
6. Be able to interpret and justify results of statistical inferences.

Description of the Lesson

Discussion begins by looking back at the first few days of the course when students simulated data to estimate whether a sample statistic might be due to either chance or to some other factor. For example, if a student was able to correctly identify Coke of Pepsi in a blind taste test vs. the student was a lucky guesser. The discussion then proceeds to when students learned how to use *Fathom* to generate P-values and confidence intervals to help in making inferences and decisions about population parameters. Now that software can be used to generate statistical results for inferences, students consider the decisions that have to be made, for example:

a. What type of analysis to run (e.g., test or estimate, one or two samples, etc.).
b. What conditions to check.
c. How to interpret the results (and also know if we made a mistake).

In the *Research Questions Involving Statistical Methods* activity, students are reminded that the computer will generate P-values for tests of significance and construct confidence intervals for population parameters, even if the conditions are not checked and met. The class discusses how one should interpret the results of a procedure where the conditions are not met. Next, students are given the following table (Table 13.2) to discuss and complete it together, which will serve as a guide for running different analyses to produce inferential statistics in *Fathom*.

Students are then given a set of research questions (as shown below in Table 13.3) and a data set to use in answering the questions, using *Fathom* software. The data set contains the body measurements for a random sample of 50 college students. First, the instructor models statistical thinking, talking out loud and demonstrating the questions and steps and interpretations involved in answering one or two of the questions on the list below. Working together, students then discuss each question, select appropriate procedures, test conditions, generate graphs and analyses, and interpret their results.

Table 13.2 A guide for running different analyses to produce inferential statistics in *Fathom*

Type of procedure	Example of research question	Fathom instructions
One sample confidence interval for proportion	What is the proportion of college students who graduate in 4 years from your school?	
One sample confidence interval for a mean	What is the average number of credits earned by students when they graduate with a bachelor's degree?	
One sample hypothesis test for a proportion	Is the proportion of students who withdraw during their first year equal to 0.15 (The proportion who withdrew 5 years ago)? Is the proportion of students who withdraw during their first year less than 0.15?	
One sample hypothesis test for a mean	Is the average number of years it takes to finish a degree equal to 5? Is the average number of years it takes to finish a degree greater than 4?	
Two sample confidence interval for the difference between two means	What is the difference in the average number of hours spent studying each week between physics majors and English majors?	
Two sample hypothesis test to compare two means	Is there a difference in the mean GPAs of first year and fourth year students?	

Table 13.3 Selecting appropriate procedures and hypotheses to given research questions

Research question	Type of procedure	Null and alternative hypothesis (if appropriate)
What proportion of students in this class has a larger arm span than height?		
What is the average hand span for students in this class?		
What is the difference in hand spans for males and females?		
Is the average height for female students greater than 163 cm?		
Is the proportion of male students who are taller than 172.72 cm different from 0.5?		
Is there a difference between males and females in head circumference?		

After students complete this work, group answers to the questions are shared and justified.

Summary

Most students studying statistics encounter great difficulty when they reach the topics of statistical inference. Some instructors have compared student response to lecturers on this topic as “the deer in the headlight” phenomena, as students seem frozen, confused, and scared when learning these difficult topics. The research literature documents the difficulty students have understanding inference, and typical misconceptions that persist regarding P-values and confidence intervals.

Although many statistics courses put statistical inference at the end of a first course in statistics, we have illustrated a research-based approach that first presents informal ideas of inference early in the class and revisits these ideas again and again, so that when the formal ideas are introduced later they are more intuitive and easier to understand. The idea of statistical hypotheses as making arguments is used to help make this difficult topic more accessible to students. At the end of the course, students are given a set of research questions and need to integrate and apply all that they have learned to determine what procedures are needed and appropriate, to provide answers, and to justify their conclusions. This process is first modeled by their instructor and then they have the opportunity to use and develop their own statistical thinking by approaching these questions as statisticians, rather than just solving a series of textbook problems for each given procedure. This approach also differs from more standard approaches because the computational procedures are not emphasized. Instead, the process of using the computer to test hypotheses and estimating parameters is stressed, along with how to do this wisely and how to justify and interpret results.

Chapter 14
Learning to Reason About Covariation[1]

Even young children have the tendency to use the covariation of events as an indicator of causality.
(Zimmerman, 2005, p. 22)

Snapshot of a Research-Based Activity on Covariation

It is the first day of a unit on bivariate data. Students consider what kind of pattern they expect to see between the number of semesters a college student has been in school and the cumulative number of credits she/he has earned. They make a conjecture about the data they would see if they took random samples of 100 college students, each group sampled from all students who are beginning their first, second, third, fourth, fifth, and sixth semesters of college. Then, working in groups, students sketch a sequence of boxplots showing what they would predict to be the distributions of cumulative college credits for these six groups of 100 randomly selected college students who have been enrolled for one to six semesters.

Students discuss their reasoning as they do this, commenting that during the first semester, students take from 10 to 18 credits, but most take about 13–15 credits. They reason about center and spread as they draw a predicted boxplot for the first semester. For each subsequent semester, they also think about expected typical values and spread around those values, drawing a total of six parallel boxplots. They share their reasoning with the class, explaining that they expect more variability over time even though they expect an average increase in cumulative credits. Next, they generate and examine graphs and summary statistics for actual random samples of data for 100 students for each of the first six semesters of their enrollment, and compare these to their conjectures. See a sample of such graphs in Fig. 14.1.

Students' attention is drawn to both center and spread of the boxplots for each semester and how they change over time. Students then discuss and write an explanation about what these graphs indicate about the relationship between length of time in school and cumulative credits earned. They are asked questions about how well they think these data generalize to all students at their institution and what types of inferences they would be willing to make based on these data. Students

[1] We gratefully acknowledge the major contribution of Dr. Andrew Zieffler to the writing of this chapter.

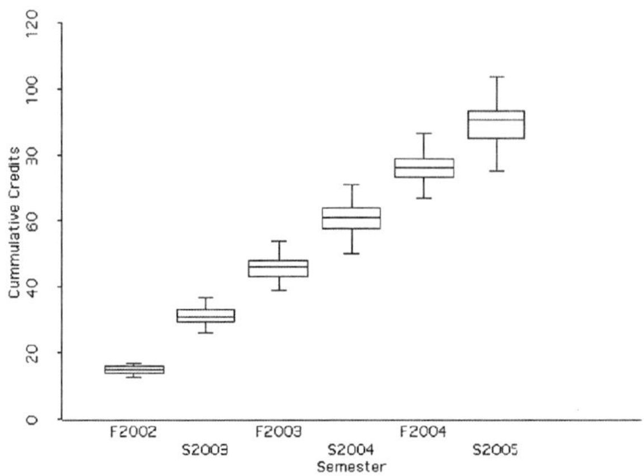

Fig. 14.1 A sequence of boxplots showing the distributions of cumulative college credits for six samples of 100 randomly selected college students, representing first through sixth semesters

then consider a variable that has an inverse relationship with number of credits, so that values of this variable get smaller as the number of credits gets bigger. They suggest, for example, the time left before graduation: that with more credits, there is a shorter time left before graduation. They are asked if this 'negative' relationship is a weak one, or if they think it is a strong one, and it is decided that the relationship is just as strong as the positive relationship with number of semesters in school.

Finally, a discussion about causation is introduced as the students are asked to discuss whether the number of semesters of being in school "causes" them to have more credits, and whether a cause and effect can be determined by examining this bivariate data set.

Rationale for This Activity

Students may have seen scatterplots before and may have ideas of trends over time, but most likely do not think about variability in bivariate data. This activity begins by having them think about variability at six intervals, as well as thinking about the linear trend, based on the centers (median values) for data in these intervals. Thus, they are guided to examine and interpret bivariate data by looking for a trend as well as scatter from the trend, by focusing on both center and spread for one variable (y) relative to values of a second, ordinal variable (x) (an approach suggested by Cook & Weisberg, 1999). This activity helps students begin to reason about the idea of a linear relationship between two variables and helps prepare them to better interpret and reason about scatterplots of data. When they examine scatterplots, they can try to remember the idea of parallel boxplots and how they represent a series of mini distributions of the y variable, each having center and spread, for the x variable, as shown in Fig. 14.1.

This lesson also brings up informally two important ideas that are often difficult for students to understand about covariation. The first is that covariation between two variables does not mean that one causes the other. The second point is the idea that a negative association between two variables does not imply a weak association.

The Importance of Understanding Covariation

Reasoning about *association* (or *relationship*) between two variables, also referred to as *covariational reasoning*, or reasoning about *bivariate data*, involves knowing how to judge and interpret a relationship between two variables. Covariational reasoning has also been defined as the cognitive activities involved in coordinating two varying quantities while attending to the ways in which they change in relation to each other (Carlson, Jacobs, Coe, Larsen, & Hsu,). This type of reasoning may take a very mathematical form (e.g., a linear function), a statistical form (reasoning about a scatterplot), or a more qualitative form (e.g., causal predictions about events, based on observed associations, such as spending more time studying seems to lead to better tests grades, as described in causal model theory in psychology). Covariational reasoning is also viewed as playing an important role in scientific reasoning (Koslowski, 1996; Schauble, 1996). Although covariation between events is a necessary but not sufficient basis for inferring a causal relationship, it is a basis for making causal inductive inferences in science (Zimmerman, 2005).

The concept of covariation may be unique in that is an important concept in the different fields of psychology, science, mathematics, and statistics, and that covariational reasoning is described somewhat differently in each discipline. Statisticians may be surprised that reasoning about covariation, which they think of as a statistical topic focusing on bivariate distributions of data, is much more complex than the familiar caution that "correlation does not imply causation," and is beyond reasoning about scatterplots, correlation, and regression analyses. Indeed, cognitive psychologists McKenzie and Mikkelsen (2007) wrote that covariational reasoning is one of the most important cognitive activities that humans perform.

It is our view that to deeply understand covariation in a statistical context, one must understand all the aspects of covariation described in the other disciplines, as well as understand aspects unique to statistics (such as model fitting and residuals) because of the inherent variability of real data. Such an understanding involves ideas of structure and strength of a bivariate relationship as well is its role in causal models and in predicting events, and the changing values of one variable relative to another, as expressed in a line. It is no wonder that students have difficulty understanding concepts related to covariation, including the correlation coefficient, r-squared, and a regression line.

Watkins et al. (2004) describe six features of a bivariate relationship that may be used to guide students as they learn to examine and reason about bivariate data. These are:

1. The individual variables and their variability.
2. The shape of a pattern revealed in a bivariate plot in terms of linearity, clusters, and outliers.

3. The trend if there is one (positive or negative).
4. The strength of the trend: strong or weak, varying or constant.
5. Whether the pattern may generalize to other situations (can be tested with inferential methods, and is also based on how the sample was obtained).
6. If there are plausible explanations for the pattern? Is it plausible that one variable may have a causal relationship or might a third, lurking variable cause the relationship.

These ideas may be developed informally and then revisited in more formal ways.

The Place of Covariation in the Curriculum

Topics involving covariation may be introduced at various times in the curriculum. In some introductory courses and textbooks, scatterplots are introduced as part of a data analysis unit. The topic of correlation and regression may be introduced before or after the topic of statistical inference, with arguments made for each placement of the topic (see Zieffler, 2006). Related ideas of contingency tables and chi-square tests often appear in a separate unit, many times at the end of a course.

We believe that the topic of covariation may be informally introduced in a unit on collecting and producing data, and that graphs of bivariate data may be informally introduced as part of data analysis, but that the formal study of bivariate data takes place after the study of statistical inference. Our reasons for this are that the study of covariation builds on and integrates many topics studied earlier in the course, and we can revisit and deepen these ideas as topics taught near or at the end of an introductory course. These ideas include distribution of data, center and spread, variability, fitting a model to data, using experiments to infer causation, using correlation to explore relationships among observational variables, and using statistical inference to run tests on correlations and regression coefficients to see if observed relationships may be generalized.

We realize that it is somewhat controversial to delay the study of scatterplots, correlation, and simple linear regression to near the end of the course. We do this based on our review of the research literature, our findings that students are able to informally reason about bivariate data well before formal instruction on this topic (see Zieffler, 2006) and based on our positive experience teaching the topic at this final point in a course. However, instructors who want to introduce some of the lessons and activities in this chapter at an earlier point in their course are welcome to do so. Finally, we want to revisit the idea mentioned in Chapter 6 about the importance of students being exposed to multivariate data sets. Once again, the data sets in this chapter are multivariate, and while students may examine pairs of variables at a time, they are also encouraged to look at the data sets in a multivariate way, disusing and integrating information about relationships between pairs of variables to better understand the variability of individual variables as well as aspects of the multivariate data set.

Review of the Literature Related to Reasoning About Covariation

Because of its important role in so many disciplines, covariational reasoning has been the focus of research in psychology, science, and mathematics education, in addition to statistics education. The research studies related to covariational understanding and reasoning are quite diverse, and vary according to the disciplinary field of the researchers. These studies are summarized in four main areas: judgment of covariation, understanding covariation as functional relationships, covariational reasoning in science education, and covariational reasoning and judgments in statistics.

Judgment of Covariation

Research by psychologists provides much of the seminal work in covariational reasoning research. Since the seminal study by Inhelder and Piaget (1958), psychologists have produced several robust findings, based on large sample sizes and use of randomized treatment designs and analyses, spanning five decades of research in this area. Regardless of the importance that covariational reasoning seems to play in the day-to-day lives of people, much of the research from the field of psychology has generally concluded that people are surprisingly poor at assessing covariation. A robust finding is that peoples' prior beliefs about the relationship between two variables have a great deal of influence on their judgments of the covariation between those variables (e.g., Jennings, Amabile, & Ross, 1982; Kuhn, Amsel, & O' Loughlin, 1988).

This finding is related to another consistent finding, that of *illusory correlation*. An illusory correlation exists when a subject believes there is a correlation between two uncorrelated events. This, for example, could encompass relying more on memory rather than examining the data/cases presented (e.g., subjects would suggest a positive relationship exists between price and quality even though the data would suggest otherwise), and viewing data/cases that confirm their expectations as more relevant than disconfirming cases (e.g., McGahan, McDougal, Williamson, & Pryor, 2000).

Additional studies have examined how people reason about covariation of data in contingency tables, indicating that people tend to not treat the four cells of a 2-by-2 contingency table as equally important. In fact, peoples' judgments seem to be most influenced by the joint presence of variables and least influenced by the joint absence of variables (e.g., Kao & Wasserman, 1993). Other studies showed that people have difficulty when a bivariate relationship is negative (e.g., Beyth-Marom, 1982), and that peoples' covariational judgment of the relationship between two variables tends to be less than optimum (i.e., smaller than the actual correlation presented in the data or graph) (e.g., Jennings et al., 1982). Still another consistent finding in these studies is that subjects have a tendency to form causal relationships based on a covariational analysis (e.g., Ross & Cousins, 1993,).

Research studies have also examined the conditions and accommodations under which people tend to make better covariational judgments. For instance, researchers have found that subjects tend to make more accurate judgments when the variables to be examined are continuous rather than dichotomous (e.g., Jennings et al., 1982), and other studies have suggested that certain accommodations such as detailed instructions (Alloy & Abramson, 1979), easy to process formats (Ward & Jenkins, 1965), subjects being told non-contingency is possible (Peterson, 1980), and low frequency of data/cases (Inhelder & Piaget, 1958) might help subjects more accurately judge covariation. Subjects have also been shown to make more accurate judgments when data are presented simultaneously rather than when it is presented one case at a time (e.g., Seggie & Endersby, 1972).

Understanding Covariation as Functional Relationships

Studies on covariational reasoning by mathematics education researchers tend to focus on lines and understanding functions, or on aspects of bivariate reasoning that might be used in algebra and calculus. For example, Carlson et al. (2002) classified college students' mental actions affecting reasoning about covariation as follows:

1. The coordination of the value of one variable with changes in the other.
2. The coordination of the direction of change of one variable with changes in the other.
3. The coordination of the amount of change of one variable with the amount of change in the other.
4. The coordination of the average rate of change of the function with uniform increments of change in the input variable.
5. The coordination of the instantaneous rate of change of the function with continuous change in the independent variable for the entire domain of the function.

Carlson's studies suggested that most students could determine the direction of change, but that many had difficulties constructing images of continuous rate of change, even after completion of a second course in calculus. The researchers noted that students have particular problems representing and interpreting graphical displays. In some cases, mathematics education researchers have suggested the need for teachers to have students think about covariation as it occurs in functions in terms of real-life dynamic events (e.g., Carlson et al., 2002).

Other researchers have pointed out that covariational reasoning is used extensively in both algebra (Nemirovsky, 1996) and calculus (Thompson, 1994). In particular, studies suggest that this type of reasoning plays a major role in students' understanding of the derivative, or rate of change (e.g., Carlson et al., 2002), and that this interpretation of covariation is slow to develop among students (e.g., Monk & Nemirovsky 1994; Nemirovsky, 1996). Studies from mathematics education have also shown that not only is students' ability to interpret graphical and functional information slow to develop, but also that students tend not to see the graph of a function as depicting covariation (Thompson, 1994).

Covariational Reasoning in Science Education

Researchers in science education have focused on either the psychological aspect of covariation in relation to identifying causal factors or on the mathematical aspects of covariation, such as reasoning about lines and functions. Some researchers have studied how students use covariational reasoning in solving science problems, finding that students tend to use sub-optimal solution strategies to determine if a relationship exists or does not exist in a set of bivariate data, and that students tend to infer a causal relationship from correlational data (e.g., Adi, Karplus, Lawson, & Pulos, 1978).

In a more recent study by Kanari and Millar (2004), students' approaches to data collection and interpretation were studied as they investigated relationships between variables, as part of students' ability to reason from data. The authors argue that it is reasoning from data that distinguishes scientific reasoning from logical reasoning. They found that students of all ages had a much lower success rate in investigations where the dependent variable did not covary with the independent variable, than in those where it did covary. They suggest that school science investigations should include both covariation and non-covariation cases to develop students' covariational reasoning.

Covariational Reasoning and Judgments in Statistics

Statistics education research has generally focused on studying students' covariational reasoning as part of instruction in statistics. The methodologies used have been primarily qualitative, following the format of design experiments (e.g., Cobb et al., 2003a) while other studies on covariational reasoning in statistics education have used methods in psychology, using a qualitative coding scheme to help categorize response data (e.g., Batanero, Estepa, & Godino, 1997; Moritz, 2004; Morris, 1997; Stockburger, 1982).

In one of the earlier studies, Stockburger (1982) asked university students enrolled in an introductory statistics course to complete many times four computer exercises, such as estimating the correlation coefficient. Stockburger found that students in general did very poorly on these exercises, and their ability to estimate correlation improved after using the computer software.

The impact of computers in developing understanding of statistical association was studied by Batanero et al. (1996, 1997). Students were asked to assess the existence of correlation between two variables given to them in a two-by-two contingency table. The researchers then identified incorrect covariational strategies employed by the students (drawing on strategies first outlined in 1958 by Inhelder & Piaget). Both studies (Batanero et al. (1996, 1997) found an overall general improvement in student strategies. They also both revealed the persistence of what they refer to as a *unidirectional misconception*. That is, students only perceive a relationship among the variables in the positive direction.

Both studies also showed that students maintained their causal misconception throughout the duration of the experiments. Both studies also showed that students had problems with several features of association such as distinguishing between the roles of independent and dependent variables and reasoning about relationships that were negative. Finally, students realized that the absolute value of the correlation coefficient was related to the magnitude of the relationship, but did not relate that idea to the spread of scatter around the regression line.

Other studies have examined students' covariational reasoning as they study regression and reported some of the difficulties associated with this topic including problems with interpretation (e.g., Sánchez, 1999), and problems with the coefficient of determination, or r^2 (Truran, 1997). Konold (2002b) presents a different view of whether or not people can make accurate covariational judgments. He suggests that people are not poor at making these judgments, but rather they have trouble decoding the ways in which these relationships are displayed (e.g., scatterplots or two-by-two contingency tables). His research has been on middle school students using *TinkerPlots* (Konold & Miller, 2005). Konold found that students were better able to make covariational judgments using a super-imposed color gradient function in *TinkerPlots*, possibly because relationships between variables are explored by use of only one dimension (e.g., horizontal axis) for one variable and color gradient for the other (see Fig. 14.2). In addition, the task is broken up into two smaller parts: First, students anticipate what will be seen and second, that they then examine the new display.

In a study of younger children, Moritz (2004) had students translate verbal statements to graphs and also translate a scatterplot into a verbal statement, and related

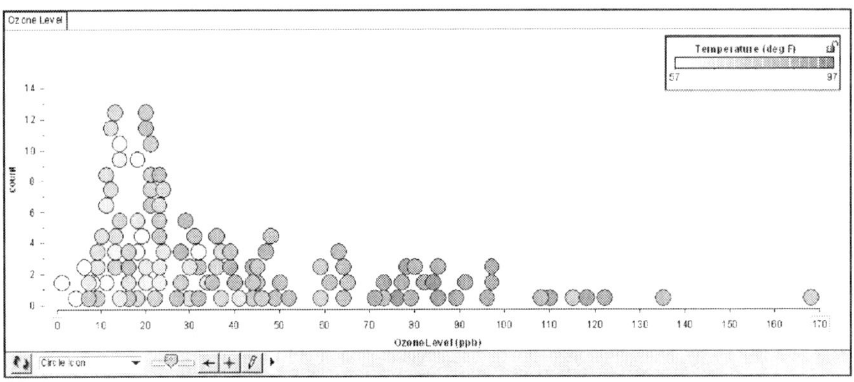

Fig. 14.2 A *TinkerPlots* graph displays how data of Ozone levels in New York City on 111 successive days in 1973 are associated with the maximum daily temperature[2]

[2] The *TinkerPlots* graph in Fig. 14.2 displays data of Ozone levels (Ground-level Ozone in parts per billion) in New York City on 111 successive days from May to September in 1973. To explore the relationship between Ozone levels and temperature (Maximum temperature in degrees Fahrenheit) we first presented the Ozone level on the x-axis, "fully separated" the cases, and then colored the icons with the grey temperature gradient. We notice how as we move up the Ozone scale we tend to go from the lower temperatures (the lighter grey) to the higher temperatures (the darker grey)

tasks. The students were given a written survey that included six or seven open-ended tasks involving familiar variables. The variables were also chosen so that students would expect a positive covariation, but the data given in the task represented a negative covariation. Moritz (2004) found many of the same student difficulties as other studies have revealed: that students often focused on isolated data points rather than on the global data set (e.g., Ben-Zvi & Arcavi, 2001). He also found that students would often focus on a single variable rather than the bivariate data, and that several students had trouble handling negative covariations when they are contradictory to their prior beliefs.

Two more recent design experiment studies investigated the role of technology in helping students reason about bivariate data, and how students differentiate between local and global variation in bivariate data. Gravemeijer (2000) results suggest that students need an idea of the global trend (prior expectation) and that students have a hard time distinguishing between arbitrary and structural covariation. He suggests that students examine and compare several univariate data sets (time series) as an introduction to examining bivariate data.

This approach was used by Cobb et al. (2003b) to help students view bivariate data as distributed in two-dimensional space, to see scatterplots as situational texts, and to track the distribution of one variable across the other (scan vertically rather than diagonally). Using the *Minitools* software (Cobb, Gravemeijer, Bowers, & Doorman, 1997) students examine the "vertical variation" across levels of x in graphs of bivariate data. Students were asked to compare differences in the distribution of the y-variable at different levels of the x-variable (see Fig. 14.3).

The results of their study suggested that the shape of a distribution is a better place to start than is variability, and that there be a continued focus on relative density and on the shape of the data within vertical slices. They also suggested that an emphasis on shape could lead to a discussion of strength and direction in a bivariate

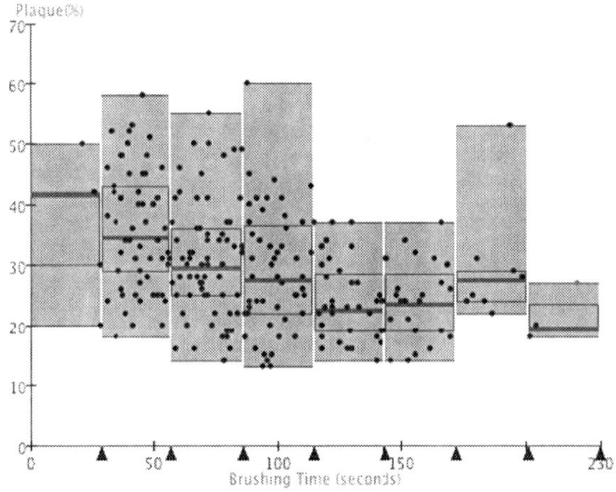

Fig. 14.3 *Minitools* software allows students to start looking at the local variation for different values on the x-axis in addition to the global trend

plot and that the focus on vertical distribution could lead to a more intuitive idea of the line of best fit.

A recent study by Zieffler (2006) examined students' development of reasoning about quantitative bivariate data during a one-semester university-level introductory statistics course. He found a surprising result that the most growth in covariational reasoning occurred early in the course, during the unit on data analysis, and before formal study of bivariate data. One plausible explanation was that students were developing their statistical reasoning in this earlier part of the course and that facilitated the growth in covariational reasoning, even before formal instruction on that topic.

Another result of his study is that students seemed to better reason and learn inference when the bivariate unit came at the end of the course after inference rather than as part of or immediately following a unit on exploring data, based on a comparison of two different sequences of course topics in his study. However, no difference was found in students' performance on an assessment of covariational reasoning.

The Role of Technology in Helping Students to Reason About Covariation

The research also suggests that particular types and uses of technology may help students make more accurate judgments of covariation (Batanero et al., 1996, 1997; Cobb et al., 2003b; Gravemeijer, 2000; Konold, 2002b; Stockburger, 1982). The appropriate use of visualizations can change students' statistical conceptualizations (Rubin, 2001) - particular software that provide flexible, dynamic capabilities and multiple, linked representations - allowing students to display, manipulate, and analyze data. This type of technology (e.g., *TinkerPlots* and *Fathom*) appears to help students understand and apply covariational reasoning (Carlson et al., 2002; Rubin, 2001). *Fathom* (Key Curriculum Press, 2006) allows additional capabilities such as visually fitting a regression line and visually showing the changing squared deviations from the line as a line is fitted to the data, as shown in Fig. 14.4.

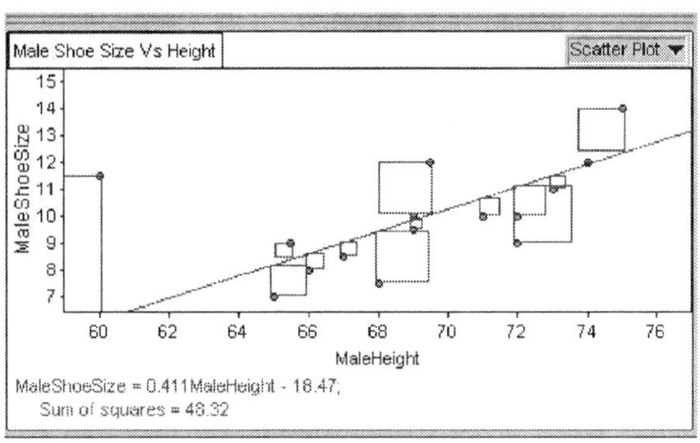

Fig. 14.4 Screen shot of *Fathom* printout showing squared residuals

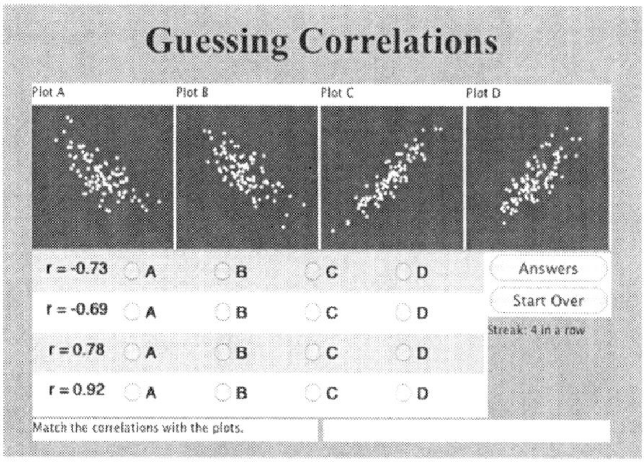

Fig. 14.5 Students guess at the appropriate correlation in the *Guessing Correlations* applet

In addition to software that allows students to explore and manipulate bivariate data, there are also Web applets that help students reason about and understand particular concepts. For example, an applet that lets students click on a graph, creating points on a scatterplot, where a changing correlation coefficient indicates the effect of each added point on the correlation (e.g., http://www.stat.vt.edu/~sundar/java/applets/CorrCreateApplet.html), and an applet that lets students visually fit a line to a bivariate plot (e.g., *Regression by Eye* applet from the *Rice Virtual Lab in Statistics* http://www.ruf.rice.edu/~lane/stat_sim/reg_by_eye/index.html).

Another tool that helps students reason about bivariate relationships is the *Guessing Correlations* applet, such as the one at http://www.stat.uiuc.edu/courses/stat100/java/GCApplet/GCAppletFrame.html. Students are shown a set of four scatterplots and asked to match them to four correlation coefficients (see Fig. 14.5). As students examine and reason about these relationships, they develop a better sense of the different degrees of covariation, and factors that make the correlation coefficient larger and smaller.

Summary of the Literature Related to Developing Covariational Reasoning

Looking at the studies across the different disciplines together, we note the following general findings:

- Students' prior beliefs about the relationship between two variables have a great deal of influence on their judgments of the covariation between those variables;
- Students often believe there is a correlation between two uncorrelated events (illusory correlation);
- Students' covariational judgments seem to be most influenced by the joint presence of variables and least influenced by the joint absence of variables;

- Students have difficulty reasoning about covariation when the relationship is negative;
- Students' covariational judgment of the relationship between two variables tends to be less than optimum (i.e. smaller than the actual correlation presented in the data or graph); and
- Students have a tendency to form causal relationships based on a covariational analysis.

These findings along with the suggestions based on design experiments lead to some implications for helping students reason about covariation.

Implications of the Research: Teaching Students to Reason About Covariation

The research studies reviewed have many implications for the teaching and assessment of bivariate data. Reasoning about covariation seems to build upon reasoning about distribution, especially the concepts of shape and variability (Cobb et al., 2003b; Gravemeijer, 2000; Konold, 2002b). It is suggested that instruction focus on relative density and on the shape of the data within vertical slices, leading to ideas of strength and direction in a bivariate plot, which could lead to a more intuitive idea of the line of best fit.

The research suggests that using data that has a context seems to be important for better student understanding of covariation (Adi et al., 1978; Cobb et al., 2003b; Gravemeijer, 2000; Moritz, 2004). This especially helps when the data are meaningful to the students (Moritz, 2004). In addition, the persistence of the positive unidirectional misconception described above implies that teachers should provide tasks that have students interpret and reason about data sets that have negative correlations and no covariation, which are less intuitive and require more thinking on the student part. An awareness of typical students' difficulties identified in the research literature can help teachers be on the lookout for these errors and misconceptions as students work together during activities and during class discussions, as well as on assessments of learning.

Progression of Ideas: Connecting Research to Teaching

Introduction to the Sequence of Activities to Develop Reasoning About Covariation

Table 14.1 shows a suggested series of ideas and activities that can be used to guide the development of students' reasoning about covariation. This sequence of activities presented in lessons are designed to lead students through a progression of ideas to help them understand the idea of covariation of data and develop covariational reasoning. This includes learning to understand and reason about graphs of bivariate relationships, interpreting and understanding features of bivariate data,

Table 14.1 Sequence of activities to develop reasoning about covariation[3]

Milestones:ideas and concepts	Suggested activities
Informal ideas prior to formal study of covariation	
• Understanding when to infer causation: from an experiment, rather than from correlated variables	• Taste Test Activity, Lesson 4, Data Unit (Chapter 6)
Formal ideas of covariation	
• Understanding ideas of trend by recognizing a positive linear pattern in the medians of vertical boxplots	• Credit Questions Activity (Lesson 1: "Reasoning about Scatterplots and Correlation")
• Understand the nature of bivariate data and the idea of covariation of two quantitative variables	• Interpreting Scatterplots Activity (Lesson 1)
• Points in a scatterplot represent pairs of data for individual cases, measured on each variable	• Interpreting Scatterplots Activity (Lesson 1)
• The idea of a linear trend in a bivariate plot	• Interpreting Scatterplots Activity (Lesson 1)
• Understanding how to distinguish between a positive trend and a negative trend, and how this differs from strength of the trend	• Interpreting Scatterplots Activity (Lesson 1)
• Reasoning about factors that could cause a linear trend in bivariate data, and that there could be a lurking variable (or a causal relationship)	• Interpreting Scatterplots Activity (Lesson 1)
• Reasoning about what would be needed to establish a causal relationship between two correlated variables	• Interpreting Scatterplots Activity (Lesson 1)
• Structure and strength in a bivariate plot: linearity, direction, closeness to the model of straight line	• Reasoning about the Correlation Coefficient Activity (Lesson 1)
• A correlation coefficient as a measure of the strength and direction of the linear relationship	• Reasoning about the Correlation Coefficient Activity (Lesson 1)
• Understating how different aspects of a bivariate data, as revealed in a scatterplot, affect the correlation coefficient	• Guessing Correlations Activity (Lesson 1)

[3] See page 391 for credit and reference to authors of activities on which these activities are based.

Table 14.1 (continued)

Milestones:ideas and concepts	Suggested activities
• Understanding how the same correlation coefficient may be obtained for very different sets of bivariate data	❖ An activity where students examine the correlation in data sets that have the same correlation coefficient, but are very different, including nonlinear patterns, such as Anscombe Data (1973). (The symbol ❖ indicates that this activity is not included in these lessons.)
• Fitting a model of a line to data.	• Diamond Rings Activity (Lesson 2: "Fitting a Line to Data")
• Strength of the relationship is based on how well the line (model) fits the data	• Diamond Rings Activity (Lesson 2)
• Linear regression as way to model a linear relationship.	• Diamond Rings Activity (Lesson 2)
• Regression model in explaining the relationships between two quantitative variables	• da Vinci and Body Measurements Activity (Lesson 2)
• What the slope and intercept mean in a bivariate data model	• da Vinci and Body Measurements Activity (Lesson 2)
• How and why better predictions have less scatter around the line (and fitted values)	• da Vinci and Body Measurements Activity (Lesson 2)
• Understanding the idea of residuals as deviations from the line (model) as providing evidence to assess how well the line provides a model for a bivariate data set	• da Vinci and Body Measurements Activity (Lesson 2)
• Understanding how to generalize bivariate relationships to a larger population	• Testing Relationships: Admissions Variables, and Baseball Variables Activities (Lesson 3: "Inferences involving Bivariate Data")
• Understanding how to interpret inferences about correlation coefficients and regression slopes	• Testing Relationships: Admissions Variables, and Baseball Variables Activities (Lesson 3)
• Understanding that it is important to consider the size of the correlation in addition to the size of the P-value: practical vs. statistical significance	• Testing Relationships: Admissions Variables, and Baseball Variables Activities (Lesson 3)
Building on formal ideas of covariation in subsequent topics	
• Knowing how to recognize when a test of a correlation coefficient or regressions slope is appropriate to answer a research question	• Research questions involving statistical methods activity (Lesson 5, Inference Unit, Chapter 13)

understanding and reasoning about the correlation coefficient, modeling bivariate data with a line, and making inferences about the nature of bivariate relationships. These activities build on foundational concepts such as distribution, variability, model, deviation, and inference.

Introduction to the Lessons

There are three lessons on covariation that focus entirely on the concepts and ideas without going into the mathematics. The activities in the lesson lead students to first think about linear trend in the presence of variability, as they move from examinations of parallel boxplots to a scatterplot to a fitted line. They develop the idea of correlation as a measure of a linear relationship between two variables, and then use the line as a model to explain and make predictions based on this relationship. Finally, students learn to apply ideas of inference to this topic by using tests of significance to test that the correlation coefficient and regression slope are different than zero.

Lesson 1: Reasoning About Scatterplots and Correlation

This lesson, as described briefly at the beginning of the chapter, guides students to develop reasoning about bivariate data by beginning with parallel boxplots and moving to scatterplots that are viewed as a series of vertical slices of data. In this way, ideas of center and spread are brought to and integrated into reasoning about bivariate distributions. Students learn how to examine and interpret scatterplots and begin to connect values of the correlation coefficient to different types of scatterplots. Causal factors are also discussed. Student learning goals for this lesson include:

1. Understand the nature of bivariate data and what that points in a scatterplot represent.
2. Understand and describe shape (form), trend (direction), and variation (strength) in a scatterplot.
3. Answer contextual questions using information obtained from a scatterplot.
4. Know that a scatterplot is the appropriate graph to create to answer certain questions about the relationship between two quantitative variables.
5. Estimate correlation coefficients from a scatterplot.
6. Understand that correlation is only about strength of linear trend.
7. Understand that a high correlation does not imply that the data are linear.
8. To be aware of lurking variables and to understand that correlation does not imply causation.

Description of the Lesson

The lesson begins with a problem involving real student data (*Credit Questions* activity). Students are told that university guidance counselors are interested in the number of credits students accumulate as they progress through their degree programs. They are interested in studying how long it takes a student to reach 60 credits (the requirement to reach Junior status), and how the length of time a college student has been in school relates to the number of credits she/he accumulates.

Students are asked to imagine six cohorts of students whose credit loads are examined, each group coming from all students in their first, second, third, fourth fifth, or sixth semesters of college. Students are asked to reason about if they expect the average number of credits that these students have accumulated would differ for students in each different semester of college. Students also reason about the variation in the total number of credits accumulated each semester of enrollment.

Next, working in groups, students make conjectures about such a data set. They discuss and then draw boxplots based on their predictions of what those distributions of cumulative credits would look like for random samples of 100 students for each of the six different cohorts. A class discussion follows as students discuss, based on their own knowledge and experience, what they expect to find in terms of number of credits across six semesters for students, focusing on center and spread for the number of credits for each subsequent semester and how the variability increases for each subsequent semester. Students are then given a data set and use it to generate actual boxplots that are used to test their conjectures. Students describe the relationship they see between length of time a college student has been in school and their cumulative number of credits, discussing trend as well as variation around the trend, in formal language.

Students are asked if they are willing to generalize about the relationship found to the population of all students at the University, based on these data. They also talk about the fact that there is a strong trend, but that one variable does not cause the other, revisiting ideas of correlation and causation from the earlier unit on data (Chapter 6). Students are asked to speculate about variables that could have an opposite or inverse relationship with these variables, distinguishing between a positive and negative trend, acknowledging that a positive or negative trend be strong or weak.

In the *Interpreting Scatterplots* activity, students are given a scatterplot representing the relationship between gestation period (length of pregnancy) and lifespan for a sample of 62 different animals, which has a moderate, positive trend. They first try to separate the points into several different vertical slices, or distributions, to determine how the centers and spread of those distributions are changing, building on the previous boxplot activity. They are asked if the data indicate a relationship between the two variables, and to describe the pattern and features of the plot. Students then examine a series of scatterplots and discuss different features that they see: lots of scatter, little scatter, linear and nonlinear patterns, positive and negative directions, different steepness of the linear pattern, outliers, clusters of points, etc. In the *Reasoning about the Correlation Coefficient* activity, students discuss different features of a scatterplot and how they are used to interpret and describe bivariate relationships. The correlation coefficient is discussed as a measure of the linearity of the pattern of points, and how it shows direction as well as strength.

The next activity (*Guessing Correlations*) has students using a Web applet to reason about the correlation coefficient and how different factors make it larger or smaller. Using the *Guessing Correlations* applet (http://www.stat.uiuc.edu/courses/stat100/java/GCApplet/GCAppletFrame.html), students go through several rounds, talking in pairs about their choices, until they are comfortable estimating the degree of linearity (correlation) in a dataset. A class discussion follows about what they learned from this activity as well as the importance of examining a scatterplot to learn about the relationship between two variables, and not just calculating a correlation coefficient. Students discuss how the correlation coefficient helps summarize the relationship between two quantitative variables much like the mean or standard deviation can be used to describe and summarize a single variable.

A wrap-up discussion has students summarize what the points in the scatterplot represent, what kind of questions can be answered by examining a scatterplot and correlation for two variables, what is considered a "low" correlation, and what information a correlation does and does not provide about two variables. Students are asked whether a high correlation indicates causation, and how they would determine casual factors.

Lesson 2: Fitting a Line to Data

This lesson focuses on using the regression line to model bivariate relationships. Students fit lines to data, discuss the interpretation of the lines in terms of direction and size of the slope. Students use a regression line to predict values of y based on values of x. They fit lines to data seeing how the sums of squares are minimized for the regression line. They also learn about the dangers of extrapolation and see how outliers may affect the fitted line. Student learning goals for this lesson include:

1. Understand the idea of a line as a model of bivariate relationship.
2. Be able to interpret the slope and y-intercept of a line of best fit in the context of data.
3. Be able to use a line to predict y from a given x.
4. Understand that a predicted value is not the same as the actual value (in most cases) and that the difference is the *residual*.
5. Be able to interpret what the residual means in the context of the data.
6. Understand that extrapolation is not appropriate.
7. Use *Fathom* to create the line of best fit from a given set of data.
8. Be able to read *Fathom* output for the line of best fit.
9. Understand the effect of an outlier on the fitted regression line.

Description of the Lesson

The lesson begins with an examination of a data set on diamond rings (*Diamond Rings* activity). Students discuss and make conjectures about what they would expect to find in terms of a relationship between the price of a diamond ring and the number of carats in the diamond. Next they generate a scatterplot using *Fathom* that contains the prices of diamond rings and the carat size of their diamond in each ring. The data look very linear, and students use *Fathom* to first estimate where a line would best fit the data. They revisit the idea of fitting a model to a data set (see Chapter 7), this time using a line as a model of the bivariate relationship. They compare the conjectured lines to the regression line ($\hat{\$} = -154.94 + 2220.66(Carat)$), which is shown in Fig. 14.6.

Students discuss if the regression line seems to be a good model and how well it fits the data. They informally discuss the line: its steepness and direction, and what that indicates about the relationship between the two variables. They also use the line to make informal predictions for different carat sizes, discussing how the actual prices for each carat size vary, and that the prediction has error involved. Students also begin to talk about the scatter of points around the line, as an indication of fit or lack of fit. They talk about residuals as the vertical deviation of individual points from the line, and see that there can be an average deviation that indicates fit, and that for this graph that would be smaller than for the graph of animal data viewed in the previous lesson, which had more scatter around a fitted line.

The second activity, *da Vinci and Body Measurements*, has students analyze body measurements to see if they conform to the "ideal" bivariate relationships described by Leonardo da Vinci, as listed below:

- Height is equal to the span of the outstretched arms.
- Kneeling height is three-fourths of the standing height.
- The length of the hand is one-ninth of the height.

Students access the body data collected earlier and begin to examine the three rules listed above. They first make and examine scatterplots for each pair of variables. Next they use *Fathom* to draw on the graphs the line of best fit. They use the graphs

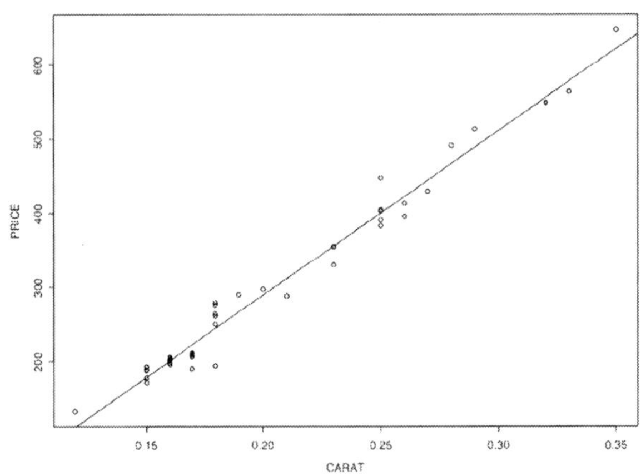

Fig. 14.6 Relationship between the price of a diamond ring (US$) and the number of carats in the diamond

and lines to make predictions of particular values of height, kneeling height, and hand length, based on measurements of arm span and height. These are compared to the ideal values based on Leonardo's "rules." The slopes of the lines are also compared to his rules to see how closely they correspond. Students are asked questions about x values that are beyond the domain of x values in the data set, leading to a discussion on extrapolating values and why that is not a good thing to do.

Students discuss how well a line models the relationships shown in each part of body data. They are guided to think about residuals as an indicator of fit, and then use the "*Show Squares*" feature in *Fathom* to incorporate the least squares criteria on their regression lines. They try using movable lines and see that the sum of these squares is bigger for all lines except the regression line.

In the wrap-up to the lesson, students are asked what they conclude about Leonardo's "rules" and to explain their response. Students are given a scatterplot and a regression line and make a prediction for a value that produces an unusual result, due to extrapolation, leading them to figure out what was wrong. Next, they are given a scatterplot of a dataset that is not linear, and they are asked whether or not a line of best fit could be used to predict a value of y in this situation. Finally, they see a demonstration that shows the effect of outliers on a regression line and reason about why they see two different effects, depending on when the outlier forces the line or is away from the line.

Lesson 3: Inference Involving Bivariate Data

Lesson 3 has students make statistical inferences, by testing hypotheses about the correlation coefficient and regression slope for a sample of bivariate data. Students learn to do this using *Fathom*, and to compare these procedures to significance tests involving a two sample test of means (see Chapter 13). Student learning goals for the lesson include:

1. To learn how to use *Fathom* to apply a test procedure involving the correlation coefficient and regression slope.
2. To continue to apply graphical and descriptive methods learned earlier in conjunction with this test.
3. To revisit and build on the ideas of statistical inference in the context of bivariate data.
4. To revisit importance of assumptions and testing conditions when making statistical inferences.

Description of the Lesson

The lesson begins with a discussion of whether or not students believe there is a relationship between the amount of time a student spends studying per week and his/her cumulative GPA (Grade Point Average). Students reason about and describe their conjectures about that relationship. Next, they examine the relationship between hours studied per week and cumulative GPA for a random sample of undergraduate students who have taken statistics in this department.

In the *Testing Relationships: Admission Variables* activity, Students use *Fathom* to test their conjectures, examining the relationship between hours studied and cumulative GPA. They produce and examine scatterplot and correlation coefficients and describe how well their conjectures matched the data and analyses.

Students are then asked if they think the relationship will hold for all undergraduate statistics students in the department, the population from which the data were sampled. They consider what they learned in the previous unit on inference (Chapter 13) and reason about how to determine if the correlation they found and regression slope are actually different than in the population. They write null and alternative hypotheses for these tests. They discuss what null distribution to compare their sample to, and how to find a P-value to determine how unlikely their sample results are (or results more extreme than theirs) if the null hypothesis is true.

Next, students use *Fathom* to run these tests and generate P-values, which they then interpret, answering the question about whether the results of the significance tests suggest that there is a positive relationship between hours studied and cumulative GPA. A discussion follows on the assumptions that are involved in making this inference.

In the *Testing Relationships: Baseball Variables* activity, students examine a second data set that contains player payrolls and won/lost records for a sample of National League Baseball teams. The research question of interest is: Is there a relationship between a baseball team's payroll and winning percentage? Students create, examine, and interpret a scatterplot to investigate the relationship between payroll and winning percentage for the sample data. They then generate a correlation coefficient, and test the hypothesis that it is significantly different from 0, and contrast the process used to conduct a hypothesis test for the correlation to the process used to conduct a hypothesis test for the mean. What is the same? What is different? A discussion follows where students compare and contrast the methods used to test hypotheses about the difference in two means and the relationship between two variables. The sampling distributions for each type of test are compared. In each case, the types of procedures, decisions, and errors are reviewed in light of a problem with a context that is provided.

This final lesson in the unit revisits the ideas of covariation, relationships, and causation. The two-sample test is revisited as a way to test for cause and effect in an experiment, while correlation and regression model covariation that may involve a causal factor, but do not provide evidence of causation.

Summary

Covariation is an important and difficult topic for students to learn and reason about. We have introduced lessons that build ideas informally, remove calculations, and focus on the concepts and reasoning. We have left out many traditional topics of bivariate data in the hope that students will study them in more detail in further coursework in statistics. Instead, we have tried to use this unit as a way to revisit and deepen understanding of previous concepts in the context of bivariate data.

Part III
Implementing Change Through Collaboration

Introduction: The Role of Collaboration in Improving Statistics Education in Learning, in Teaching, and in Research

The more voices we allow to speak about one thing, the more eyes, different eyes we can use to observe one thing, the more complete will our concept of this thing, our objectivity, be.
(Nietzsche)

The two chapters in Part III conclude our book by focusing on one of the most important ways to make positive changes happen in education, via collaboration. Chapter 15 discusses the value and use of collaboration in the statistics classroom to facilitate and promote student learning. Chapter 16 focuses on collaboration among teachers and among classroom researchers. The first part of this chapter makes the case for collaboration among teachers of statistics as a way to implement and sustain instructional changes and as a way to implement a Statistical Reasoning Learning Environment (SRLE, Chapter 3). The second part of the chapter describes collaboration among teachers as researchers in order to generate new methods to improve teaching and learning and to contribute to the knowledge base in statistics education. As an introduction to Part III, we begin by describing the practice of statistics as a collaborative endeavor.

Collaboration is a fundamental aspect of human activity and society, which refers to all processes wherein people work together with others to achieve a common goal. Collaboration applies both to the work of individuals as well as to larger collectives and societies. Research into the general nature and processes of collaboration has intensified with the growing importance of collaboration in many fields, and the advent of the Internet (collaborative editing, computer-mediated communication, etc.).

Most statistical work is collaborative. Statisticians need to be able to work smoothly on teams and to communicate effectively with their collaborators, who may have little or no background in statistics. This is true across academe, business and industry, where statisticians offer statistical consulting for various projects in different disciplines. In this role, they provide guidance in thinking about and making decisions regarding the statistical aspects of research projects at various stages including design, data collection as well as data analysis. The contributions of these statisticians are greatest when statistical thought is integrated with contextual knowledge in the content areas of the research project. In a true collaboration, the statistician collaborates on the overall problem (not just the statistical questions) as an equal researcher, sharing responsibility for project success or failure, as well as for publication and patenting. Nearly all graduate and even undergraduate programs in statistics today prepare their students to be statistical consultants, and some also prepare their students in communication skills, realizing statistical practice requires high level skills in teamwork, collaboration, and communication.

If statistics instruction should resemble statistical practice, even for nonstatistics majors, then students should learn about and experience collaboration, teamwork, and develop their communication skills as part of their learning. However, there are additional important reasons to use collaboration in the classroom, which are described in Chapter 15. Chapter 16 examines the positive impact of collaboration among teachers of statistics, in teaching as well as in research.

The goal of these final chapters is to convince readers that collaboration is an essential way to bring about instructional change, to create new knowledge, and most important of all, to improve student learning of statistics.

Chapter 15
Collaboration in the Statistics Classroom[1]

A crucial difference exists between simply putting students in groups to learn and in structuring cooperation among students.

(*Johnson, Johnson, & Smith, 1991, p. iv*)

Overview

This chapter attempts to make a convincing case for the use of cooperative learning experiences for students in the statistics classroom. It builds on the premise that statistics instruction ought to resemble statistical practice, an inherently cooperative enterprise. Statisticians typically need to work on teams and communicate effectively with their collaborators, who may have little or no background in statistics. Today, nearly all graduate and even undergraduate programs in statistics prepare their students to be statistical consultants, and some also prepare their students in communication skills, realizing statistical practice requires high level skills in teamwork, collaboration, and communication.

We present definitions and examples of cooperative learning, the theory and research that support the use of cooperative learning, and examples and practical tips for successful implementation of this instructional method. While this chapter focuses on cooperative learning in the classroom, the following chapter (Chapter 16) focuses on collaboration among statistics educators and researchers. The argument is made that collaboration among both parties – i.e., among students and among educators and researchers – ultimately enhances and sustains the other.

Cooperative Learning in the Statistics Classroom

Collaboration is not only just an end goal of statistics instruction, but also a means to help students learn statistics. Indeed, educators, psychologists, and statisticians alike have all called for students to have opportunities to work together as they learn statistics (e.g., Garfield, 1993; Hogg, 1992; Lovett & Greenhouse, 2002). The recently adopted GAISE guidelines (see Chapter 3 and Franklin & Garfield,

[1] This chapter is based on the article: Roseth, C. J., Garfield, J. B., & Ben-Zvi, D. (2008). Collaboration in learning and teaching statistics. *Journal of Statistics Education, 16*(1). Online: http://www.amstat.org/publications/jse/v16n1/roseth.html.

2006) make similar recommendations explicit, stating that, "As a rule, teachers of statistics should rely much less on lecturing, and much more on the alternatives such as projects, lab exercises, and group problem solving and discussion activities" (http://www.amstat.org/education/gaise/GAISECollege.htm).

While increasing collaboration and active learning are relatively simple ideas, implementing such methods is not. Statistics educators may reasonably ask how these teaching methods should be translated to the statistics classroom. As highlighted by the 7th International Conference on Teaching Statistics (ICOTS), working cooperatively in statistics education involves obvious benefits and challenges, with the latter raising questions about the degree to which cooperative methods actually translate to statistics education (Osvaldo Fabbroni, Chang Lee, Lievesley, Martin-Guzman, Morettin, & Rossman, 2007). This chapter addresses this issue by providing specific examples of how statistics educators may apply a cooperative framework to classroom teaching, student assessment, and teacher collaboration. Thus, the purpose of this chapter is a practical one, connecting a theoretical framework to guidelines and materials for statistics teachers.

This first section focuses on practical tips and materials for successfully implementing cooperative learning methods in the statistics classroom. We begin, however, by differentiating the terms peer learning, active learning, cooperative learning, and group work. We also introduce social interdependence theory (Deutsch, 1949, 1962; Johnson & Johnson, 1989, 2005), the guiding theoretical framework for much of the research on the effects of cooperation.

Definition of Terms

How is Cooperative Work Different from Group Work?

To use cooperative learning effectively, statistics teachers must realize that not all groups are cooperative groups. Study groups, lab groups, and discussion groups may be groups, but they are not necessarily cooperative. In fact, while some kinds of groups facilitate student learning and increase the quality of life in the classroom, other types may hinder student learning and create disharmony and dissatisfaction in the classroom (Fiechtner & Davis, 1992).

How is Cooperative Learning Different from Active Learning?

Cooperative learning and active learning are often used interchangeably to describe instructional methods that allow students to solve problems, participate in activities, and discuss content with students. While cooperative learning is a form of active learning, however, active learning is not necessarily cooperative. An important distinction is that cooperative learning methods capitalize on the motivational and epistemic processes that occur *between* individuals rather than students' epistemic curiosity, work ethic, or the provocative nature of a given activity. From the cooperative learning perspective, engagement and interest are primarily derived from the way peers' individual goals are linked to each other. Simply put, knowing that your peers' success depends on your own – that you "sink or swim" together – is a powerful motivator (Kohn, 1986).

An example helps to illustrate this distinction. Consider an activity focusing on understanding the standard deviation (for the original lesson plan, see http://www.causeweb.org/repository/StarLibrary/activities/delmas2001). In this activity, student pairs must decide whether one histogram has a larger standard deviation than another histogram, or if the two histograms have the same standard deviation. This lesson is clearly active in that students are not passive recipients of the teacher's knowledge. Simply asking students to complete the activity, however, – even asking students to work *together* in completing the activity – does not make the activity cooperative. Indeed, it is possible that some students may choose *not* to engage in the activity, preferring instead to let their partner do all of the work.

Now, consider how the histogram activity may be structured cooperatively. Student pairs are again presented with several pairs of histograms and must decide whether one histogram has a larger standard deviation than another histogram, or if the two histograms have the same standard deviation. The key instructional step is then explaining the cooperative goal structure. Specifically, students are told that after completing the histogram activity, individual students will then form new groups of two and compare their answers. The new pairs must reach consensus about their answers and, most importantly, *both individuals* must be able to explain what characteristics of the graphs support their answer. The instructor will randomly select one individual from the new pairs to explain their answer for a given problem.

The cooperative structure described above builds on *social interdependence theory* (Deutsch, 1949, 1962; Johnson & Johnson, 1989, 2005), the basic premise of which is that the way in which interdependence is structured moderates how individuals interact which, in turn, determines outcomes. Cooperative goal structures result in promotive interaction (such as mutual help and assistance, sharing resources and information, acting in trustworthy and trusting ways), competitive goal structures result in oppositional interaction (such as obstructing of each other's goal achievement efforts, hiding resources and information from each other, acting in distrustful and distrusting ways), and the absence of goal structures results in the absence of interaction. The basic model proposed by social interdependence theory may be represented as follows:

Goal structures (Interdependence) → Interaction patterns → Situational outcomes

What is Different About the Cooperatively Structured Histogram Activity?

When individuals perceive that they can reach their goals if and only if the other individuals with whom they are cooperatively linked also reach their goals, social interdependence theory predicts that individuals tend to seek outcomes that are beneficial to all those with whom they are cooperatively linked. Thus, in the histogram activity, individuals know that, after working together, they must individually present their answers to another student. Students also know that, after reaching consensus with their second partner, they are also individually responsible for justifying their answer to the instructor. In short, whether or not individual students complete the histogram activity successfully depends on realizing the right answer with his or her partners. It is in this way that cooperative goal structures tend to result in promotive interaction while competitive and individualistic goal structures result in oppositional or no interaction, respectively. Readers interested in the empirical support of social interdependence theory are referred to several meta-analyses studies (e.g., Johnson & Johnson, 1989; Johnson, Maruyama, Johnson, Nelson, & Skon, 1981; Roseth, Johnson, & Johnson, 2008). Readers are also referred to related meta-analyses on cooperative effects, albeit from slightly different theoretical orientations (e.g., Ginsburg-Block, Rohrbeck, & Fantuzzo, 2006; Rohrbeck, Ginsburg-Block, Fantuzzo, & Miller, 2003; Slavin, 1980, 1983, 1995).

From a cooperative learning perspective, *any* activity may be motivating and interesting if cooperative structures link students' goals, behavior, and outcomes. In this book, we use *cooperative learning* as a term for all forms of peer learning in which students work together to maximize their own and each other's learning (Johnson, Johnson, & Holubec, 1998a). Thus, by cooperative learning, we also include collaborative learning, peer tutoring, cross-age tutoring, and other teaching strategies aimed at structuring the way students interact with each other as they learn. For excellent reviews of these methods, see O'Donnell (2006), Boud, Cohen, and Sampson (2001), and Falchikov (2001). Also, for research focusing on cooperative learning in statistics education, see Chick and Watson (2002), Courtney et al. (1994), Giraud (1997), Keeler and Steinhorst (1995), Magel (1998), Perkins and Saris (2001), and Potthast (1999).

Implementing Cooperative Learning

The following sections highlight materials and provide practical tips for effectively using cooperative learning in the statistics classroom. These recommendations are meant to address common concerns about using group work, some of which make instructors reluctant to adopt more student-centered methods or, alternatively, to stop using them after initial difficulties. Our hope is to provide practitioners with the tools and understanding needed to capitalize on the benefits of cooperative learning.

We begin by providing a general introduction to the steps typically involved in a cooperative lesson. We then provide examples of how these methods may be used in the statistics classroom, focusing especially on the use of cooperative learning assessment procedures. We emphasize three statistics activities that are also part of a cooperative learning module available online at the Consortium for the Advancement of Undergraduate Statistics Education (CAUSE, http://serc.carleton.edu/sp/cause/cooperative/example.html).

Conducting a cooperative lesson typically involves four steps: (a) making pre-instructional decisions about the lesson, (b) explaining the task and cooperative structure to students, (c) monitoring and, if necessary, intervening with each learning group, and after the lesson, (d) processing and evaluating student achievement (Johnson et al., 1998a; see also Davis, 1993). While detailed instructions for accomplishing each of these steps is beyond the scope of this book, we offer a few helpful hints.

Making pre-instructional decisions. Planning a lesson begins with specifying the academic objectives. In cooperative learning, it is also recommended that educators specify the social skill objectives that specify what interpersonal and small group skills are to be emphasized during the lesson. The importance of social skills cannot be overstated as, increasingly, large numbers of children, adolescents, and young adults do not possess the social skills necessary to establish and maintain positive relationships with their peers. Further, research suggests that educators must confront social norms making academic disinterest increasingly acceptable (Steinberg, Brown, & Dornbusch, 1996). Instructors must create the conditions in which students feel safe to say things like, "That's interesting. Tell me more." Thus, social skill objectives may enhance student participation in statistics activities.

Explaining the task and cooperative structure to students. The second and arguably most important step in a cooperative lesson is telling your students what to do and how to do it. It is here that instructors must (a) assign students to groups, (b) specify the criteria for successfully completing the activity, and (c) structure the cooperative goals linking individual students.

- Group size. Cooperative learning groups typically range in size from two to five. In general, remember that as the size of the learning group increases, so too do the resources needed to help the group succeed (Johnson & Johnson, 2006). Thus, smaller groups are typically more successful than larger groups of students, especially when time is short and/or students lack the skills to work with several peers at once. Larger groups also decrease individual member's opportunity to contribute; correspondingly, smaller groups make it more difficult for students to hide and not contribute their share of the work. Smaller groups also make it easier to identify any difficulties students may have in working together.
- Individual accountability. Instructors must be clear about the criteria by which student performance will be evaluated. Following Brophy (2004), effective criterion make goals *immediate* (here and now rather than for the distant future), *specific* ("answer all questions with no more than one mistake" rather than "do your best"), and *challenging* (i.e., difficult but reachable). For example, instead of saying "compare the two distributions with your neighbor," it is much more effective to say: "With your neighbor, identify three similarities and three differences between the two distributions. After 5 min I will call on one student to share their answers with the class."
- Make positive interdependence explicit. Above all, cooperative goal structures must be made explicit. For example, when using a problem set to review for a quiz, cooperation may be structured by saying: "Each of you has two

responsibilities. First, you are each responsible for learning how to solve the problems and, so doing, preparing for the quiz. Second, you are responsible for making sure that every member of your group also knows how to solve the problem. Each member of the group will receive 5 extra points on their individual quiz if all their group members score above 85%."

Monitoring and intervening during the cooperative activity. Once an activity begins, instructors must monitor and, when necessary, intervene in students' group work. This is not the time to get a cup of coffee. To the contrary, cooperative learning requires that teachers observe interaction among group members, assessing both academic progress and the appropriate use of interpersonal and small group skills. This is also the time when the window to student's reasoning and thinking begins to open. Student discourse provides a uniquely powerful way for educators to understand how students understand a given concept (Case, 1992).

Group processing. Finally, time should be given for students to process, reflect, and evaluate what was helpful and unhelpful in completing the assignment. For example, instructors may ask students to identify at least two behaviors that were helpful to the group and at least two behaviors that would help the group perform even more effectively in the future (Davis, 1993). Not only do these discussions help to clarify whether learning objectives were achieved, they also help to reinforce classroom norms, values, and behavioral expectations. Statements like, "Raise your hand if everyone in your group participated in your discussion" makes it clear that, in this classroom everyone's participation is expected and valued.

Group processing may occur at two levels. Each learning group may discuss what they accomplished or, in whole-class processing, the teacher may give the class feedback with students sharing specific examples from their own groups. For more on group processing, readers are referred to Johnson et al. (1991, 1998), Rau and Heyl (1990), and Walvoord (1986).

Practical Tips

In this section, we suggest some practical tips for statistics instructors willing to try cooperative learning in their classes.

Assigning students to groups. For brief exercises, students may work effectively with their friends. For extended cooperative activities however, letting students assemble into groups of their own choosing is typically not successful (Fiechtner & Davis, 1992).

Random groups, especially for one-time projects, are easy to assemble. They can be set up before class using a roster or they can be formed spontaneously by having students count off. For example, in a group of 40 students where you want groups of four, ask students to count off from 1 to 10, repeating this four times. Then, ask that all the "1's" get together in a group, the "2's" in a second group, etc. This method works well in creating heterogeneous groups as it rearranges students who started the class sitting together.

Using base groups to check in and track preparation for class. In addition to using different, informal learning groups to complete activities during class, students may also be assigned to a consistent base group. These students can be given the task of checking in with each other at the beginning of class, to ask each other questions, compare notes, and discuss their preparation for class. In our own teaching, we have even had students in a base group give themselves a daily rating on how well prepared they are for class. The base group then calculates the group mean and charts it over time. By looking at the group average fluctuate over the semester, students recognize that they are accountable to their peers and gain awareness of their study habits. Students also gain additional experience looking and interpreting real data.

Getting groups started on an activity. We suggest that students' first interaction with each other be structured. Never let groups suffer through the awkward silence of "getting started." Social roles inevitably take over in such instances – e.g., the "talkers" always talk, and the "quiet" students always remain quiet. Cooperative groups work best when social roles are trumped by interdependent roles and goal structures. An easy way to structure the way groups get started is to say something like, "You're probably wondering who will start? The person born closest (or farthest) from this room should begin." Using questions like this provides a quick way for students to interact personally without being too distracting or too "touchy-feely."

Importantly, "structured starts" can also be used post hoc. Anytime instructors hear silence after beginning an activity is an ideal time to say, "You're probably wondering who will go first..." Structuring the start, even after the fact, helps students begin their work as quickly and as effectively as possible.

Assigning roles to group members. Another practical tip is to assign roles to students that give them a specific task in the group. Assigning roles can help to avoid traditional social roles (e.g., female students serving as the note-taker or male students serving as the group spokesman), and they can also be used to develop thinking and reasoning skills. Roles can also help "level" perceptions of different status among group members (see Cohen, 1994; Cohen & Lotan, 1997). Following Garfield (1993) and Johnson et al. (1998), examples of such critical-thinking roles include the following:

- Summarizing out loud: A student summarizes out loud what has just been read or discussed as completely as possible and without referring to notes or the original material. Students vocalize their understanding in order to make the implicit reasoning process overt and thus open to correction and discussion.
- Seeking Accuracy: A student seeks accuracy by correcting a member's summary, adding important information he or she did not include, and pointing out the ideas or facts that were summarized incorrectly (e.g., "I'm not sure if that is right. I thought it was...").
- Seeking Elaboration: A student relates material being learned to earlier learned material and to other things they know (e.g., "This is like what we studied last week..." or, "How does this relate to...?").

- Extending Answers: A student extends other members' answers or conclusions by adding further information or implications (e.g., "Here is something else to consider..." or, "I also think that...").
- Probing Questions: A student asks questions of the group that lead to deeper understanding of synthesis (e.g., "Would it work in this situation...?," "What else makes you believe...?" or, "Would this also work...?").

Helping students build their teamwork skills. The greater the students' teamwork skills, the higher will be the quality and quantity of their learning. Operationally, teamwork skills are defined by specifying which behaviors are and are not appropriate within learning groups. It is not enough to tell students what skills you wish to see them use during the lesson. Teachers must also explain exactly what is expected by modeling the skill in front of the class. Students must have a clear idea of what the skill sounds and looks like. For example, consider what it looks and sounds like when students do not pay attention. Not paying attention often looks like slouching, reading the newspaper, completing the Sudoku, checking e-mail, sleeping, etc. Not paying attention can also sound like finger tapping, turning pages, gossiping with a peer, typing, snoring, etc. In our experience, demonstrating such behaviors can offer a moment of levity in the classroom, with humor making it "safe" for students to engage in more productive behaviors that, in other contexts, may not be socially acceptable among their peers. Practical examples of social skills that enhance critical thinking were provided above (see "assigning roles to group members").

Helping a group that is not working well together. Not every group will work well together. Students vary in their preparation, motivation, and preferences for learning. If a group does not seem to be working well together it is important to try to figure out the cause. Is one student unprepared and perhaps not willing to help? Are students sitting silently because they don't understand where to begin? The teacher can help by going to the group and asking a question or offering a prompt to get them going. If one student seems to be a problem, the teacher can invite that student into the hallway and find out what seems to be going wrong. If a student says that the rest of the group members are not contributing or that they themselves have not prepared well, the instructor can offer advice to the student or the group, as appropriate.

Here again we emphasize that forming different, random groups for each activity can prevent the problem of some groups working consistently better than others. Simply put, random assignment ensures that individual differences in achievement, interpersonal skills and the like are evenly distributed across learning groups. Over time, these differences are also distributed across all learning groups, ensuring that – again, over time – any effects of between-group differences are minimized. Importantly, randomization of student groups also helps students to get to know as many of their classmates as possible, a key step in turning a classroom into a caring, supportive learning community. Consider the following student's comment on the introductory statistics course evaluation:

> I have never had a professor or instructor that was this effective ... I always felt valued, always ... we would work in groups frequently which always helped ... Big picture: he

> provided a unique learning environment that will help a variety of learning styles ... Working in groups has made a big impact. Having built rapport with the other students quickly in the semester helped. Having a base group and assigned seating really helped me not feel so lost. Having to report to what level we did our homework to our base groups, made each of us accountable and probably made each of us come to class more and do homework more. Overall, I can't compare his style to any other professor or instructor I have ever had here...

Assessment and Cooperative Learning

Assessment provides yet another opportunity to capitalize on the effects of cooperative activities. For example, in the GIG procedure (Group preparation, Individual assessment, Group assessment; Johnson & Johnson, 2002), students first meet in their cooperative learning groups to discuss study questions and to come to consensus about the answers. The cooperative goal is to ensure that all group members understand how to answer the study questions correctly. Next, students take the assessment (e.g., quiz, test, etc.) individually, the task (obviously) being to answer each question correctly. Last, after all group members have finished the individual portion of the assessment, the group as a whole then meets again to complete the group portion of the test.

There are a myriad of ways of using the GIG procedure. As suggested earlier, extra credit may be awarded if every member of the learning group scores above a certain threshold on the individual assessment. Other times, students may meet in their cooperative learning groups and retake the entire test, the cooperative goal being to reach consensus on the answers and to ensure that all members can explain the answer. Calling on individual members of a group is, again, one way to ensure that individuals are accountable for contributing to the group goal.

In our own undergraduate statistics courses, we use a procedure whereby students complete the individual portion of a quiz at home. Individual questions emphasize statistical literacy, thus providing each student with a structured review of the material. The next day, students turn in their individual quiz and meet in cooperative learning groups to complete the group portion of the quiz. For the group portion, students are randomly assigned to groups of three, thus ensuring that different students work together during every assessment. Group quiz questions emphasize statistical reasoning and thinking, thus going beyond the individual portion of the quiz and capitalizing on the power of group discourse to realize deeper understanding of the concepts.

In our version of group assessment, every individual member of the group writes their own answers to each question. Students are told that one, randomly chosen member of each group will have their quiz graded, and that their score will be given to the group as a whole. Thus, we structure the sink-or-swim requirement, with the cooperative goal of getting the best possible grade encouraging discourse, helping, and individual accountability. Group members also review each other's quiz answers until every member is satisfied with each other's answers. If students cannot agree on an answer, instructors intervene and/or encourage them to compare their answer with an adjacent group. As in other cooperative activities, these assessments serve as windows to students' reasoning and thinking. The group quizzes also reinforce and deepen students' understanding of the course's more complex topics. It is our experience that an incredible amount of learning occurs during these group assessments, with students often leaving class smiling and saying, "I really understand this now!"

Using Cooperative Learning in Large Classes

A frequently asked question about implementing cooperative groups in statistics classes is how to do this in a large class. This is a reasonable concern, as the logistics of managing large groups does require some additional planning. We emphasize however that cooperative learning can be used in classes of almost any size. In fact, many faculty have commented that cooperative learning is *especially* important in large classes, where getting students involved can be especially challenging (e.g., MacGregor, Cooper, Smith, & Robinson, 2000). Few if any students are willing to ask questions and volunteer answers in front of hundreds of classmates, but there is nothing threatening about speaking with two or three people in a small group. Thus, Felder (2001) has argued that the only difference between a large and small class is the number of small groups that you have working at one time.

In statistics classes, Magel (1998) and Harkness (2005) write about their experiences using small groups for learning and assessment in large statistics classes, both finding very positive results. Also interesting are the reports of eight engineering faculty who use cooperative learning in classes ranging from 20 to 400 students (see http://clte.asu.edu/active/implework.htm). While noting that adjustments for class size have to be made, these faculty also describe very positive experiences. Some tips for implementing cooperative groups in large classes include:

1. Find good ways to quickly assign students to groups, such as giving them a group number as they walk in the classroom, and then showing on an overhead the location of where each group number will meet that day.
2. Assign regular seats to students and then assign long term base groups based on seating location, so groups of 2–4 students sitting near each other form a group. Allow students time to discuss content, make conjectures, and question each other's conclusions or assertions during specific times in the class. The Think-Pair-Share method works particularly well for such discussions (Ledlow, 2001).
3. After a group activity, call on randomly selected students from a few groups to report on the results of their group activity. Let students know in advance that this will occur, so they know they each have to be accountable for the results of the group activity.
4. Establish a signal for quickly getting groups to stop talking and listen to the instructor – e.g., use a whistle or turn the lights off and on.

5. Use undergraduate teaching assistants and undergraduate peer teachers to facilitate small group work. Have these facilitators walk around during class activities, listen to student discussion and answer questions.

For more suggestions for managing cooperative learning activities in large classes, see Smith (2000), MacGregor et al. (2000), and the Website, "Implementing Cooperative Learning – Working with Large Classes" (http://clte.asu.edu/active/implework.htm).

Online Materials for Statistics Educators

While many activities can be easily structured to facilitate cooperative learning, we highlight three statistics activities that are already structured for cooperative learning. They are available online at the Consortium for the Advancement of Undergraduate Statistics Education (CAUSE; see http://serc.carleton.edu/sp/cause/cooperative/example.html).

- Body Measures: Exploring Distributions and Graphs – Using cooperative learning methods, this lesson introduces distributions for univariate data, emphasizing how distributions help us visualize central tendencies and variability. Students collect real data on head circumference and hand span, then describe the distributions in terms of shape, center, and spread. The lesson moves from informal to more technically appropriate descriptions of distributions.
- Histogram Sorting – Using cooperative learning methods, this activity provides students with 24 histograms representing distributions with differing shapes and characteristics. By sorting the histograms into piles that seem to go together, and by describing those piles, students develop awareness of the different versions of particular shapes (e.g., different types of skewed distributions, or different types of normal distributions), and that not all histograms are easy to classify. Students also learn that there is a difference between models (normal, uniform) and characteristics (skewness, symmetry, etc.).
- Understanding the standard deviation: What makes it larger or smaller? – Using cooperative learning methods, this activity helps students develop a better intuitive understanding of what is meant by variability in statistics. Emphasis is placed on the standard deviation as a measure of variability. This lesson also helps students to discover that the standard deviation is a measure of the density of values about the mean of a distribution. As such, students become more aware of how clusters, gaps, and extreme values affect the standard deviation.

Summary

The purpose of cooperative learning is ultimately to improve the learning of every individual student. Using these methods, students learn statistics in ways that not only enhance their statistical reasoning and communication skills, but also give them practice in working collaboratively, which models the collaborative nature of real statistical work. Cooperative learning groups also enhance critical thinking, conceptual understanding, and other higher order skills.

Group work does not necessarily constitute *cooperative* group work. Thus, statistics educators must understand the theoretical underpinnings of these methods and become experienced through practicing this method. An important way to support instructors in this process is through collaborative teaching groups. Indeed, it is our belief that successfully using cooperative learning depends, at least in part, on successful collaboration among statistics educators and researchers, the subject of the next chapter.

Chapter 16
Collaboration in Teaching and Research

What is good for students is even better for faculty.
(D. W. Johnson, R. T. Johnson, & Smith, 1991, p. 115)

Overview

While cooperative learning has been widely promoted for use in the classroom, it has been much less visible as a method of faculty development and instructional change. The same authors who have widely disseminated research on the positive effects of collaboration in student learning, state that "the research that validates the use of cooperative learning in the classroom also validates the use of cooperative faculty teams at the department or college level" (Johnson et al., 1991, p. 115). These authors note that most faculty are not used to, nor skilled in working in groups with their peers. However, the use of cooperative faculty groups can be a powerful way to bring about needed changes in teaching and curriculum, such as those outlined in Chapter 3 of this book. This chapter describes and advocates the use of two different, but related forms of faculty collaborations: in a teaching group and in a research group. There is also a description of Japanese Lesson study and its use to help faculty collaborate in the improvement of teaching and the production of research-based lessons.

Collaboration in Teaching Statistics

Most people who teach statistics (and this is also true in other disciplines) do so in relative isolation, even if they have colleagues who also teach statistics. What we do in our classrooms is often known only by us and by our students. Academic freedom allows many college faculty to develop and teach courses without any collaboration with other colleagues, and rarely do we have our colleagues sit in our classes to even see how we are teaching. In this chapter, we propose a major alternative to the teaching of statistics as a solo activity. We build an argument for forming a collaborative teaching group with other faculty either at the same school or at comparable schools, to discuss and share teaching experiences and ideas and to provide support with teaching challenges.

In an article devoted to this topic, Rumsey (1998) described cooperative teaching of statistics as "an environment where teachers share ideas and experiences, support each other, and work together toward a common goal of quality teaching. Individual accountability is reflected by the performance and effectiveness of each instructor."

What Collaboration in Teaching Provides

We see six main reasons why forming a collaborative teaching group can help us become better teachers of statistics.

1. When we collaborate with other teachers, we can accomplish more and at a higher level than working alone (Rumsey, 1998). For example, we can produce better materials, assessments, and teaching techniques by building on the diverse backgrounds and experiences of the teachers in the group.
2. Collaboration promotes reflection on our teaching by verbalizing and justifying what we believe and practice, which also leads us to question our beliefs and practices.
3. Collaboration can motivate and support us in making changes that may be daunting to try on our own. It can provide an environment to reflect on these changes and move forward, rather than abandoning efforts when they are not immediately successful.
4. Collaboration provides a mechanism to develop and maintain a level of consistency from section to section within the same course. For example, collaboratively developing and using a common syllabus, teaching materials and assessment materials facilitates consistency, and developing these as a group with discussion about what is important for students to learn can ensure that they are high quality and represent a shared vision.
5. Collaboration provides a sense of community: Working together toward a mutual goal also results in emotional bonding where group members develop positive feelings toward the group and commitment toward working together. Rumsey (1998) notes that discussions and group decision making about teaching, testing and grading, as well as soliciting and providing feedback to peers, creates an atmosphere of teamwork and community that can improve and enhance our work environment and our job satisfaction.
6. Collaboration provides support and guidance for new teachers: New instructors can benefit from the support and experience a more positive beginning to their career teaching statistics.

Cooperative teaching groups can expand our knowledge and awareness of other perspectives on teaching and learning. Verbally sharing and discussing ideas with colleagues can help us better articulate our goals for students, help us reflect on activities in ways that allow us to improve them, and provide insights into ways to improve teaching. For example, in a discussion with colleagues about the sequence of activities used in the distribution unit (Chapter 8), it became clear that the order should be reversed, an approach we probably would not have considered without this discussion.

While most teachers may keep their teaching experiences to themselves, especially when things do not go well in a particular class, a cooperative teaching group provides support in challenging situations. Such a group can also provide advice and support in dealing with difficult situations (e.g., technology problems or a student having problems using the software).

The Role of Collaborative Teaching in Supporting Innovation and Change

A very important type of support offered by a cooperative teaching group is support for a teacher trying to move from a lecture style of teaching to a more active learning approach, especially a teacher who has felt comfortable lecturing and is not initially comfortable in a student-centered classroom. Colleagues can offer tips, advice and empathy while supporting their peer in making this difficult change. This may prevent the instructor from saying (as we have heard others say) "I have tried cooperative groups, but it didn't work for me so I am going back to the lecture method I'm used to."

As readers may have recognized in reading Chapter 3 in this book, developing a *Statistical Reasoning Learning Environment* (SRLE) is a radically different approach to teaching than what is being done in most introductory statistics classes today. We suspect that some instructors who may be interested in trying this approach may find it daunting to undertake on their own and may worry that their colleagues or department will not be supportive of their efforts. Therefore, we encourage the use of a cooperative teaching group to discuss the desired changes, what it would take to implement them, and how to provide the necessary support. It might be that one member of the group begins one change in the desired direction and reports on this to the group for feedback, or that several members of the group together try to implement a change and discuss their efforts and results. We strongly believe that this type of collaborative teaching group can play a major role in supporting and sustaining instructional innovation and curricular change.

It is important to note that beginning a collaborative teaching group takes commitment. Rumsey (1998) also points out that the main disadvantage of cooperative teaching is that it demands time, and in fact may require a large initial time commitment. However, she also points out that once the group is established and new materials and methods are in place, the individual time commitment diminishes and takes on more of a maintenance role. She also notes that any attempt to make changes in teaching requires a large amount of effort on the part of the instructor, but that a collaborative teaching group can minimize the amount of effort needed.

How Does a Collaborative Teaching Group Work?

We provide six examples of different types of collaborative teaching groups. The first four are more structured around specific courses at a single institution. The last two are more informal and include faculty from multiple institutions.

Introductory Statistics at Kansas State

Rumsey (1998) described weekly teaching meetings of a faculty member and graduate teaching assistants. The primary goals were to offer a forum to confirm statistical concepts, to discuss and present ways to teach the concepts in the spirit of general education, and to provide a testing ground for new ideas. Each meeting included an overview of the material for the upcoming week (based on a common syllabus). They discussed the teaching philosophy, which was to move from the traditional approach (hand computations and formulas, the flow-chart approach to working problems, daily lectures, and small, contrived datasets and examples) toward an environment of discovery, hands-on activities, critical thinking, and making connections to the students' everyday lives and professional careers through relevant, real-world examples. Discussion topics included textbook selection, writing good assessment items, implementing teaching techniques such as group work or leading good discussions, and ways to bring more relevance into the classes. After the first semester of cooperative teaching, the format and goals of the group were altered as needs changed and more resources had been collected or developed.

Rumsey (1998) felt positive about the use of cooperative teaching in helping the department move to a more "general education" approach to teaching statistics than a traditional, more mathematical approach. She wrote:

> The cooperative teaching approach allows us to implement the pedagogical themes of general education in a way that minimizes the overall amount of additional time and effort. This is accomplished through teamwork and cooperation among instructors as they develop and share ideas, receive and offer feedback, and work together toward the common goal of teaching introductory statistics in the best way possible. We are very happy with the progress we have made in establishing a teaching resource notebook and a weekly teaching meeting structure; ideas are collectively developed, presented, tested, and written down in an organized form for easy retrieval and use. Our instructors have the responsibility and the freedom to develop their own teaching styles and leadership qualities; this is an important investment in the future of statistics education.

A Cooperative Model at Auckland, New Zealand

Wild (2006) describes using quality management ideas and teamwork to implement systems to deliver continual improvement in the Auckland University large first-year service courses despite a continual turnover of personnel. This collaborative team serves over 3,500 students per year. It has a small stable core of people who do much of the teaching and almost all of the development. Their aim in collaborating with each other is to find "ways to seal enhancements permanently into the system as a whole so that the courses improved each time they were taught, regardless of who taught them."

The group of teachers works collectively on a common product used by all. They share their best ideas and favorite tricks with the rest of the team, so everyone can take advantage of them in order to try to capture as much as possible of what makes the best teachers good and transfer that to everyone in the group; this way everyone in the group can learn from and take advantage of these best practices. The teamwork produced a very flexible, well-integrated learning environment with high-quality activities and supporting materials such as animated Lecture Slides with narrated soundtrack, extensive Computer Manuals and Tutorials with narrated movies, Applets, video clips, a large online test bank with extensive feedback, online Forums that get tens of thousands of hits each month, extensive use of online surveys, and instant-messaging discussion groups.

Wild (2006) notes that it became clear that when the system retains the most important contributions from those who move on, teacher turnover is an advantage rather than a burden. It continually revitalizes the team with fresh ideas, enthusiasm, and creativity. There has been recognition of the team's success in the form of several university teaching awards and a national teaching excellence award. The team members also find time for other cooperative activities such as being involved in the development of the national curriculum for statistics in schools and running nation-wide outreach activities like the New Zealand CensusAtSchool project (http://www.censusatschool.org.nz/) and teacher workshops.

Wild (2006) describes some of the advantages of the collaborative teamwork:

> The team is very productive because they are no longer doing their preparation work in parallel reinventing their own slightly different versions of the wheel. They are working closely together, are regularly in discussion and sparking ideas off one another ... This way of working is particularly valuable where there is turnover or where you have, from time to time, to use inexperienced people and want to avoid significant drops in the quality of teaching and assessment experienced by students. But there also are a myriad of unexpected subtle ways in which teamwork can increase quality and creativity. Much of it stems from the way in which smart, committed people who are teaching the same material, regularly discussing their experiences and polishing a common set of resources, continually learn from one another and their environment. It is superb for teacher development.

Introductory Graduate and Undergraduate Statistics Courses at the University of Minnesota

The Department of Educational Psychology has always provided statistics courses for students in the department as well as students across the University of Minnesota. In 1998, the department chair asked one of the authors (Joan Garfield) to take charge of the introductory graduate and undergraduate statistics courses, which at that time were challenging. Various faculty and graduate students taught the course, and the courses were taught in inconsistent ways. Many students did not complete the courses and had to retake them, and complaints about instructors were frequently made to the department chair. Garfield was given *carte blanche* to redesign the courses and work with graduate students to prepare them to teach a newer, more effective introductory course. Thus, the original objectives were to improve the introductory course, to ensure that the multiple sections were taught in an effective and consistent way, and to prepare graduate students to teach these courses.

The course revisions built on current theories of learning – actively engaging students in constructing knowledge, using cooperative groups to enhance student learning, and utilizing state of the art technology to analyze data and to illustrate abstract concepts. In addition, a variety of high-quality assessment methods were introduced to evaluate student learning and for course evaluation and improvement.

As the new courses were being designed and implemented, the department constructed a state of art computer lab to support the teaching of these courses. This lab allowed the integration of multimedia software as well as statistical software into these classes. A collaborative teaching group was established that consisted of Garfield and the graduate students who were either serving as Teaching Assistants (TAs) or course instructors. All of these students were required to attend weekly meetings, and all future course instructors were required to first serve as a TAs and attend every class session to see how activities and technology were used, and how discussions and collaborative learning were facilitated.

The weekly meetings were used to discuss the course content, to share activities used to develop concepts, to have TAs share problems they noticed in their grading, to discuss ways to help students better understand these areas, and to make sure that the courses were being taught in fairly consistent manners. Each instructor used the same textbook and covered the same content, but wrote their own syllabus and exams. However, copies of materials were widely shared so that new instructors did not have to start from scratch. Garfield was also able to help alert the instructors to difficult concepts and misconceptions students have in each topic area, often justifying choices of activities, technology, and content. For example, she needed to explain why software used for teaching (e.g., Data Desk) was different than software used for research (e.g., SPSS). She provided supervision for the instructors by observing them teach and providing feedback on their teaching. By observing them at different points in time, she could monitor changes and improvements in their teaching. She also had the instructors give out midterm feedback forms, which were discussed as a group noting similarities and sometimes differences from section to section. This model seemed to help graduate students prepare to teach the introductory course, and some positive outcomes were soon noted:

1. Fewer complaints about instructors
2. Fewer incompletes and dropouts
3. Increased demand for the courses
4. Higher mean ratings on course evaluations
5. Increased interest from graduate students to serve as TAs in preparation **to** teach the introductory courses.

One of the first graduate students to become a TA and teach this course was Michelle Everson, who became such a popular and effective teacher, she was hired as a full time Instructor to oversee the introductory graduate course and coordinate the collaborative group of teachers and TAs for this course. She now meets regularly with the instructional staff for this class, prepares and revises course materials, observes the teachers, collects and monitors midterm feedback from students in the course as well as end of course evaluations, and provides feedback to the graduate students teaching these courses. Dr. Everson has developed an innovative online version of the introductory courses (see Everson, 2006) and is preparing graduate students to teach this online course as well as the in class version of the course. She meets with the online instructors and TAs in a separate collaborative teaching group.

Graduate students who wish to teach the undergraduate course continue to serve as TAs first, sitting on the entire course for at least one semester. The course instructors vary from graduate students to PhDs who teach the course on an adjunct basis. Over the past few years, this group has decided to change textbooks, software, and sequences of course content, all as a result of group discussions.

Collaboration in a Complete and Radical Revision of the Undergraduate Course

The current version of the undergraduate introductory course that served as the model for lessons described in this book was the result of an intense, two-year teaching collaboration. This collaboration began in the summer, as a faculty member met with two graduate students weekly, to completely redesign the course. The impetus for this change was the desire to use a new textbook that seemed more aligned with their curricular goals (*Statistics in Action* by Watkins et al., 2004), the desire to use *Fathom* software (Key Curriculum Press, 2006), the desire to better utilize the classroom computer lab, and the goal of focusing on the big ideas of statistics rather than procedures. They also wanted to develop a course aligned with the GAISE guidelines (see Chapter 1; Franklin & Garfield, 2006). The way to do this transformation was through intense, ongoing, collaboration. Both graduate students were experienced high school teachers who had also taught the previous version of this course at the college level.

The group developed a lesson plan format (as a result of the Japanese Lesson study project, described later in this chapter) and decided to use it in developing lessons for each day of the course. The plan was to divide the course into topic areas, and each took some of these areas. They developed lesson plans and produced accompanying student handouts. They tried to focus on the big ideas, create good discussion questions to engage students and have them make and test conjectures using data, and gather and use data from the students.

Each week that summer, the group met for several hours to review lesson plans and student activities, discuss, argue, and revise them. During the fall semester, the group continued to meet weekly to debrief how the lessons went and to look ahead at the lessons to come. The group also discussed assessments used in the course in terms of their development, use, and the feedback provided. Although it was a challenging experience for the two graduate students, they were also excited by the positive results they were seeing. The collaborative group provided support, encouragement, and positive feedback.

The collaborative teaching group resulted in a unique and innovative course that none of the participants could have designed alone, provided a mechanism to teach this course in a reflective way that allowed them to evaluate and improve it, and developed materials that could be shared with the wider statistics education community (as part of an NSF grant and this book). The lessons have continued to change as the new members are added each year and as some students graduate and leave. However, each year a group of faculty and students continues to meet weekly to discuss the activities, design new assessments, and make revisions to the course materials.

The two types of collaborative groups used in this department as described above had similarities and differences. However, they both led to many positive outcomes, which include:

1. Development of a course that focuses on the big ideas, incorporates innovative technology, and has students actively engaged in learning statistics.
2. Courses that are taught in a fairly consistent way across sections. In the undergraduate course, the same syllabus and assessment are used across sections.
3. Student satisfaction with the undergraduate and graduate introductory statistics courses has been consistently higher. The doctoral students teaching the courses typically receive positive or excellent student teaching evaluations. Several have won graduate student teaching awards for excellence in teaching.
4. Increased numbers of sections of each course due to high student demand, from 6 a year to almost 20 sections a year (with up to 35 students per section).
5. More diversity in the areas of specialization of the graduate students in this program (e.g., Learning and Cognition, Quantitative Methods, and School Psychology) as students from different areas express their desire to participate in teaching and assisting in teaching statistics. Therefore, this is also creating a diverse group of excellent teachers of statistics.
6. Two fulltime instructors (former graduate students in the department) now supervise each of the statistics courses and meet with each instructional team, as described above. They also observe and supervise the new group of doctoral students teaching the courses.

A Collaborative Group across Institutions

A more informal group of teachers of statistics was formed in fall 2005 in the Twin Cites area of Minnesota. This group, called *Stat Chat* is described as an informal but informative monthly get-together of local statistics educators (see http://www.macalester.edu/~kaplan/statchat/index.html). The group is convened by three faculty and draws about 12–20 participants at each meeting. After a year of loosely organized meetings (e.g., various presentations of current projects and teaching methods), the meetings now have a different theme each month (e.g., online assessment, simulation software, connecting to other disciplines). Meetings follow a structure that includes dinner and "Data for Dessert," (a presentation of a dataset or method for collecting data that might be useful for a class).

The main part of the meeting consists of a presentation (e.g., implementing the GAISE guidelines in a course) or a series of small presentations (e.g., different uses of online assessments) interspersed with discussion. Another format is to have a general topic (e.g., using Bayesian statistics in an introductory statistics class) and no presentations, just discussion. The benefits are that participants share ideas and materials, raise issues that lead to reflections on teaching practice, and suggest solutions to practical problems that arise in teaching (e.g., how to deal with cheating in an online testing format).

Meeting in the evening on one designated day per month (e.g., the last Tuesday of each month) seems to help people commit to coming to the meetings. Materials are posted to the Website from each presentation, and additional materials may be added that are discussed or seem relevant. Anyone in the group is invited to suggest a topic, give a presentation, or lead a discussion.

A Virtual Collaborative Teaching Group

Another option for faculty who do not have colleagues or graduate students available to form a collaborative group is to form a group with colleagues at other institutions. One informal collaboration among teachers of statistics at different institutions is the *Isolated Statisticians Group* (see http://www.lawrence.edu/fast/jordanj/isostat.html). This is a collection of academic statisticians, each of whom is usually the only statistician (or one of two) in a mathematics department. Their Website states: "In such circumstances it is difficult to have meaningful discussions with departmental colleagues about the fine points of teaching or practicing statistics." While this group has many members, a few active participations on the email listserv raise questions or provide suggestions in responses to questions, mostly about teaching statistics (e.g., how to choose a good text for a mathematics statistics course, a good Web applet to use for a particular topic, how to help students overcome a particular misconception about randomization).

Many group members meet at the annual mathematics and statistics conferences. This loose collaboration provides some support but many of the members are silent, so there is less of a sense of developing shared materials and approaches than in a formal group. With today's technological advances, it seems possible that a collaborative group could be formed by statistics teachers at different schools, which could be maintained using Web conference calls, collaborative editing Websites (such as Wikis), and emails.

How to Get Started: Forming a Collaborative Teaching Group

Cooperative activity among faculty should be as carefully structured as is cooperative interactions among students in classrooms (Johnson et al., 1998b). These authors provide many practical suggestions for what they term *Collegial Support Groups*, groups to support colleagues in implementing cooperative learning in the classroom. However, their suggestions generalize to cooperative teaching groups such as those described earlier. Johnson et al. (1998b) suggest three key activities of such a group:

1. Frequent professional discussions about student learning
2. Discussing, planning, and developing curriculum materials
3. Teaching together and/or observing each other teach.

These authors also discuss how such a group can provide support and leadership in implementing cooperative learning strategies in one's classes, an experience that can be challenging and difficult for many instructors used to teaching in more traditional, teacher-centered formats.

Based on her experience at Kansas State, Rumsey (1998) offered some practical tips for forming and using a cooperative teaching group to promote instructional change: These include creating an agreed upon structure, being clear and realistic about expectations of group members, being flexible and creative while at the same time, being organized, and offering support and guidance in a structured environment. Weekly teaching meetings help greatly in this regard.

We think that there are many ways to start such a group. In the examples described above or by Rumsey, there was often a faculty member who started the group and required graduate teaching assistants to participate. It is also possible to start such a group with faculty colleagues and invite interested students to join. It helps to have one person who is wiling to schedule and run the meetings and coordinate the group, although tasks can often be divided among, and selected by, group members. We encourage members of the group to observe each other's classes to see how the same lesson can be taught. It is very illuminating to then compare the different outcomes of the same lesson in two different settings by two different teachers.

Summary: Collaborative Groups to Implement Change in Teaching

The social psychologist Kurt Lewin is credited with the saying that "the way to change individuals is by changing groups" (D. Johnson, personal communication, December 14, 2006). Johnson, an international expert on collaboration, sees collaborative teaching groups as a method to engender real and sustained changes in teaching. He also suggested that such a group can serve as a way for novice teachers to become aware of and comfortable using approaches to teaching that are different than their own training. Consider a new PhD in statistics who has only experienced traditional methods of teaching and is now at an institution where active learning is encouraged. Never having experienced active learning, this person may be reluctant to try something other than lectures. A collaborative teaching group can help the new teacher by sharing experiences, inviting the person to see their classes, and providing guidance and support.

While many academics may initially feel most comfortable thinking about and making decisions about teaching on their own, we strongly urge them to consider forming a collaborative group with other statistics teachers. We realize that there is an investment in time involved, but even devoting one hour per week to such a group can have major, positive benefits. The positive benefits of participating in a collaborative group include promoting and sustaining innovation and changes in teaching and can lead to improved statistical teaching, and, we believe, to improved student learning.

Collaboration in Producing New Knowledge: Collaborative Classroom-Based Research

Another important area of collaboration is in classroom-based research. Members of collaborative classroom research group may be statistics faculty but may also include colleagues in other disciplines such as psychology, measurement, and/or mathematics education.

What is a Collaborative Classroom Research?

Classroom research, also referred to as *action research*, is an increasingly popular approach to educational research, where research is conducted in the classroom, focusing on problems and questions that arise from teaching. In contrast to more traditional forms of scientific or lab-based research, classroom research does not attempt to answer questions definitively nor to find and generalize solutions to problems. While classroom research is often described in the context of elementary and secondary education, it is recommended for use by postsecondary instructors as a tool for studying and improving their classes (Cross & Steadman, 1996).

Classroom research can also be done collaboratively, which brings in more viewpoints, perspectives, and instructional settings. delMas et al. (1999) developed a model of collaborative classroom research for statistics education. They outline four stages, structured around the following questions:

1. *What is the problem? What difficulties are students having (across the different classes taught by the group of teachers) learning a particular topic or learning from a particular type of instructional activity?* The identification of the problem emerges from the teachers' experience in the classroom, through observing students, reviewing student work, and reflecting on this information. Once a problem begins to be identified, published research is studied to better understand the problem, to see what has already been learned, and to understand what might be causing the difficulty.
2. *What technique might be developed and used in each of the teacher's classrooms to address the learning problem?* A new instructional technique may be designed and implemented in class, a modification may be made to an existing technique, or alternative materials may be used, to help address the learning problem.
3. *What types of evidence might be gathered in each classroom to help evaluate whether the new technique or materials is effective*? Types of assessments and data need to be gathered and evaluated, that will provide feedback on effectiveness of the technique or materials.
4. *What should be done next in each class setting, based on what was learned?* Once a change has been made, and data have been gathered and used to evaluate the impact of the change, the researchers consider how the technique or materials might be further modified to improve student learning. They also address the question of how should new data be gathered and evaluated.

For each stage, the statistics instructors would discuss and come to consensus on responses to these questions, leading them to carry out a focused research project within their own classrooms, allowing them to jointly create materials, design methods, and compare results. These stages are carried out in an iterative cycle, as new information gathered in each round often leads back to modifying and trying out new materials and methods.

The authors point out that while the results of many classroom research studies may not be viewed as suitable for dissemination because they focus on a particular class setting and are not generalizable, nevertheless the results of a carefully designed collaborative classroom research project can yield valuable insights into the teaching and learning of statistics. They concluded that collaborative classroom research is an exciting and productive model for research in statistics education and encouraged other statistics faculty to try out this model in their own classrooms as a way to better understand and improve student learning of statistics. (Note: the third lesson in the Sampling Unit, Chapter 12, resulted from their research findings.)

For more information on the results of this collaborative research group as they studied students' reasoning about sampling distributions and how that was impacted by an instructional material and *Sampling SIM* software, see Chance et al. (2004). Lunsford et al. (2006) extended this research and method as they investigated student understanding of sampling distributions, finding similar results but with a different population of students.

Japanese Lesson Study as One Form of Collaborative Classroom Research

Japanese Lesson Study (JLS) is a method used by teachers to collaboratively develop "research lessons" that are used by teachers to carefully and systematically study how to achieve a particular learning goal (Bass, Usiskin, & Burrill, 2002; Hiebert, Morris, & Glass, 2003; Fernandez, Cannon, & Chokshi, 2003). JLS has been studied and written about extensively and is the focus of many new research projects in mathematics and science education (e.g., Fernandez, 2002; Lewis, 2000; Lewis & Tsuchida, 1998). Rather than offer a new technique for teaching students, it offers a set of concrete steps that teachers can take, over time, to improve their teaching (Stigler & Hiebert, 1999).

These lessons embody theories about how to help students reach particular learning goals (Hiebert, Gallimore, & Stigler, 2002). They are classroom lessons taught to a regular class of students but that have special features linked to the research process. According to Lewis and Tsuchida (1998), JLS lessons are focused on important learning goals, carefully planned in collaboration with other teachers, observed by teachers, recorded, discussed, and then revised. They consist of research lessons developed over time by a study group and contain descriptions of the learning goals, the rationale for the lesson design, descriptions of activities, anticipated responses of students, and suggested responses by the teachers. These lessons may be disseminated widely to other teachers, provide a professional knowledge base for teachers, and thereby contribute to the improvement of teaching and learning.

Roback, Chance, Legler, & Moore (2006) summarize Curcio (2002), in outlining several important aspects of a lesson study group:

1. *Collaborative planning*. It is recommended (Fernandez, 2002) that groups of 4–6 teachers come together for 10–15 hours over 3–4 weeks to carefully plan a single specific lesson (as opposed to a longer unit of material) that will address one or more overarching goals.
2. *Teaching and observing*. One member of the group teaches the lesson as designed, while the other group members and outsiders observe the class, taking detailed notes regarding the reactions and engagement of the students.
3. *Analytic reflection*. The teacher, other group members, and observers gather soon after the lesson has been taught to share thoughts and insights, and to evaluate the success of the lesson in meeting its objectives.
4. *Ongoing revision*. Based on experience and evidence, the lesson is often revised and taught again, and the process is repeated.

Japanese Lesson Study in Statistics

There have been a few uses of JLS method in the context of a college statistics course. One group took place in the Department of Educational Psychology at the University of Minnesota, with two faculty members, one full time instructor, and several graduate students (Garfield et al., 2007). The second group was based at St. Olaf College in Minnesota, with faculty from several different colleges (Roback et al., 2006).

The University of Minnesota JLS Group

This group decided to try to adapt JLS to college statistics classes as a way to examine and develop students' statistical thinking and reasoning about variability. They saw JLS as a form of design experiment (Cobb et al., 2003a; The Design-Based Research Collective, 2003) in that JLS could be used to enrich their understanding of students' reasoning and learning (e.g., lesson designs are based on hypothetical learning trajectories and developed through iterative design cycles that compare teacher/researcher conjectures to dense descriptive observations of the learning environment). Their main research question became: Can we construct activities through a JLS process that will engage students, elicit and build on their informal intuitions about variability, and help develop their formal reasoning about variability? A secondary research goal was to see how the JLS experience impacted novice and experienced teachers of statistics (Garfield, delMas, & Chance, 2005).

The JLS group at the University of Minnesota began in September, 2003, and originally consisted of two faculty, one full-time instructor, and four doctoral students who teach their own sections of introductory statistics. It took five sessions to discuss the learning goal and design a lesson that was ready to be taught. The group developed a plan that involved immersing students in a real data set (data on graduating seniors) and required them to use technology to test their conjectures about the data. After developing the lesson as a group, one person taught the lesson in a class, and the others observed, taking notes on their observations of how students were engaged, their conjectures, reasoning, and misconceptions. The group also collected student worksheets and minute papers written by the students to provide data on the impact of the lesson.

At the next group meeting, the lesson was debriefed and critiqued. During spring semester, the group continued to meet every other week and developed a new version of this lesson, based on a new set of data, collected from students in all sections of the introductory statistics classes on the first day of spring semester. One of the TAs taught the revised lesson, which was broken into two parts. The group observed the lesson, and later discussed and critiqued it, and then developed the second part, which was taught during the next to last week of the semester. Again, the lesson was later discussed and critiqued and further revisions were made.

As a result of the JLS group, two research-based lessons were developed, that included a student activity sheet, a data set, teacher notes, and observation notes. These two lessons appeared to lead students to a deeper understanding of statistical measures of variability and their complexities as well as limitations. This evidence was based on written student assessments as well as observations of students working in groups during the lessons.

In addition to the production of innovative lesson plans that seemed to be effective in moving students toward statistical thinking by developing their reasoning about variability, the JLS experience had important effects on both the novice and experienced teachers. See Garfield et al. (2007) for more details on this lesson and the research results.

Impact on Experienced Teachers

Collaborative discussions and planning of lessons appeared to be very productive for teachers of statistics. It forced questions about the relationship of an activity to important learning goals, and it challenged the group to predict how students might respond and how to respond to their responses. It also helped all participants to clarify and deepen their own understanding of the concept of variability and develop a better knowledge of how students come to learn and reason about variability.

The group also felt that their experience in the JLS group over the year deepened their own understanding of the complexities of the concept of variability and helped them to better understand students' difficulties with this concept and the related statistical measures. They also benefited from the experience of observing the teaching of their research lessons, activities that allowed students to construct knowledge and engage in statistical thinking, rather than just follow procedures to a predictable outcome.

Impact on Novice Teachers

The JLS group was ostensibly used to develop a lesson in statistics, but this experience was also designed to impact novice teachers by modeling the development of lessons that engage students in active learning, stimulate good discussions, and lead to important learning goals. The lessons developed provided an alternative to lecturing, working out problems, and offering explanations, which is the dominant method that all of the graduate students (novice teachers) had experienced as students of statistics.

By the end of the year, these TAs appeared to believe that a well-designed activity to help students construct important ideas is an important and valuable instructional method and that developing such an activity with colleagues, discussing, trying it, watching it, evaluating, and revising it, is a useful method of ongoing professional development.

Because of the conversations and group discussions, the TAs appear to have carried over some of the techniques used in the research lesson in their courses. They recognized the power of an activity that engages students in reasoning about statistics and were impressed with the end results of the lesson and students' understanding. This was quite a contrast to their own experience as students of statistics.

The St. Olaf College JLS Group

A second JLS group consisted of faculty from different colleges met at St. Olaf College in 2004. They wrote about their experience getting started (Roback et al., 2006):

> We began our undertaking intrigued by what we knew of Japanese Lesson Study, but extremely "green" with respect to its implementation. Not only were we newcomers to the ideas shaping lesson study, but we could find nothing in the literature to guide the implementation of lesson study specifically at the college level, especially in an upper-level statistics course. Thus, we embarked on a pilot implementation – a preliminary attempt to assess the feasibility of Japanese Lesson Study principles in upper-level undergraduate statistics courses. We hoped to gain insight into concrete benefits and potential pitfalls.

In the end, this group focused on developing a lesson for a Mathematical Statistics course at St. Olaf that was taught in the spring semester. The course has a prerequisite of Probability Theory, so it was targeted toward juniors and seniors who were mathematics majors or statistics concentrators with no previous course in applied statistics. They spent their first meeting watching a videotape overview of Japanese Lesson Study (Curcio, 2002) and discussing the lesson study philosophy and process. The goal of their second meeting was to brainstorm about big goals and content for a research lesson. The discussion was wide-ranging, and it consumed much of the next few meetings as they discussed the important ideas they wanted students to remember from this statistics class. Eventually, a lesson was collaboratively developed on goodness-of-fit tests and sampling distributions.

Specific objectives used to develop and later evaluate the lesson were:

1. engaging students in an active way with lesson material;
2. having students apply statistical thinking to develop a test statistic;
3. having students suggest the need to examine an empirical sampling distribution (assuming the null is true) to decide if an observed test statistic value is surprising;
4. introducing the theory of the chi-square statistic, distribution, and goodness-of-fit test; and
5. extending goodness-of-fit tests from the categorical to the discrete to the continuous case.

The lesson was taught in the next-to-last week of the semester, immediately after a unit on regression analysis and inference. Only one of the group members was able to observe the class and take notes on his observations. Roback et al. (2006) provide complete details of this lesson. In reflecting on their use of the lesson study experience, they wrote:

> Is lesson study a worthwhile endeavor at the undergraduate level, and, in particular, for an upper-level course such as Mathematical Statistics? Our pilot experience with lesson study principles suggests that the answer is yes, with proper preparation, faculty commitment, and realistic expectations.

They also described specific benefits as:

1. Focused and energized collaboration.
2. Insight into student learning.
3. Development of a strong lesson plan.
4. Facilitation of pedagogical research.

The authors also offer suggestions for other colleagues interested in forming a lesson study group for a college statistics course, noting that focusing intensely on a single lesson can provide an "achievable and generalizable" means for examining an entire course as a whole. They note that this process required commitment and time, but that this results in a valuable and worthwhile experience "which has had a lasting impact on our teaching beyond the single lesson on which we collaborated."

Summary of Japanese Lesson Study in Statistics

Japanese Lesson Study may be viewed as a special type of collaborative teaching group, as well as a unique form of collaborative classroom research. By bringing together a set of teachers who teach the same course and having them together develop a research-based lesson to achieve an important learning goal, these groups are contributing to the knowledge base for teaching statistics (Stigler & Hiebert, 1999). Indeed, it was the JLS experience at the University of Minnesota that led to the creation of the series of lessons that are now part of this book and will contribute to the knowledge base for statistics education. We encourage teachers of statistics to consider forming collaborative groups, and to try develop a "research" lesson that involves students in constructing knowledge of an important statistical idea. We believe that going through the process of designing, teaching, critiquing, revising, and re-teaching this lesson will not only contribute important new knowledge to teaching statistics, but will also ground this knowledge in practical classroom context. Thus, we see collaborative classroom research as a vital way to connect research and teaching practice.

Summary of Collaboration among Teachers and Researchers

> *Ultimately, our success in teaching for understanding depends on our design skills: our abilities to deign activities and assessments that naturally raise questions and new ideas instead of telling students what we know, and assuming they understand.*
>
> (Wiggins & McTighe, 1998, p. 175)

We see collaboration as the mechanism to achieve what Wiggins and McTighe describe in the above quote. Collaborative learning moves to a more student-centered approach where students are learning from experience and from each other, rather than "receiving" knowledge from the teacher. Teachers working together collaboratively can help lead and support each other in moving to a more student-centered classroom that uses collaboration in effective ways. And collaborative classroom research can help develop and identify effective activities and assessments to contribute to the knowledge base on teaching and learning statistics. It is our belief that all of these forms of collaboration are forms of "communities of practice"[1] (Lave & Wenger, 1991), in that all involve learning as part of a group (whether students learning from each other, teachers learning from each other, or researchers learning from their collaborative project as well as from each other). These different types of collaboration are all needed to improve student learning of statistics, so we end our book with a plea for statistics teachers to build on the collaborative foundations of statistical work and try one or more of these methods in the coming year.

[1] The concept of a *community of practice* (often abbreviated as CoP) refers to the process of social learning that occurs when people who have a common interest in some subject or problem collaborate over an extended period to share ideas, find solutions, and build innovations.

References

Abelson, R. P. (1995). *Statistics as principled argument*. Hillsdale, NJ: Erlbaum.

Aberson, C. L., Berger, D. E., Healy, M. R., Kyle, D., & Romero, V. L. (2000). Evaluation of an interactive tutorial for teaching the Central Limit Theorem. *Teaching of Psychology, 27*, 289–291.

Abrahamson, D., Janusz, R. M., & Wilensky, U. (2006, March). There once was a 9-block ... – A middle-school design for probability and statistics. *Journal of Statistics Education, 14*(1). Retrieved July 15, 2007, from http://www.amstat.org/publications/jse/v14n1/abrahamson.html

Adams, V. M. (1989). Affective issues in teaching problem solving: A teacher's perspective. In D. B. McLeod, & B. M. Adams (Eds.), *Affect and mathematical problem solving*. New York: Springer-Verlag.

Adi, H., Karplus, R., Lawson, A., & Pulos, S. (1978). Intellectual development beyond elementary school VI: Correlational reasoning. *School Science & Mathematics, 78*(8), 675–683.

Ainley, J., Nardi, E., & Pratt, D. (2000). The construction of meanings for trend in active graphing. *International Journal of Computers for Mathematical Learning, 5*, 85–114.

Alldredge, J. R., & Som, N. A. (2002). Comparison of multimedia educational materials used in an introductory statistical methods course. In B. Phillips (Ed.), *Proceedings of the sixth international conference on the teaching of statistics* (ICOTS-6), Cape Town, South Africa [CD-ROM]. Voorburg, The Netherlands: International Statistical Institute. Retrieved September 28, 2007, from http://www.stat.auckland.ac.nz/~iase/publications/1/6c4_alld.pdf

Allen, K., Stone, A., Rhoads, T. R., & Murphy, T. J. (2004). The statistics concepts inventory: Developing a valid and reliable instrument. *Proceedings, American Society for Engineering Education Annual Conference and Exposition*. Retrieved September 20, 2007, from https://engineering.purdue.edu/ SCI/pubs/ASEE%202004%20SCI.pdf

Alloy, L. B., & Abramson, L. Y. (1979). Judgment of contingency in depressed and nondepressed students: Sadder but wiser? *Journal of Experimental Psychology: General, 108*, 441–485.

American Association for the Advancement of Science (1993). *Benchmarks for science literacy* (Project 2061). New York: Oxford University Press. Retrieved October 21, 2007, from http://www.project2061.org/publications/ bsl/online/bchin.htm

Angelo, T., & Cross, K. P. (1993). *A handbook of classroom assessment techniques for college teachers* (2nd ed.). San Francisco: Jossey-Bass.

Anscombe, F. J. (1973). Graphs in statistical analysis. *American Statistician, 27*, 17–21.

Bakker, A. (2004a). *Design research in statistics education: On symbolizing and computer tools* (Doctoral Dissertation). Utrecht, The Netherlands: CD Beta Press. Retrieved October 21, 2007, from http://www.stat.auckland.ac.nz/~iase/publications/dissertations/04.Bakker.Dissertation.pdf

Bakker, A. (2004b, November). Reasoning about shape as a pattern in variability. *Statistics Education Research Journal, 3*(2), 64–83. Retrieved July 15, 2007, from http://www.stat.auckland.ac.nz/~iase/serj/SERJ3(2)_Bakker.pdf

Bakker, A., Biehler, R., & Konold, C. (2004). Should young students learn about Boxplots? In G. Burrill & M. Camden (Eds.), *Curricular development in statistics education, IASE 2004 roundtable on curricular issues in statistics education*, Lund Sweden. Voorburg, The Netherlands: International Statistical Institute.

Bakker, A., Derry, J., & Konold, C. (2006). Technology to support diagrammatic reasoning about center and variation. In A. J. Rossman & B. L. Chance (Eds.), *Proceedings of the seventh international conference on the teaching of statistics* (ICOTS-7), Salvador, Bahia, Brazil, [CD-ROM]. Voorburg, The Netherlands: International Statistical Institute. Retrieved July 15, 2007, from http://www.stat.auckland.ac.nz/~iase/publications/17/2D4_BAKK.pdf

Bakker, A., & Gravemeijer, K. P.E. (2004). Learning to reason about distributions. In D. Ben-Zvi & J. Garfield (Eds.), *The challenge of developing statistical literacy, reasoning, and thinking* (pp. 147–168). Dordrecht, The Netherlands: Kluwer Academic Publishers.

Bakker, A., & Gravemeijer, K. P.E. (2006). An historical phenomenology of mean and median. *Educational Studies in Mathematics, 62*(2), 149–168.

Barnett, V. (Ed.) (1982). *Teaching statistics in schools throughout the world.* Voorburg, The Netherlands: International Statistical Institute.

Bass, H., Usiskin, Z., & Burrill, G. (2002). *Studying classroom teaching as a medium for professional development: Proceedings of a U. S. – Japan workshop.* Washington, DC: National Academies Press.

Batanero, C. (2000). Controversies around the role of statistical tests in experimental research. *Mathematical Thinking and Learning*, *2*(1–2), 75–97.

Batanero, C., Estepa, A., & Godino, J. D. (1997). Evolution of students' understanding of statistical association in a computer based teaching environment. In J. B. Garfield & G. Burrill (Eds.), *Research on the role of technology in teaching and learning statistics: Proceedings of the 1996 IASE roundtable conference* (pp. 191–205). Voorburg, The Netherlands: International Statistical Institute. Retrieved September 28, 2007, from http://www.dartmouth.edu/~chance/teaching_aids/IASE/15.Batanero.pdf

Batanero, C., Estepa, A., Godino, J. D., & Green, D. R. (1996). Intuitive structures and preconceptions about association in contingency tables. *Journal for Research in Mathematics Education, 27*(2), 151–169.

Batanero, C., & Sánchez, E. (2005). What is the nature of high school student's conceptions and misconceptions about probability? In G. Jones (Ed.), *Exploring probability in school: Challenges for teaching and learning* (pp. 260–289). Dordrecht, The Netherlands: Kluwer Academic Publisher.

Batanero, C., Tauber, L., & Sánchez, V. (2004). Student's reasoning about the normal distribution. In D. Ben-Zvi & J. Garfield (Eds.), *The challenge of developing statistical literacy, reasoning, and thinking* (pp. 257–276). Dordrecht, The Netherlands: Kluwer Academic Publishers.

Bean, J. (2001). *Engaging ideas: The professor's guide to integrating writing, critical thinking, and active learning in the classroom.* San Francisco: Jossey-Bass.

Ben-Zvi, D. (2000). Toward understanding the role of technological tools in statistical learning. *Mathematical Thinking and Learning, 2*(1–2), 127–155.

Ben-Zvi, D. (2004a). Reasoning about data analysis. In D. Ben-Zvi & J. Garfield (Eds.), *The challenge of developing statistical literacy, reasoning, and thinking* (pp. 121–145). Dordrecht, The Netherlands: Kluwer Academic Publishers.

Ben-Zvi, D. (2004b, November). Reasoning about variability in comparing distributions. *Statistics Education Research Journal, 3*(2), 42–63. Retrieved July 15, 2007, from http://www.stat.auckland.ac.nz/~iase/serj/SERJ3(2)_BenZvi.pdf

Ben-Zvi, D. (2006). Scaffolding students' informal inference and argumentation. In A. J. Rossman & B. L. Chance (Eds.), *Proceedings of the seventh international conference on the teaching of statistics* (ICOTS-7), Salvador, Bahia, Brazil, [CD-ROM]. Voorburg, The Netherlands: International Statistical Institute. Retrieved July 15, 2007, from http://www.stat.auckland.ac.nz/~iase/publications/17/2D1_BENZ.pdf

Ben-Zvi, D. (2007, October). Using wiki to promote collaborative learning in statistics education. *Technology Innovations in Statistics Education Journal 1*(1). Retrieved October 15, 2007, from http://repositories.cdlib.org/uclastat/cts/tise/vol1/iss1/art4/

Ben-Zvi, D., & Amir, Y. (2005, July). How do primary school students begin to reason about distributions? In K. Makar (Ed.), *Reasoning about distribution: A collection of current research studies*. Proceedings of the fourth international research forum on statistical reasoning, thinking, and literacy (SRTL-4), University of Auckland, New Zealand, 2–7 July, 2005. Brisbane: University of Queensland.

Ben-Zvi, D., & Arcavi, A. (2001). Junior high school students' construction of global views of data and data representations. *Educational Studies in Mathematics, 45*(1–3), 35–65.

Ben-Zvi, D., & Garfield, J. (2004a). Statistical literacy, reasoning, and thinking: Goals, definitions, and challenges. In D. Ben-Zvi & J. Garfield (Eds.), *The challenge of developing statistical literacy, reasoning, and thinking* (pp. 3–16). Dordrecht, The Netherlands: Kluwer Academic Publishers.

Ben-Zvi, D., & Garfield, J. (Eds.) (2004b). *The challenge of developing statistical literacy, reasoning, and thinking*. Dordrecht, The Netherlands: Kluwer Academic Publishers.

Ben-Zvi, D., Garfield, J., & Zieffler, S. A. (2006). Research in the statistics classroom: Learning from teaching experiments. In G. F. Burrill (Ed.), *Thinking and reasoning about data and chance: Sixty-eighth NCTM yearbook* (pp. 467–481). Reston, VA: National Council of Teachers of Mathematics.

Ben-Zvi, D., Gil, E., & Apel, N. (2007, August). What is hidden beyond the data? Helping young students to reason and argue about some wider universe. In D. Pratt & J. Ainley (Eds.), *Reasoning about informal inferential statistical reasoning: A collection of current research studies*. Proceedings of the fifth international research forum on statistical reasoning, thinking, and literacy (SRTL-5), University of Warwick, UK, August, 2007.

Beyth-Marom, R. (1982). Perception of correlation reexamined. *Memory & Cognition, 10*, 511–519.

Biehler, R. (1990). Changing conceptions of statistics: A problem area for teacher education. In A. Hawkins (Ed.), *Training teachers to teach statistics* (pp. 20–29). Voorburg, The Netherlands: International Statistical Institute.

Biehler, R. (1991). Computers in probability education. In R. Kapadia & M. Borovcnik (Eds.), *Chance encounters: Probability in education* (pp. 169–211). Amsterdam: Kluwer.

Biehler, R. (1993). Software tools and mathematics education: The case of statistics. In C. Keitel & K. Ruthven (Eds.), *Learning from computers: Mathematics education and technology* (pp. 68–100). Berlin: Springer.

Biehler, R. (1997a). Software for learning and doing statistics. *International Statistical Review, 65*, 167–189.

Biehler, R. (1997b). Students' difficulties in practicing computer-supported data analysis: Some hypothetical generalizations from results of two exploratory studies. In J. Garfield & G. Burrill (Eds.), *Research on the role of technology in teaching and learning statistics*, (pp. 169–190). Voorburg, The Netherlands: International Statistical Institute. Retrieved September 28, 2007, from http://www.dartmouth.edu/~chance/teaching_aids/IASE/14.Biehler.pdf

Biehler, R. (1998). Students – statistical software – statistical tasks: A study of problems at the interfaces. In L. Pereira-Mendoza (Ed.), *Proceedings of the fifth international conference on the teaching of statistics* (ICOTS-5), Singapore (pp. 1025–1031). Voorburg, The Netherlands: International Statistical Institute. Retrieved September 28, 2007, from http://www.stat.auckland.ac.nz/ ~iase/publications/2/Topic7zg.pdf

Biehler, R. (2001). *"Girls (tend to) watch less television than boys" – Students' hypotheses and data exploration strategies in group comparison tasks*. Talk presented at LOGOS #10, Mathematics Education Unit, Department of Mathematics, The University of Auckland, New Zealand.

Biehler, R. (2003). Interrelated learning and working environments for supporting the use of computer tools in introductory classes. *Proceedings of the IASE satellite conference on statis-*

tics education and the internet, August 2003, Berlin, [CD-ROM]. Voorburg, The Netherlands: International Statistical Institute.

Biehler, R., & Steinbring, H. (1991). Entdeckende statistik, stengel-und-blätter, boxplots: Konzepte, begründungen und erfahrungen eines unterrichtsversuches [statistics by discovery, stem-and-leaf, boxplots: Basic conceptions, pedagogical rationale, and experiences from a teaching experiment]. *Der Mathematikunterricht, 37*(6), 5–32.

Biggs, J. B., & Collis, K. F. (1982). *Evaluating the quality of learning: The SOLO taxonomy*. New York: Academic Press.

Bloom, B. S. (Ed.) (1956). *Taxonomy of educational objectives: The classification of educational goals: Handbook I, cognitive domain*. New York: Longmans, Green.

Boud, D., Cohen, R., & Sampson, J. (2001). *Peer learning in higher education: Learning from and with each other*, (pp. 1–17). London: Kogan Page.

Box, G. E.P. (1979). Robustness in scientific model building. In R. L. Launer & G. N. Wilkinson (Eds.), *Robustness in statistics* (pp. 201–236). New York: Academic Press.

Bransford, J., Brown, A. L., & Cocking, R. R. (Eds.) (2000). *How people learn: Brain, mind, experience, and school*. Washington, DC: National Academy Press.

Bright, G. W., & Friel, S. N. (1998). Graphical representations: Helping students interpret data. In S. P. Lajoie (Eds.), *Reflections on statistics: Agendas for learning, teaching, and assessment in K-12* (pp. 63–88). Mahwah, NJ: Lawrence Erlbaum Associates.

Brophy, J. (2004). *Motivating students to learn, (2nd ed.)*. Mahwah, NJ: Lawrence Erlbaum.

Brown, A. L. (1975). The development of memory: Knowing, knowing about knowing, and knowing how to know. In H. W. Reese (Ed.), *Advances in child development and behavior* (Vol. 10). New York: Academic Press.

Brown, A. L., & Campione, J. C. (1994). Guided discovery in a community of learners. In K. McGilly (Ed.), *Classroom lessons: Integrating cognitive theory and classroom practice* (pp. 229–270). Cambridge, MA: MIT Press.

Cai, J. (1998). Exploring students' conceptual understanding of the averaging algorithm. *School Science and Mathematics, 2*, 93–98.

Callingham, R. A. (1997). Teachers' multimodal functioning in relation to the concept of average. *Mathematics Education Research Journal, 9*(2), 205–224.

Canada, D. (2004). *Pre-service elementary teachers' conceptions of variability*. Unpublished doctoral dissertation, Portland State University, Portland, OR. Retrieved October 21, 2007, from http://www.stat.auckland.ac.nz/ ~iase/publications/dissertations/04.Canada. Dissertation.pdf

Canada, D. (2006, May). Elementary pre-service teachers' conceptions of variation in a probability context. *Statistics Education Research Journal, 5*(1), 36–64. Retrieved July 15, 2007, from http://www.stat.auckland.ac.nz/ ~iase/serj/SERJ5(1)_Canada.pdf

Cardenas, K. (1998, May/June). Technology in today's classroom: It slices and it dices, but does it serve us well? *Academe, 84*(3), 27–29.

Carlson, M., Jacobs, S., Coe, E., Larsen, S., & Hsu, E. (2002). Applying covariational reasoning while modeling dynamic events. *Journal for Research in Mathematics Education, 33*(5), 352–378.

Carver, R. P. (1978). The case against statistical significance testing. *Harvard Educational Review, 48*(3), 378–399.

Case, R. (1992). *The mind's staircase: Exploring the conceptual underpinnings of children's thought and knowledge*. Mahwah, NJ: Lawrence Erlbaum.

Chance, B. L. (2000). Experiences with authentic assessment techniques in an introductory statistics course. In T. L. Moore (Ed.) *Teaching Statistics: Resources for Undergraduate Instructors* (pp. 209–218). Washington DC: Mathematical Association of America.

Chance, B. L. (2002, November). Components of statistical thinking and implications for instruction and assessment. *Journal of Statistics Education, 10*(3). Retrieved July 15, 2007, from http://www.amstat.org/publications/ jse/v10n3/chance.html

Chance, B. L., delMas, R., & Garfield, J. (2004). Reasoning about sampling distributions. In D. Ben-Zvi & J. Garfield (Eds.), *The challenge of developing statistical literacy, reasoning, and thinking* (pp. 295–323). Dordrecht, The Netherlands: Kluwer Academic Publishers.

Chance, B. L., & Rossman, A. J. (2001). Sequencing topics in introductory statistics: A debate on what to teach when. *The American Statistician, 55*, 140–144.

Chance, B. L., & Rossman, A. J. (2005). *Investigating statistical concepts, applications, and methods*. Duxbury Press.

Chance, B. L., & Rossman, A. J. (2006). Using simulation to teach and learn statistics. In A. J. Rossman & B. L. Chance (Eds.), *Proceedings of the seventh international conference on the teaching of statistics*. (ICOTS-7), Salvador, Bahia, Brazil, [CD-ROM]. Voorburg, The Netherlands: International Statistical Institute. Retrieved July 15, 2007, from http://www.stat.auckland.ac.nz/ ~iase/publications/17/7E1_CHAN.pdf

Chick, H. L., & Watson, J. M. (2002). Collaborative influences on emergent statistical thinking – a case study. *Journal of Mathematical Behavior, 21*, 371–400.

Chu, S. (2001, July). Pricing the C's of diamond stones. *Journal of Statistics Education, 9*(2). Retrieved December 6, 2006, from http://www.amstat.org/publications/jse/v9n2/datasets.chu.html

Ciancetta, M. (2007). *Undergraduate and graduate statistics students' reasoning when comparing distributions*. Unpublished doctoral dissertation, Portland State University, Portland, OR.

Clark, J., Karuat, G., Mathews, D., & Wimbish, J. (2003). *The fundamental theorem of statistics: Classifying student understanding of basic statistical concepts*. Unpublished paper. Retrieved October 21, 2007, from http://www1.hollins.edu/faculty/clarkjm/stat2c.pdf

Clement, J. (1987). Overcoming students' misconceptions in physics: The role of anchoring intuitions and analogical validity. *Proceedings of the second international seminar, misconceptions and educational strategies in science and mathematics*. Ithaca, NY: Cornell University.

Cobb, G. W. (1992). Report of the joint ASA/MAA committee on undergraduate statistics. In the *American Statistical Association 1992 proceedings of the Section on Statistical Education*, (pp. 281–283). Alexandria, VA: American Statistical Association.

Cobb, G. W. (1993, July). Reconsidering statistics education: A national science foundation conference. *Journal of Statistics Education, 1*(1), 1–28. Retrieved November 6, 2006, from http://www.amstat.org/publications/jse/v1n1/cobb.html

Cobb, G. W. (1998). *The objective-format question in statistics: Dead horse, old bath water, or overlooked baby?* Paper presented at the annual meeting of the American Educational Research Association (AERA). Retrieved April 26, 2008, from http://www.jmu.edu/assessment/resources/resource_files/cobb_1998.pdf

Cobb, G. W. (2005). *The introductory statistics course: A saber tooth curriculum*? Presentation at United States Conference on Teaching Statistics. Columbus, Ohio. Retrieved October 21, 2007, from http://www.causeweb.org/uscots/ uscots05/plenary/USCOTS_Dinner_Cobb.ppt

Cobb, G. W. (2007, October). The introductory statistics course: A Ptolemaic curriculum? *Technology Innovations in Statistics Education, 1*(1). Retrieved October 21, 2007, from http://repositories.cdlib.org/uclastat/cts/tise/vol1/iss1/art1/

Cobb, G. W., & Miao, W. (1998). Bears in space: Activities to introduce basic ideas of design. In L. Pereira-Mendoza (Ed.), *Proceedings of the fifth international conference on the teaching of statistics* (ICOTS-5), Singapore (pp. 214–220). Voorburg, The Netherlands: International Statistical Institute. Retrieved September 28, 2006, from www.stat.auckland.ac.nz/~iase/publications/2/Topic2n.pdf

Cobb, G. W., & Moore, D. S. (1997). Mathematics, statistics, and teaching. *American Mathematical Monthly, 104*, 801–823.

Cobb, P. (1994). Where is the mind? Constructivist and sociocultural perspectives on mathematical development. *Educational Researcher, 23*(7), 13–20.

Cobb, P. (1999). Individual and collective mathematical development: The case of statistical data analysis. *Mathematical Thinking and Learning, 1*(1), 5–43.

Cobb, P., Confrey, J., diSessa, A., Lehrer, R., & Schauble, L. (2003a). Design experiments in educational research. *Educational Researcher, 32*(1), 9–13.

Cobb, P., Gravemeijer, K. P. E., Bowers, J., & Doorman, M. (1997). *Statistical Minitools* [applets and applications]. Nashville & Utrecht: Vanderbilt University, TN & Freudenthal Institute, Utrecht University. Available from: http://peabody.vanderbilt.edu/depts/tandl/mted/Minitools/Minitools.html

Cobb, P., & McClain, K. (2004). Principles of instructional design for supporting the development of students' statistical reasoning. In D. Ben-Zvi & J. Garfield (Eds.), *The challenge of developing statistical literacy, reasoning, and thinking* (pp. 375–396). Dordrecht, The Netherlands: Kluwer Academic Publishers.

Cobb, P., McClain, K., & Gravemeijer, K. P.E (2003b). Learning about statistical covariation. *Cognition and Instruction, 21*, 1–78.

Cobb, P., Yackel, E., & Wood, T. L. (1992). A constructivist alternative to the representational view of mind in mathematics education. *Journal for Research in Mathematics Education, 23*(1), 2–33.

Cohen, D. K., & Ball, D. L. (1999). *Instruction, capacity, and improvement*. Philadelphia, PA: Consortium for Policy Research in Education, University of Pennsylvania (CPRE RR-43).

Cohen, E. G. (1994). *Designing groupwork: strategies for heterogeneous classrooms*. NY: Teachers College Press.

Cohen, E. G., & Lotan, R. A. (Eds.) (1997). *Working for equity in heterogeneous classrooms: Sociological theory in practice*. NY: Teachers College Press.

Collins (2003). *Collins English dictionary* (6Rev Ed edition). Collins.

Collins, A., Brown, J. S., & Newman, S. E. (1989). Cognitive apprenticeship: Teaching the crafts of reading, writing, and mathematics. In L. B. Resnick (Ed.), *Knowing, learning, and instruction: Essays in honour of Robert Glaser*, (pp. 453–494). Hillsdale, NJ: Erlbaum.

Collins, L., & Mittag, K. C. (2005, May). Effect of calculator technology on student achievement in an introductory statistics. *Statistics Education Research Journal, 4*(1), 7–15. Retrieved October 15, 2007, from http://www.stat.auckland.ac.nz/ ~iase/serj/SERJ4(1)_Collins_Mittag.pdf

Cook, R. D., & Weisberg, S. (1999). *Applied regression including computing and graphics*. (Wiley Series in Probability and Statistics). New York: Wiley-Interscience.

Cooper, H., Robinson, J. C., & Patall, E. A. (2006). Does homework improve academic achievement? A synthesis of research, 1987–2003. *Review of Educational Research, 76*(1), 1–62.

Courtney, D. P. Courtney, M., & Nicholson, C. (1994). The effect of cooperative learning as an instructional practice at the college level. *College Student Journal, 28*, 471–477.

Cox, R. C. (1965). Item selection techniques and evaluation of instructional objectives. *Journal of Educational Measurement*, *2*(2), 181–185.

Crooks, T. J. (1988, Winter). The impact of classroom evaluation practices on students. *Review of Educational Research, 58*(4), 438–481.

Cross, K. P., & Steadman, M. H. (1996). *Classroom research: Implementing the scholarship of teaching*. San Francisco, CA: Jossey–Bass.

Cryer, J. (2001). *Problems with using Microsoft Excel for statistics*. Presented at the American Statistical Association (ASA) Joint Statistical Meeting, Atlanta, Georgia. Retrieved April 26, 2008, from http://www.cs.uiowa.edu/ ~jcryer/JSMTalk2001.pdf

Curcio, F. R. (2002). *A user's guide to Japanese lesson study: Ideas for improving mathematics teaching*. Reston, VA: National Council of Teachers of Mathematics.

Dauphinee, T. L., Schau, C., & Stevens, J. J. (1997). Survey of attitudes toward statistics: factor structure and factorial invariance for women and men. *Structural Equation Modeling, 42*, 129–141.

Davis, B. G. (1993). *Tools for teaching*. San Francisco: Jossey-Bass Publishers.

de Lange, J., Burrill, G., Romberg, T. A., & van Reeuwijk, M. (1993). *Learning and testing mathematics in context: The case of data visualization*. Scotts Valley, CA: Wings for Learning.

de Saint-Exupéry, A. (2000). *The little prince*. Harcourt children's books.

De Veaux, R. D., Velleman, P. F., & Bock, D. E. (2005). *Stats: Data and models*. Boston: Pearson Education, Inc.

delMas, R. C. (2001a). *Sampling SIM* (version 5) [Computer software]. Retrieved November 12, 2006, from: http://www.tc.umn.edu/~delma001/stat_tools

delMas, R. C. (2001b). What makes the standard deviation larger or smaller? STAR library. Retrieved October 21, 2007, from http://www.causeweb.org/repository/StarLibrary/activities/delmas2001/

delMas, R. C. (2002, November). Statistical literacy, reasoning, and learning: A commentary. *Journal of Statistics Education, 10*(3). Retrieved November 6, 2006, from http://www.amstat.org/publications/jse/v10n3/delmas_intro.html

delMas, R. C. (2004). A comparison of mathematical and statistical reasoning. In D. Ben-Zvi & J. Garfield (Eds.), *The challenge of developing statistical literacy, reasoning, and thinking* (pp. 79–96). Dordrecht, The Netherlands: Kluwer Academic Publishers.

delMas, R. C. (2005). *Activities to promote an understanding of variability*. Paper presented at the Annual Meeting of the American Mathematics Association of Two-Year Colleges (AMATYC). San Diego, CA, November 11.

delMas, R. C., Garfield, J., & Chance, B. L. (1999, November). A model of classroom research in action: Developing simulation activities to improve students' statistical reasoning. *Journal of Statistics Education, 7*(3). Retrieved December 18, 2006, from http://www.amstat.org/publications/jse/secure/v7n3/delmas.cfm

delMas, R. C., Garfield, J., & Chance, B. L. (2004). *Using assessment to study the development of students' reasoning about sampling distributions*. Paper presented at the Annual meeting of the American Educational Research Association, San Diego. Retrieved April 26, 2008, from http://www.gen.umn.edu/ faculty_staff/delmas/AERA_2004_samp_dist.pdf

delMas, R. C., Garfield, J., & Ooms, A. (2005, July). *Using assessment items to study students' difficulty with reading and interpreting graphical representations of distributions*. Presented at the fourth international research forum on statistical reasoning, thinking, and literacy (SRTL-4), July 6, 2005. Auckland, New Zealand.

delMas, R. C., Garfield, J., Ooms, A., & Chance, B. L. (2007, November). Assessing students' conceptual understanding after a first course in statistics. *Statistics Education Research Journal 6*(2), 28–58. Retrieved April 20, 2008, from http://www.stat.auckland.ac.nz/~iase/serj/SERJ6(2)_delMas.pdf

delMas, R. C., & Liu, Y. (2005, May). Exploring students' conceptions of the standard deviation. *Statistics Education Research Journal, 4*(1), 55–82. Retrieved November 14, 2006, from http://www.stat.auckland.ac.nz/ ~iase/serj/SERJ4(1)_delMas_Liu.pdf

delMas, R. C., Ooms, A., Garfield, J., & Chance, B. L. (2006). Assessing students' statistical reasoning. In A. J. Rossman & B. L. Chance (Eds.), *Proceedings of the seventh international conference on the teaching of statistics* (ICOTS-7), Salvador, Bahia, Brazil, [CD-ROM]. International Statistical Institute. Retrieved July 15, 2007, from http://www.stat.auckland.ac.nz/~iase/publications/17/6D3_DELM.pdf

Derry, S. J., Levin, J. R., Osana, H. P., Jones, M. S., & Peterson, M. (2000). Fostering students' statistical and scientific thinking: Lessons learned from an innovative college course. *American Educational Research Journal, 37*, 747–773.

Deutsch, M. (1949). A theory of cooperation and competition. *Human Relations, 2*, 129–151.

Deutsch, M. (1962). Cooperation and trust: Some theoretical notes. In M. Jones (Ed.), *Nebraska symposium on motivation*. Lincoln: University of Nebraska Press.

Dietz, E. J. (1993). A cooperative learning activity on methods of selecting a sample. *The American Statistician, 47*(2), 104–108.

Dietz, E. J., Gould, R., Hartlaub, B., & Rossman, A. J. (2005). *Nuts and bolts of classroom assessment*. Invited panel presentation, Joint Statistical Meetings, Minneapolis.

Doerr, H. M., & Zangor, R. (2000, Feb.). Creating meaning for and with the graphing calculator. *Educational Studies in Mathematics, 41*(2), 143–163.

Dunham, P. H., & Dick, T. P. (1994). Connecting research to teaching: research on graphing calculators. *The Mathematics teacher, 87*(6), 440–445.

Earley, M. A. (2001). *Improving statistics education through simulations: The case of the sampling distribution*. Paper presented at the Annual Meeting of the Mid-Western Educational Research Association. Chicago, Illinois, October 24–27.

Elbow, P. (1997). High stakes and low stakes in assigning and responding to writing. In M. D. Sorcinelli & P. Elbow (Eds.), *Writing to learn: Strategies for assigning and responding to writing across the disciplines* (pp. 5–13). San Francisco: Jossey-Bass.

Elmore, P. B., & Vasu, E. S. (1986). Model of statistics achievement using spatial ability, feminist attitudes and mathematics-related variables as predictors. *Educational and Psychological Measurement, 46*, 215–222.

Erickson, T. (2001). *Data in depth: Exploring mathematics with Fathom*. Emeryville, CA: Key Curriculum Press.

Erickson, T. (2002). *Fifty Fathoms: Statistics demonstrations for deeper understanding*. Oakland, CA: EEPS Media.

Everson, M. (2006). Group discussion in online statistics courses. *E-learn Magazine*, (4/14/06). Available from: http://www.elearnmag.org/subpage. cfm?section=case_studies&article=35–1

Everson, M., Zieffler, S. A., & Garfield, J. (in press). Implementing new reform guidelines in teaching introductory college statistics courses. *Teaching Statistics*.

Fairbrother, R. W. (1975). The reliability of teachers' judgment of the abilities being tested by multiple choice items. *Educational Research, 17Z*(3), 202–210.

Falchikov, N. (2001). *Learning together: Peer tutoring in higher education*. London: Routledge Falmer.

Falk, R. (1981). The perception of randomness. In C. Comiti & G. Verguaud (Eds.), *Proceedings of the fifth international conference for the psychology of mathematics education* (pp. 222–229). Grenoble, France.

Falk, R. (1988). Conditional probabilities: Insights and difficulties. In R. Davidson & J. Swift (Eds.), *Proceedings of the second international conference on the teaching of statistics* (ICOTS-2) (pp. 292–297). Victoria BC, Canada: University of Victoria. Retrieved April 20, 2008, from http://www.stat.auckland.ac.nz/ ~iase/publications/icots2/Falk.pdf

Falk, R., & Greenbaum, C. W. (1995). Significance tests die hard: The amazing persistence of a probabilistic misconception. *Theory and Psychology, 5*, 75–98.

Falk, R., & Konold, C. (1992). The psychology of learning probability. In F. S. Gordon & S. P. Gordon (Eds.), *Statistics for the twenty-first century* (pp. 151–164). Mathematical Association of America.

Falk, R., & Konold, C. (1994). Random means hard to digest. *Focus on Learning Problems in Mathematics, 16*, 2–12.

Falk, R., & Konold, C. (1997). Making sense of randomness: Implicit encoding as a basis for judgment. *Psychological Review, 104*, 301–318.

Falk, R., & Konold, C. (1998). Subjective randomness. In S. Kotz, C. B. Read & D. L. Banks (Eds.), *Encyclopedia of statistical sciences* (pp. 653–659). NY, NY: J. Wiley & Sons.

Fast, G. R. (1997). *Using analogies to produce long term conceptual change: Overcoming high school mathematics students' probability misconceptions*. Paper presented at the Annual Meeting of the American Educational Research Association, Chicago, IL. Retrieved April 20, 2008, from http://eric.ed.gov/ERICDocs/data/ericdocs2sql/content_storage_01/0000019b/80/16/92/52.pdf

Feenberg, A. (1999, Summer). Whither educational technology? *Peer Review, 1*(4).

Felder, R. (2001). Interview transcript retrieved October 2, 2007, from Foundation Coalition, Active/Cooperative Learning Website: http://clte.asu.edu/active/implework.htm

Fennema, E. (1989). The study of affect and mathematics: A proposed generic model for research. In McLeod, D. B. & Adams, B. M. (Eds.), *Affect and mathematical problem solving* (pp. 205–219). New York: Springer-Verlag.

Fernandez, C. (2002). Learning from Japanese approaches to professional development: The case of lesson study. *Journal of Teacher Education, 53*(5), 390–405.

Fernandez, C., Cannon, J., & Chokshi, S. (2003). A U. S.-Japan lesson study collaboration reveals critical lenses for examining practice. *Teaching and Teacher Education, 19*(2), 171–185.

Fiechtner, S. B., & Davis, E. A. (1992). Why some groups fail: A survey of students' experiences with learning groups. In A. Goodsell, M. Maher, V. Tinto, & Associates (Eds.), *Collaborative learning: A sourcebook for higher education*. University Park: National Center on Postsecondary Teaching, Learning, and Assessment, Pennsylvania State University.

Fillebrown, S. (1994, November). Using projects in an elementary statistics course for non-science majors. *Journal of Statistics Education, 2*(2). Retrieved November 6, 2006, from http://www.amstat.org/publications/jse/v2n2/fillebrown.html

Finch, S., & Cumming, G. (1998). Assessing conceptual change in learning statistics. In L. Pereira-Mendoza (Ed.), *Proceedings of the fifth international conference on the teaching of statistics* (ICOTS-5), Singapore (pp. 897–904). Voorburg, The Netherlands: International Statistical Institute. Retrieved on September 28, 2007, from http://www.stat.auckland.ac.nz/~iase/publications/2/Topic7n.pdf

Finney, S. J. (2000). *The meaning of success for students in statistical methods courses: A phenomenological study*. Paper presented at the annual meeting of the American Educational Research Association, New Orleans.

Fischbein, E. (1975). *The intuitive sources of probabilistic thinking in children*. Dordrecht, The Netherlands: D. Reidel.

Fisher, R. A. (1925). *Statistical methods for research workers*. Oliver and Boyd, Edinburgh.

Flavell, J. H. (1976). Metacognitive aspects of problem-solving. In L. B. Resnick (Ed.), *The nature of intelligence*. Hillsdale, NJ: Erlbaum.

Flores, A. (2006). Using graphing calculators to redress beliefs in the "law of small numbers". In G. F. Burrill (Eds.), *Thinking and reasoning about data and chance: Sixty-eighth NCTM Yearbook* (pp. 291–304). Reston, VA: National Council of Teachers of Mathematics.

Fong, G. T., Krantz, D. H., & Nisbett, R. E. (1986). The effects of statistical training on thinking about everyday problems. *Cognitive Psychology, 18*, 253–292.

Forster, P. A. (2007, Mar). Technologies for teaching and learning trend in bivariate data. *International Journal of Mathematical Education in Science and Technology, 38*(2), 143–161.

Franklin, C. A. (2006). The statistical education of grades Pre-K-2 Teachers: A shared responsibility. In G. F. Burrill (Ed.), *Thinking and reasoning about data and chance: Sixty-eighth NCTM Yearbook* (pp. 335–344). Reston, VA: National Council of Teachers of Mathematics.

Franklin, C. A., & Garfield, J. (2006). The guidelines for assessment and instruction in statistics education (GAISE) project: Developing statistics education guidelines for pre K-12 and college courses. In G. F. Burrill, (Ed.), *Thinking and reasoning about data and chance: Sixty-eighth NCTM Yearbook* (pp. 345–375). Reston, VA: National Council of Teachers of Mathematics. Available from: http://www.amstat.org/Education/gaise/GAISECollege.htm, and http://www.amstat.org/education/gaise/GAISEPreK-12.htm

Franklin, C. A., & Mewborn, D. S. (2006). The statistical education of grades Pre-K-2 teachers: A shared responsibility. In G. F. Burrill (Ed), *Thinking and reasoning with data and chance: Sixty-eighth NCTM yearbook*. Reston, VA: National Council of Teachers of Mathematics (NCTM).

Freund, J. E., & Perles, B. M. (1987). A new look at quartiles of ungrouped data. *The American Statistician, 41*(3), 200–203.

Friel, S. N. (2007). The research frontier: Where technology interacts with the teaching and learning of data analysis and statistics. In G. W. Blume & M. K. Heid (Eds.), *Research on technology and the teaching and learning of mathematics: Cases and perspectives* (Vol. 2, pp. 279–331). Greenwich, CT: Information Age Publishing, Inc.

Friel, S. N., Curcio, F. R., & Bright, G. W. (2001). Making sense of graphs: Critical factors influencing comprehension and instructional implications. *Journal for Research in Mathematics Education, 32*(2), 124–158.

GAISE (2005a). *Guidelines for assessment and instruction in statistics education (GAISE) report: A curriculum framework for PreK-12 statistics education*. The American Statistical Association (ASA). Retrieved December 4, 2006, from http://www.amstat.org/education/gaise/GAISEPreK-12.htm

GAISE (2005b). *Guidelines for assessment and instruction in statistics education (GAISE) college report*. The American Statistical Association (ASA). Retrieved December 4, 2006, from http://www.amstat.org/education/gaise/GAISECollege.htm

Gal, I. (Ed.) (2000). *Adult numeracy development: Theory, research, practice*. Cresskill, NJ: Hampton Press.

Gal, I. (2002). Adults' statistical literacy: Meaning, components, responsibilities. *The International Statistical Review, 70*(1), 1–25.

Gal, I., & Garfield, J. (Eds.) (1997). *The assessment challenge in statistics education*. Amsterdam: IOS Press. Retrieved April 27, 2008, from http://www.stat.auckland.ac.nz/~iase/publications/assessbk/

Gal, I., & Ginsburg, L. (1994, November). The role of beliefs and attitudes in learning statistics: Towards an assessment framework. *Journal of Statistics Education*, *2*(2). Retrieved November 6, 2006, from http://www.amstat. org/publications/jse/v2n2/gal.html

Gal, I., Rothschild, K., & Wagner, D. A. (1989). *Which group is better? The development of statistical reasoning in elementary school children.* Paper presented at the meeting of the Society for Research in Child Development, Kansas City. Retrieved April 20, 2008, from http://eric.ed.gov/ERICDocs/ data/ericdocs2sql/content_storage_01/0000019b/80/ 1f/d9/82.pdf

Gal, I., Rothschild, K., & Wagner, D. A. (1990). *Statistical concepts and statistical heuristics in school children: convergence or divergence?* Paper Presented at the Annual Meeting of the American Educational Research Association. Boston, Massachusetts, USA.

Gani, J. (1979). The role of ISI in the field of statistical education: 1949–1979. *International Statistical Review*, *47*(3), 237–244.

Garcia, G. E., & Pearson, P. D. (1990). *Modifying reading instruction to maximize its effectiveness for all students* (Report No. 489). Urbana-Champaign, IL: Center for the Study of Reading.

Garfield, J. (1993, July). Teaching statistics using small-group cooperative learning. *Journal of Statistics Education*, *1*(1). Retrieved December 30, 2006, from http://www.amstat.org/ publications/jse/v1n1/Garfield.html

Garfield, J. (1994, July). Beyond testing and grading: Using assessment to improve student learning. *Journal of Statistics Education*, *2*(1). Retrieved November 6, 2006, from http://www.amstat.org/publications/jse/v2n1/garfield.html

Garfield, J. (1995). How students learn statistics. *International Statistical Review*, *63*, 25–34.

Garfield, J. (1999). *Thinking about statistical reasoning, thinking, and literacy*. Paper presented at the first international research forum on statistical reasoning, thinking, and literacy (STRL-1), Kibbutz Be'eri, Israel.

Garfield, J. (2000). Beyond testing and grading: New ways to use assessment to improve student learning. In T. L. Moore (Ed.) T*eaching statistics: Resources for undergraduate instructors* (pp. 201–208). Washington DC: Mathematical Association of America.

Garfield, J. (2002a). Histogram sorting. *Statistics teaching and resource library* (STAR). Retrieved December 24, 2006, from http://www.causeweb.org/ repository/StarLibrary/activities/

Garfield, J. (2002b, November). The challenge of developing statistical reasoning. *Journal of Statistics Education, 10*(3). Retrieved July 15, 2007, from http://www.amstat.org/ publications/jse/v10n3/garfield.html

Garfield, J. (2003, May). Assessing statistical reasoning. *Statistics Education Research Journal*, *2*(1), 22–38. Retrieved December 26, 2006, from http://www.stat.auckland.ac.nz/ ~iase/serj/SERJ2(1)

Garfield, J. (Ed.). (2005). *Innovations in teaching statistics* (MAA Notes number 65). Washington, DC: Mathematical Association of America.

Garfield, J. (2006). Collaboration in statistics education research: Stories, reflections, and lessons learned. In A. J. Rossman and B. L. Chance (Eds.), *Proceedings of the seventh international conference on the teaching of statistics* (ICOTS-7), Salvador, Bahia, Brazil, [CD-ROM]. Voorburg, The Netherlands: International Statistical Institute. Retrieved December 26, 2006, from http://www.stat.auckland.ac.nz/~iase/publications/17/PL2_GARF.pdf

Garfield, J., & Ahlgren, A. (1988). Difficulties in learning basic concepts in probability and statistics: Implications for research. *Journal for Research in Mathematics Education, 19*(1), 44–63.

Garfield, J., & Ben-Zvi, D. (2004). Research on statistical literacy, reasoning, and thinking: issues, challenges, and implications. In D. Ben-Zvi & J. Garfield (Eds.), *The challenge of developing statistical literacy, reasoning, and thinking* (pp. 397–409). Dordrecht, The Netherlands: Kluwer Academic Publishers.

Garfield, J., & Ben-Zvi, D. (2005, May). A framework for teaching and assessing reasoning about variability. *Statistics Education Research Journal*, *4*(1), 92–99. Retrieved December 26, 2006, from http://www.stat.auckland.ac.nz/ ~iase/serj/SERJ4(1)_Garfield_BenZvi.pdf

Garfield, J., & Burrill, G. (Eds.). (1997). *Research on the role of technology in teaching and learning statistics*. Voorburg, The Netherlands: International Statistical Institute. Proceedings of the IASE roundtable conference, University of Granada, Spain, 23-27 July, 1996. Retrieved April 20, 2008, from http://www.dartmouth.edu/~chance/teaching_aids/IASE/IASE.book.pdf

Garfield, J., & Chance, B. L. (2000). Assessment in statistics education: Issues and challenges. *Mathematics Thinking and Learning, 2,* 99–125.

Garfield, J., Chance, B. L., & Snell, J. L. (2000). Technology in college statistics courses. In D. Holton et al. (Eds.), *The teaching and learning of mathematics at university level: An ICMI study* (pp. 357–370). Dordrecht, The Netherlands: Kluwer Academic Publishers. Retrieved December 4, 2006, from http://www.dartmouth.edu/~chance/teaching_aids/books_articles/technology.html

Garfield, J., delMas, R., & Chance, B. L. (2000). Tools for teaching and assessing statistical inference (NSF project). Available: http://www.tc.umn.edu/ ~delma001/stat_tools/

Garfield, J., delMas, R., & Chance, B. L. (2005). *The impact of Japanese lesson study on teachers of statistics*. Paper presented at the Joint Statistical Meetings, Minneapolis, MN.

Garfield, J., delMas, R., & Chance, B. L. (2007). Using students' informal notions of variability to develop an understanding of formal measures of variability. In M. C. Lovett & P. Shah (Eds.), *Thinking with data* (pp. 117–148). Mahwah, NJ: Lawrence Erlbaum Associates.

Garfield, J., Hogg, B., Schau, C., & Whittinghill, D. (2002, July). First courses in statistical science: The status of educational reform efforts. *Journal of Statistics Education, 10*(2). Retrieved November 6, 2006, from http://www.amstat.org/publications/jse/v10n2/garfield.html

Gelman, A., & Nolan, D. (2002). *Teaching statistics: A bag of tricks*. New York: Oxford University Press.

Gigerenzer, G. (1996). On narrow norms and vague heuristics: A reply to Kahneman and Tversky (1996). *Psychological Review, 103*(3), 592–596.

Ginsburg-Block, M. D., Rohrbeck, C. A., & Fantuzzo, J. W. (2006). A meta-analytic review of social, self-concept, and behavioral outcomes of peer-assisted learning. *Journal of Educational Psychology, 98,* 732–749.

Giraud, G. (1997, November). Cooperative learning and statistics instruction. *Journal of Statistics Education 5*(3). Retrieved December 30, 2006, from http://www.amstat.org/publications/jse/v5n3/giraud.html

Glanz, J. (2003). *Action research: An educational leader's guide to school improvement*. Norwood, MA: Christopher-Gordon Publishers.

Gnanadesikan, M., Scheaffer, R. L., & Swift, J. (1987). *The art and techniques of simulation*. Palo Alto, CA: Dale Seymour Publications.

Good, P. (2006). *Resampling methods* (3rd ed.). Birkhauser.

Goodchild, S. (1988). School pupils' understanding of average. Teaching Statistics, 10(3), 77–81.

Gordon, F., & Gordon, S. (Eds.) (1992). *Statistics for the twenty-first century* (MAA Notes Volume 26). Washington DC: Mathematical Association of America.

Gould, R. (2004, November). Variability: One statistician's view. *Statistical Education Research Journal, 3*(2), 7–16. Retrieved November 14, 2006, from http://www.stat.auckland.ac.nz/~iase/serj/SERJ3(2)_Gould.pdf

Graham, A. (1987). *Statistical investigations in the secondary school*. Cambridge, UK: Cambridge University Press.

Graham, A. (2006). *Developing thinking in statistics*. London: Paul Chapman Publishing.

Gravemeijer, K. P.E. (2000). *A rationale for an instructional sequence for analyzing one- and two-dimensional data sets*. Paper presented at the annual meeting of the American Educational Research Association, Montreal, Canada.

Gravemeijer, K. P.E. (2002). Developing research, a course in elementary data analysis as an example. In F. Lin (Eds.), *Common sense in mathematics education* (pp. 43–68). Taipei, Taiwan: National Taiwan Normal University.

Groth, R. E. (2002). Characterizing secondary students' understanding of measures of central tendency and variation. In D. S. Mewborn, P. Sztajn, D. Y. White, H. G. Wiegel, R. L. Bryant & K. Nooney (Eds.), *Proceedings of the twenty-fourth annual meeting of the North American chapter of the international group for the psychology of mathematics education* (Vol. 1, pp. 247–257). Columbus, OH: ERIC Clearinghouse for Science, Mathematics, and Environmental Education.

Groth, R. E. (2003). High school students' levels of thinking in regard to statistical study design. *Mathematics Education Research Journal, 15*, 252–269.

Groth, R. E. (2005). An investigation of statistical thinking in two different contexts: Detecting a signal in a noisy process and determining a typical value. *Journal of Mathematical Behavior, 24*, 109–124.

Groth, R. E., & Bergner, J. A. (2005, November). Preservice elementary school teachers' metaphors for the concept of statistical sample. *Statistics Education Research Journal, 4*(2), 27–42. Retrieved July 15, 2007, from http://www.stat.auckland.ac.nz/~iase/serj/SERJ4(2)_groth_bergner.pdf

Groth, R. E., & Bergner, J. A. (2006). Preservice elementary teachers' conceptual and procedural knowledge of mean, median, and mode. *Mathematical Thinking and Learning, 8*, 37–63.

Haladyna, T. M., Downing, S. M., & Rodriguez, M. C. (2002). A review of multiple-choice item-writing guidelines for classroom assessment. *Applied Measurement in Education, 15*(3), 309–334.

Haller, H., & Krauss, S. (2002). Misinterpretations of significance: A problem students share with their teachers? *Methods of Psychological Research, 7*(1), 1–20.

Halloun, I., & Hestenes, D. (1985). The initial knowledge state of college physics students. *American Journal of Physics, 53*(11), 1043–1055.

Hammerman, J. K., & Rubin, A. (2003). Reasoning in the presence of variability. In C. Lee (Ed.), *Reasoning about variability: A collection of current research studies*, Proceedings of the third international research forum on statistical reasoning, thinking, and literacy (SRTL-3). July 23-28 2003, The University of Nebraska-Lincoln, [CD-ROM].

Hammerman, J. K., & Rubin, A. (2004, November). Strategies for managing statistical complexity with new software tools. *Statistics Education Research Journal, 3*(2), 17–41. Retrieved December 4, 2006, from http://www.stat.auckland.ac.nz/~iase/serj/ SERJ3(2)_Hammerman_Rubin.pdf

Hancock, C., Kaput, J. J., & Goldsmith, L. T. (1992). Authentic inquiry with data: Critical barriers to classroom implementation. *Educational Psychologist, 27*(3), 337–364.

Hardiman, P., Well, A. D., & Pollatsek, A. (1984). Usefulness of a balance model in understanding the mean. *Journal of Educational Psychology, 76*, 792–801.

Harkness, W. (2005). Becoming an active learning facilitator in the statistics classroom. In J. B. Garfield (Ed.), *Innovations in teaching statistics*, MAA Notes # 65 (pp. 93–100). Washington, DC: Mathematical Association of America.

Hawkins, A. S. (1997). Myth-conceptions. In J. Garfield & G. Burrill (Eds.), *Research on the role of technology in teaching and learning statistics* (pp. 1–14). Voorburg, The Netherlands: International Statistical Institute. Retrieved April 20, 2008, from http://www.dartmouth.edu/~chance/teaching_aids/IASE/1.Hawkins.pdf

Hawkins, A. S., & Kapadia, R. (1984). Children's conception of probability-A psychological and pedagogical review. *Educational Studies in Mathematics, 15*, 349–377.

Hayden, R. (1989). Using writing to improve student learning of statistics. *The WAC Journal (Writing Across the Curriculum), 8*, 198–205.

Healy, L. (2006). Constructing simulations to express developing statistical knowledge. In A. J. Rossman & B. L. Chance (Eds.), *Proceedings of the seventh international conference on the teaching of statistics* (ICOTS-7), Salvador, Bahia, Brazil, [CD-ROM]. Voorburg, The Netherlands: International Statistical Institute. Retrieved July 15, 2007, from http://www.stat.auckland.ac.nz/ ~iase/publications/17/7G1_HEAL.pdf

Heaton, R. M., & Mickelson, W. T. (2002). The learning and teaching of statistical investigation in teaching and teacher education. *Journal of Mathematics Teacher Education, 5*, 35–59.

Hershkowitz, R., Dreyfus, T., Schwarz, B., Ben-Zvi, D., Friedlander, A., Hadas, et al. (2002). Mathematics curriculum development for computerized environments: A designer – researcher – teacher – learner activity. In L. D. English (Ed.), *Handbook of international research in mathematics education: Directions for the 21st century* (pp. 657–694). London: Laurence Erlbaum Associates Publishers.

Herweyers, G. (2002) Simulating experiments with the graphic calculator Ti83plus. In B. Phillips (Ed.), *Proceedings of the sixth international conference on the teaching of statistics* (ICOTS-6), Cape Town, South Africa, [CD-ROM]. Voorburg, The Netherlands: International Statistical Institute. Retrieved October 14, 2007, from http://www.stat.auckland.ac.nz/~iase/publications/1/7d2_herw.pdf

Hiebert, J., Gallimore, R., & Stigler, J. W. (2002). A knowledge base for the teaching profession: What would it look like and how can we get one? *Educational Researcher, 31*(5), 3–15.

Hiebert, J., Morris, A. K., & Glass, B. (2003). Learning to learn to teach: An "experiment" model for teaching and teacher preparation in mathematics. *Journal of Mathematics Teacher Education, 66*, 201–222.

Hirsch, L. S., & O'Donnell, A. M. (2001, July). Representativeness in statistical reasoning: Identifying and assessing misconceptions. *Journal of Statistics Education, 9*(2). Retrieved July 15, 2007, from http://www.amstat.org/publications/ jse/v9n2/hirsch.html

Hogg, R. V. (1992). Towards lean and lively courses in statistics. In F. Gordon & S. Gordon (Eds.), *Statistics for the twenty-first century* (pp. 3–13). Washington, DC: Mathematical Association of America.

Holcomb, J. (2004). *Assessing student learning in introductory statistics: an authentic assessment approach.* Presented at ARTIST Roundtable Conference on Assessment in Statistics, Lawrence University.

Holcomb, J., & Spalsbury, A. (2005, November). Teaching students to use summary statistics and graphics to clean and analyze data. *Journal of Statistics Education, 13*(3). Retrieved December 6, 2006, from http://www.amstat. org/publications/jse/v13n3/datasets.holcomb.html

Horvath, J. K., & Lehrer, R. (1998). A model-based perspective on the development of children's understanding of chance and uncertainty. In S. P. Lajoie (Ed.), *Reflections on statistics: Agendas for learning, teaching and assessment in K-12* (pp. 121–148). Mahwah, NJ: Lawrence Erlbaum Associates.

Hubbard, R. (1997, March). Assessment and the process of learning statistics. *Journal of Statistics Education, 5*(1). Retrieved November 6, 2006, from http://www.amstat.org/publications/ jse/v5n1/hubbard.html

Huberty, C. J., Dresden, J., & Bak, B. (1993). Relations among dimensions of statistical knowledge. *Educational and Psychological Measurement, 53*, 523–532.

Huck, S., Cross, T. L., & Clark, S. B. (1986). Overcoming misconceptions about *z*-scores. *Teaching Statistics, 8*(2), 38–40.

Huffman, D. Goldberg, F., & Michlin, M. (2003). Using computers to create constructivist learning environments: Impact on pedagogy and achievement. *Journal of Computers in Mathematics and Science Teaching, 22*(2), 151–168.

Hunt, D. N. (1996). Teaching statistics with Excel 5.0. *Maths & Stats, 7*(2), 11–14.

Inhelder, B., & Piaget, J. (1958). *The growth of logical thinking from childhood to adolescence.* London: Routledge & Kegan Paul.

Jacobs, V. R. (1999). How do students think about statistical sampling before instruction? *Mathematics Teaching in the Middle School, 5*(4), 240–246.

Jennings, D., Amabile, T., & Ross, L. (1982). Informal covariation assessment: Data-based versus theory-based judgments. In D. Kahneman, P. Slovic & A. Tversky (Eds.), *Judgment under uncertainty: Heuristics and biases* (pp. 211–230). Cambridge, England: Cambridge University Press.

Johnson, D. W., & Johnson, F. P. (2006). *Joining together: Group theory and group skills,* (9th ed.). Boston: Allyn & Bacon.

Johnson, D. W., & Johnson, R. T. (1989). *Cooperation and competition: Theory and research.* Edina, MN: Interaction Book Company.

Johnson, D. W., & Johnson, R. T. (1996). *Meaningful and manageable assessment through cooperative learning.* Edina, Minnesota: Interaction Book Company.

Johnson, D. W., & Johnson, R. T. (2002). *Meaningful assessment: A manageable and cooperative process.* Boston, MA: Allyn & Bacon.

Johnson, D. W., & Johnson, R. T. (2005). New developments in social interdependence theory. *Genetic, Social, and General Psychology Monographs, 131,* 285–358.

Johnson, D. W., Johnson, R. T., & Holubec, E. (1998a). *Cooperation in the classroom* (6th ed.). Edina, MN: Interaction Book Company.

Johnson, D. W., Johnson, R. T., & Smith, K. A. (1991). Cooperative learning: Increasing college faculty instructional productivity. ASHE-ERIC Higher Education Report, No. 4, Washington DC, The George Washington University School of Education and Human Development. Retrieved April 26, 2008, from http://www.eric.ed.gov/ERICDocs/data/ericdocs2sql/content_storage_01/0000019b/80/23/f1/c1.pdf

Johnson, D. W., Johnson, R. T., & Smith, K. A. (1998b). *Active learning: Cooperation in the college classroom* (2nd ed.).Edina, MN: Interaction Book Company.

Johnson, D. W., Maruyama, G., Johnson, R. T., Nelson, D., and Skon, L. (1981). The effects of cooperative, competitive, and individualistic goal structures on achievement: A meta-analysis. *Psychological Bulletin, 89,* 47–62.

Jones, G. A. (2005). Reflections. In G. A. Jones (Ed.), *Exploring probability in school: Challenges for teaching and learning* (pp. 367–372). New York: Springer.

Jones, G. A., Langrall C. W., & Mooney E. S. (2007). Research in probability: Responding to classroom realities. In F. K. Lester (Ed.), *The Second Handbook of Research on Mathematics* (pp. 909–956). Reston, VA: National Council of Teachers of Mathematics (NCTM).

Jones, G. A., Langrall, C. W., Mooney, E. S., & Thornton, C. A. (2004). Models of development in statistical reasoning. In D. Ben-Zvi & J. Garfield (Eds.), *The Challenge of Developing Statistical Literacy, Reasoning, and Thinking* (pp. 97–117). Dordrecht, The Netherlands: Kluwer Academic Publishers.

Jones, G. A., Langrall, C. W., Thornton, C. A., & Mogill, A. T. (1999). Students' probabilistic thinking in instruction. *Journal for Research in Mathematics Education, 30,* 487–519.

Jones, G. A., Thornton, C. A., Langrall, C. W., Mooney, E. S., Perry, B., & Putt, I. J. (2000). A framework for characterizing children's statistical thinking. *Mathematical Thinking and Learning*, 2(4), 269–307.

Jones, P., Lipson, K., & Phillips, B. (1994). A role for computer intensive methods in introducing statistical inference. In L. Brunelli & G. Cicchitelli (Eds.), *Proceedings of the first scientific meeting of the International Association for Statistical Education* (IASE), 23-24 August, 1993 (pp. 1255–1263). Perugia, Italy: University of Perugia. Retrieved April 26, 2008, from http://www.stat.auckland.ac.nz/ ~iase/publications/proc1993/255-263rec.pdf

Jordan, J. (2004, August). *The use of writing as both a learning and an assessment tool.* Paper presented at the ARTIST Roundtable Conference on Assessment in Statistics. Retrieved November 6, 2006, from http://www.rossmanchance.com/ artist/proceedings/jordan.pdf

Kader, G., & Perry, M. (1994). Learning statistics with technology. *Mathematics Teaching in the Middle School, 1*(2), 130–136.

Kahneman, D., Slovic, P., & Tversky, A. (Eds.). (1982). *Judgment under uncertainty: Heuristics and biases.* New York: Cambridge University Press.

Kanari, Z., & Millar, R. (2004). Reasoning from data: How students collect and interpret data in science investigations. *Journal of Research in Science Teaching, 41*(7), 748–769.

Kao, S. F., & Wasserman, E. A. (1993). Assessment of an information integration account of contingency judgment with examination of subjective cell importance and method of information presentation. *Journal of Experimental Psychology: Learning, Memory, and Cognition, 19,* 1363–1386.

Kaplan, D. (2007, October). Computing and introductory statistics. *Technology Innovations in Statistics Education, 1*(1), Article 5. Retrieved April 26, 2008, from http://repositories.cdlib.org/uclastat/cts/tise/vol1/iss1/art5

Kaplan, J. J. (2006). *Factors in statistics learning: Developing a dispositional attribution model to describe differences in the development of statistical proficiency.*

Unpublished Ph.D. Dissertation, University of Texas, Austin. Retrieved April 26, 2008, from http://www.stat.auckland.ac.nz/~iase/publications/ dissertations/06.Kaplan.Dissertation.pdf

Keeler, C. M., & Steinhorst, R. K. (1995, July). Using small groups to promote active learning in the introductory statistics course: A report from the field. *Journal of Statistics Education, 3*(2). Retrieved December 30, 2006, from http://www.amstat.org/publications/jse/v3n2/keeler.html

Kelly, A. E., & Lesh, R. A. (Eds.). (2000). *Handbook of research design in mathematics and science education.* Mahwah, NJ: Erlbaum.

Key Curriculum Press (2006). *Fathom dynamic data*™ software (Version 2.03) [Computer software]. Emeryville, CA: Key Curriculum Press. Available: http://www.keypress.com/fathom

Kirschner, P. A., Buckingham-Shum, S. J., & Carr, C. S. (2003). *Visualizing argumentation: Software tools for collaborative and educational sense-making*. London, England: Springer Verlag.

Kleiman, G. M. (2004). Myths and realities about technology in k-12 schools: Five years later. *Contemporary Issues in Technology and Teacher Education, 4*(2), 248–253.

Koehler, M. H. (2006). Using graphing calculator simulations in teaching statistics. In G. F. Burrill (Ed.), *Thinking and reasoning about data and chance: Sixty-eighth NCTM yearbook* (pp. 257–272). Reston, VA: National Council of Teachers of Mathematics (NCTM).

Kohn, A. (1986). *No contest: The case against competition.* Boston: Houghton Mifflin.

Konold, C. (1989a). Informal conceptions of probability. *Cognition and Instruction, 6*, 59–98.

Konold, C. (1989b). An outbreak of belief in independence? In the *eleventh annual meeting of the North American chapter of the international group for the psychology of mathematics education* (Vol. 2, pp. 203–209). New Brunswick, NJ: Center for Mathematics, Science and Computer Education. Retrieved April 26, 2008, from http://eric.ed.gov/ERICDocs/data/ericdocs2sql/content_storage_01/0000019b/80/1e/b2/ac.pdf

Konold, C. (1991). Understanding students' beliefs about probability. In E. V. Glasersfeld (Ed.), *Radical constructivism in mathematics education* (pp. 139–156). Amsterdam: Kluwer Academic Publishers. Retrieved April 26, 2008, from http://eric.ed.gov/ERICDocs/data/ericdocs2sql/content_storage_01/ 0000019b/80/1e/5d/94.pdf

Konold, C. (1994a). Teaching probability through modeling real problems. *Mathematics Teacher, 87*, 232–235.

Konold, C. (1994b). Understanding probability and statistical inference through resampling. In L. Brunelli & G. Cicchitelli (Eds.), *Proceedings of the first scientific meeting of the International Association for Statistical Education* (IASE), 23–24 August, 1993, (pp. 199–211). Perugia, Italy: University of Perugia. Retrieved April 26, 2008, from http://www.stat.auckland.ac.nz/~iase/publications/proc1993/199-211rec.pdf

Konold, C. (1995, March). Issues in assessing conceptual understanding in probability and statistics. *Journal of Statistics Education, 3*(1). Retrieved July 15, 2007, from http://www.amstat.org/publications/jse/v3n1/konold.html

Konold, C. (2002a). *Hat plots?* Unpublished manuscript, University of Massachusetts, Amherst.

Konold, C. (2002b). Teaching concepts rather than conventions. *New England Journal of Mathematics, 34*(2), 69–81. Retrieved April 26, 2008, from http://www.umass.edu/srri/serg/papers/KonoldNEJM.pdf

Konold, C., & Higgins, T. (2002). Highlights of related research. In S. J. Russell, D. Schifter & V. Bastable (Eds.), *Developing mathematical ideas: Working with data* (pp. 165–201). Parsippany, NJ: Seymour. (Staff-development materials for elementary teachers.). Available from: http://www.umass.edu/srri/serg/ papers/KonoldHiggins.pdf

Konold, C., & Higgins, T. (2003). Reasoning about data. In J. Kilpatrick, W. G. Martin & D. E. Schifter (Eds.), *A research companion to principles and standards for school mathematics* (pp. 193–215). Reston, VA: National Council of Teachers of Mathematics (NCTM).

Konold, C., Higgins, T., Russell, S. J., & Khalil, K. (2003). *Data seen through different lenses*. Unpublished Manuscript, Amherst, MA. Retrieved September 28, 2007, from http://www.umass.edu/srri/serg/papers/Konold-Higgins, et al.pdf

Konold, C., Kazak, S., Dahlstrom-Hakki, I., Lehrer, R., & Kim, M. J. (2007, August). *Students building data models*. Paper presented at the fifth international research forum on statistical reasoning, thinking, and literacy (SRTL-5). The University of Warwick, UK.

Konold, C., & Miller, C. D. (1994). *DataScope* [Computer software, Version 1.4]. Santa Barbara: Intellimation Library for the Macintosh. Available: http://www.umass.edu/srri/serg/software/download-chanceplus/datascopedl.htm

Konold, C., & Miller, C. D. (2005). *TinkerPlots: Dynamic data exploration* (Version 1.0) [Computer software]. Emeryville, CA: Key Curriculum Press. Available: http://www.keypress.com/tinkerplots

Konold, C., & Pollatsek, A. (2002). Data analysis as the search for signals in noisy processes. *Journal for Research in Mathematics Education, 33*(4), 259–289.

Konold, C., Pollatsek, A., Well, A. D., & Gagnon, A. (1997). Students analyzing data: Research of critical barriers. In J. Garfield & G. Burrill (Eds.), *Research on the role of technology in teaching and learning statistics* (pp. 151–168). Voorburg, The Netherlands: International Statistical Institute. Retrieved September 28, 2007, from http://www.dartmouth.edu/~chance/teaching_aids/IASE/13.Konold.pdf

Konold, C., Robinson, A., Khalil, K., Pollatsek, A., Well, A. D., Wing, R., & Mayr, S. (2002). Students' use of modal clumps to summarize data. In B. Phillips (Ed.), *Proceedings of the sixth international conference on the teaching of statistics* (ICOTS-6), Cape Town, South Africa, [CD-ROM]. Voorburg, The Netherlands: International Statistical Institute. Retrieved September 28, 2007, from http://www.stat.auckland.ac.nz/~iase/publications/1/8b2_kono.pdf

Koslowski, B. (1996). *Theory and evidence: The development of scientific reasoning (Learning, development, & conceptual change)*. Cambridge, MA: MIT Press.

Kuhn, D., Amsel, E., & O'Loughlin, M. (1988). *The development of scientific thinking skills*. Orlando, FL: Academic Press.

Kuhn, J. R. D. (2003, November). Graphing calculator programs for instructional data diagnostics and statistical inference. *Journal of Statistics Education, 11*(2). Retrieved October 8, 2007, from www.amstat.org/publications/jse/v11n2/kuhn.html

Lajoie, S. P. (1999). Understanding of statistics. In E. Fennema & T. A. Romberg (Eds.), *Mathematics classrooms that promote understanding* (pp. 109–132). Mahwah, NJ: Lawrence Erlbaum Associates.

Landwehr, J. M., & Watkins, A. E. (1986). *Exploring data* (The Quantitative Literacy Series). Palo Alto, CA: Dale Seymour.

Lane, D. M., & Peres, S. C. (2006). Interactive simulations in the teaching of statistics: Promise and pitfalls. In A. J. Rossman and B. L. Chance (Eds.), *Proceedings of the seventh international conference on the teaching of statistics* (ICOTS-7), Salvador, Bahia, Brazil, [CD-ROM]. Voorburg, The Netherlands: International Statistical Institute. Retrieved July 15, 2007, from http://www.stat.auckland.ac.nz/~iase/ publications/17/7D1_LANE.pdf

Lane, D. M., & Tang, Z. (2000). Effectiveness of simulation training on transfer of statistical concepts. *Journal of Educational Computing Research, 22*(4), 383–396.

Lane-Getaz, S. J. (2006). What is statistical thinking, and how is it developed? In G. F. Burrill (Ed.), *Thinking and reasoning about data and chance: Sixty-eighth NCTM Yearbook* (pp 273–289). Reston, VA: National Council of Teachers of Mathematics.

Lave, J., & Wenger, E. (1991). *Situated learning: Legitimate peripheral participation*. Cambridge, England: University of Cambridge Press.

Leavy, A. (2006, November). Using data comparisons to support a focus on distribution: Examining pre-service teachers' understandings of distribution when engaged in statistical inquiry. *Statistics Education Research Journal, 5*(2), 89–114. Retrieved July 15, 2007, from http://www.stat.auckland.ac.nz/ ~iase/serj/SERJ5(2)_Leavy.pdf

Lecoutre, M. P. (1992). Cognitive models and problem spaces in "purely random" situations. *Educational Studies in Mathematics, 23*(6), 557–568.

Ledlow, S., (2001). *Using think-pair-share in the college classroom*. Retrieved October 2, 2007, from Arizona State University, Center for Learning and Teaching Excellence Website: http://clte.asu.edu/active/usingtps.pdf

Lee, C. (2000). *Developing student-centered learning environments in the technology era – the case of introductory statistics*. Presented at the American Statistical Association (ASA) Joint Statistical Meeting, Indianapolis, Indiana.

Lee, C., Zeleke, A., & Wachtel, H. (2002). Where do students get lost? The concept of variation. In B. Phillips (Ed.), *Developing a statistically literate society: Proceedings of the sixth international conference on the teaching of statistics* (ICOTS-6), Cape Town, South Africa, [CD-ROM]. Voorburg, The Netherlands: International Statistical Institute. Retrieved September 28, 2007, from http://www.stat.auckland.ac.nz/~iase/publications/1/10_08_le.pdf

Lee, T. J. (2005). Fool me twice, shame on me. *Mathematics Teacher, 98*(8), 560–563.

Lehrer, R., & Schauble, L. (2000). Inventing data structures for representational purposes: Elementary grade students' classification models. *Mathematical Thinking and Learning*, *2*, 49–72.

Lehrer, R., & Schauble, L. (2002). Distribution: A resource for understanding error and natural variation. In B. Phillips (Ed.), *Proceedings of the sixth international conference on the teaching of statistics* (ICOTS-6), Cape Town, South Africa, [CD-ROM]. Voorburg, The Netherlands: International Statistical Institute. Retrieved September 28, 2007, from http://www.stat.auckland.ac.nz/ ~iase/publications/1/8b3_lehr.pdf

Lehrer, R., & Schauble, L. (2004). Modeling natural variation through distribution. *American Educational Research Journal*, *41*(3), 635–679.

Lehrer, R., & Schauble, L. (2007). Contrasting emerging conceptions of distribution in contexts of error and natural variation. In M. C. Lovett & P. Shah (Eds.), *Thinking with data* (pp. 149–176). Mahwah, NJ: Lawrence Erlbaum Associates.

Leon, M. R., & Zawojewski, J. S. (1993). *Conceptual understanding of the arithmetic mean*. Paper presented at the Annual Meeting of the American Educational Research Association, Atlanta GA. Retrieved April 26, 2008, from http://eric.ed.gov/ERICDocs/data/ericdocs2sql/content_storage_01/0000019b/80/13/0f/94.pdf

Lesser, L. (2007, January). Technology tips: Using graphing calculators to do statistics: A pair of problematic pitfalls. *Mathematics Teacher, 100*(5), 375.

Levin, J., & Waugh, M. (1998). Teaching teleapprenticeships: Frameworks for integrating technology into teacher education. *Interactive Learning Environments, 6*(1–2), 39–58.

Lewis, C. (2000). *Lesson study: The core of Japanese professional development.* Invited address to the Special Interest Group on Research in Mathematics Education, Annual meeting of the American Educational Research Association (AERA), New Orleans. *Retrieved April 26, 2008, from* http://lessonresearch.net/aera2000.pdf

Lewis, C., & Tsuchida, I. (1998). A lesson is like a swiftly flowing river: How research lessons improve Japanese education. *American Educator, 22*(4), 12–17, 50–52.

Lipson, K. (2002). The role of computer based technology in developing understanding of the concept of sampling distribution. In B. Phillips (Ed.), *Developing a statistically literate society: Proceedings of the sixth international conference on the teaching of statistics* (ICOTS-6), Cape Town, South Africa, [CD-ROM]. Voorburg, The Netherlands: International Statistical Institute. Retrieved September 28, 2007, from http://www.stat.auckland.ac.nz/~iase/publications/1/6c1_lips.pdf

Lipson, K., Kokonis, S., & Francis, G. (2003). Investigation of students' experiences with a Web-based computer simulation. *Proceedings of the IASE satellite conference on statistics education and the Internet*, August 2003, Berlin, [CD-ROM]. Voorburg, The Netherlands: International Statistical Institute. Retrieved November 29, 2006, from http://www.stat.auckland.ac.nz/~iase/publications/6/Lipson.pdf

Liu, Y. (2005). *Teachers' understandings of probability and statistical inference and their implications for professional development.* Unpublished Ph.D. Dissertation. Available from: http://www.stat.auckland.ac.nz/ ~iase/publications/dissertations/05.liu.Dissertation.pdf

Lovett, M. C. (2001). A collaborative convergence on studying reasoning processes: A case study in statistics. In S. Carver & D. Klahr (Eds.), *Cognition and instruction: Twenty-five years of progress* (pp. 347–384). Mahwah, NJ: Erlbaum.

Lovett, M. C., & Greenhouse, J. B. (2002) Applying cognitive theory to statistics instruction. *The American Statistician*, *54*(3), 196–206.

Lunsford, M. L., Rowell, G. H., & Goodson-Espy, T. (2006, November). Classroom research: Assessment of student understanding of sampling distributions of means and the central limit theorem in post-calculus probability and statistics classes. *Journal of Statistics*

Education, 14(3). Retrieved December 28, 2006, from http://www.amstat.org/publications/jse/v14n3/lunsford.html

MacGregor, J., Cooper, J. L., Smith, K. A., & Robinson, P. (2000). *Strategies for energizing large classes: From small groups to learning communities*. San Francisco: Jossey-Bass.

Magel, R. C. (1998, November). Using cooperative learning in a large introductory statistics class. *Journal of Statistics Education, 6*(3). Retrieved December 30, 2006, from http://www.amstat.org/publications/jse/v6n3/magel.html

Makar, K. (Ed.) (2005, July). Reasoning about distribution: A collection of current research studies. *Proceedings of the fourth international research forum on statistical reasoning, thinking, and literacy* (SRTL-4), University of Auckland, New Zealand, 2–7 July. Brisbane: University of Queensland.

Makar, K., & Confrey, J. (2002). Comparing two distributions: Investigating secondary teachers' statistical thinking. In B. Phillips (Ed.), *Proceedings of the sixth international conference on the teaching of statistics* (ICOTS-6), Cape Town, South Africa, [CD-ROM]. Voorburg, The Netherlands: International Statistical Institute. Retrieved April 20, 2008, from http://www.stat.auckland.ac.nz/~iase/ publications/1/10_18_ma.pdf

Makar, K., & Confrey, J. (2004). Secondary teachers' statistical reasoning in comparing two groups. In D. Ben-Zvi & J. Garfield (Eds.), *The challenge of developing statistical literacy, reasoning, and thinking* (pp. 353–374). Dordrecht, The Netherlands: Kluwer Academic Publishers.

Makar, K., & Confrey, J. (2005, July). Using distributions as statistical evidence in well-structured and ill-structured problems. In K. Makar (Ed.), *Reasoning about distribution: A collection of current research studies*. Proceedings of the fourth international research forum on statistical reasoning, thinking, and literacy (SRTL-4), University of Auckland, New Zealand, 2–7 July. Brisbane: University of Queensland.

Mallows, C. (1998). The zeroth problem. *The American Statistician, 52*, 1–9.

Mandler, G. (1984). *Mind and body*. New York: Norton.

Marlena, H. (2007). What students choose to do and have to say about use of multiple representations in college algebra. *Journal of Computers in Mathematics and Science Teaching, 26*(1), 27–54.

Mason, R. L., McKenzie, J. D., Jr., & Ruberg, S. J. (1990). A brief history of the American Statistical Association, 1839–1989. *American Statistician, 44*(2), 68–73.

Mathews, D., & Clark, J. (2003). *Successful students' conceptions of mean, standard deviation and the central limit theorem*. Unpublished paper. Retrieved October 20, 2007, from http://www1.hollins.edu/faculty/clarkjm/stats1.pdf

Mayer, R. E. (2001) *Multimedia learning*. New York: Cambridge University Press.

McClain, K., & Cobb, P. (2001). Supporting students' ability to reason about data. *Educational Studies in Mathematics, 45*(1–3), 103–129.

McCullough, B. D., & Wilson, B. (1999). On the accuracy of statistical procedures in Microsoft Excel 97. *Computational Statistics and Data Analysis, 31*, 27–37.

McGahan, J. R., McDougal, B., Williamson, J. D., & Pryor, P. L. (2000). The equivalence of contingency structure for intuitive covariation judgments about height, weight, and body fat. *Journal of Psychology, 134*, 325–335.

McKeachie, W. J., Pintrich, P. R., Lin, Y. G., Smith, D. A., & Sharma, R. (1990). *Teaching and learning in the college classroom: A review of the research literature* (2nd ed.). Ann Arbor, MI: University of Michigan, National Center for Research to Improve Postsecondary Teaching and Learning.

McKeachie, W. J., & Svinicki, M. (2006). *Teaching tips: Strategies, research, and theory for college and university teachers* (12th ed.). Boston: Houghton Mifflin.

McKenzie, C. R.M., & Mikkelsen, L. A. (2007). A Bayesian view of covariation assessment. *Cognitive Psychology, 54*(1), 33–61.

McLeod, D. B. (1989). The role of affect in mathematical problem solving. In D. B. McLeod, & B. M. Adams, (Eds.), *Affect and mathematical problem solving* (pp. 20–36). New York: Springer-Verlag.

McLeod, D. B. (1992). Research on affect in mathematics education: A reconceptualization. In D. A. Grows (Ed.), *Handbook of research on mathematics teaching and learning* (pp. 575–596). New York: Macmillan.

Means, B., Penuel, W., & Padilla, C. (2001). *The connected school: Technology and learning in high school.* San Francisco, CA: Jossey-Bass.

Meletiou, M., & Lee, C. (2002, December). Teaching students the stochastic nature of statistical concepts in an introductory statistics course. *Statistics Education Research Journal, 1*(2), 22–37. Retrieved July 15, 2007, from http://www.stat.auckland.ac.nz/~iase/serj/SERJ1(2).pdf

Metz, K. E. (1999). Why sampling works or why it can't: Ideas of young children engaged in research of their own design. In R. Hitt & M. Santos (Eds.), *Proceedings of the twenty-first annual meeting of the North American chapter of the international group for the psychology of education* (pp. 492–498). Columbus, OH: ERIC Clearinghouse of Science, Mathematics, and Environmental Education.

Mevarech, Z. R. (1983). A deep structure model of students' statistical misconceptions. *Educational Studies in Mathematics, 14*, 415–429.

Meyer, O., & Lovett, M. C. (2002). Implementing a computerized tutor in a statistical reasoning course: getting the big picture. In B. Phillips (Ed.), *Proceedings of the sixth international conference on the teaching of statistics* (ICOTS-6), Cape Town, South Africa, [CD-ROM]. Voorburg, The Netherlands: International Statistical Institute. Retrieved September 28, 2007, from http://www.stat.auckland.ac.nz/~iase/publications/1/10_16_me.pdf

Mickelson, W. T., & Heaton, R. M. (2004). Primary teachers' statistical reasoning about data. In D. Ben-Zvi & J. Garfield (Eds.), *The challenge of developing statistical literacy, reasoning, and thinking* (pp. 295–324). Dordrecht, The Netherlands: Kluwer Academic Publishers.

Miller, J. (2000). *The quest for the constructivist statistics classroom: Viewing practice through constructivist theory*. Unpublished Ph.D. Dissertation, Ohio State University. Retrieved December 4, 2006, from http://www.stat.auckland.ac.nz/~iase/publications/dissertations/00.Miller.Dissertation.pdf

Mills, J. D. (2002, March). Using computer simulation methods to teach statistics: A review of the literature. *Journal of Statistics Education, 10*(1). Retrieved December 28, 2006, from http://www.amstat.org/publications/jse/v10n1/mills.html

Mills, J. D. (2004). Learning abstract statistics concepts using simulation. *Educational Research Quarterly, 28*(4), 18–33.

Mittag, K. C., & Thompson, B. (2000). A national survey of AERA members' perceptions of statistical significance tests and other statistical issues. *Educational Researcher, 29*(4), 14–20.

Mokros, J., & Russell, S. J. (1995). Children's concepts of average and representativeness. *Journal for Research in Mathematics Education, 26*(1), 20–39.

Monk, S., & Nemirovsky, R. (1994). The case of Dan: Student construction of a functional situation through visual attributes. In E. Dubinsky, J. Kaput & A. Schoenfeld (Eds.), *Research in collegiate mathematics education* (Vol. 1, pp. 139–168). Providence, RI: American Mathematics Society.

Mooney, E. S. (2002). A framework for characterizing middle school students' statistical thinking. *Mathematical Thinking and Learning, 4*(1), 23–63.

Moore, D. S. (1990). Uncertainty. In L. A. Steen (Ed.), *On the shoulders of giants: A new approach to numeracy* (pp. 95–137). Washington, DC: National Academy of Sciences.

Moore, D. S. (1992). Teaching statistics as a respectable subject. In F. Gordon & S. Gordon (Eds.), *Statistics for the twenty-first century* [MAA Notes #26], (pp. 14–25). Washington, DC: Mathematical Association of America.

Moore, D. S. (1997). New pedagogy and new content: The case of statistics. *International Statistical Review, 65*(2), 123–137.

Moore, D. S. (1998). Statistics among the liberal arts. *Journal of the American Statistical Association, 93*, 1253–1259.

Moore, D. S. (1999). Discussion: What shall we teach beginners? *International Statistical Review, 67*, 250–252.

Moore, D. S. (2004). *The basic practice of statistics* (3rd ed.). New York: W. H. Freeman.

Moore, D. S. (2005). Learning from data. In R. Peck, G. Casella, G. Cobb, R. Hoerl, D. Nolan, R. Starbuck & H. Stern (Eds.), *Statistics: A guide to the unknown* (4th ed.). Belmont, CA: Thomson Publishers.

Moore, D. S., Cobb, G. W., Garfield, J., & Meeker, W. Q. (1995, August). Statistics education fin de siècle. *The American Statistician, 49*(3), 250–260.

Moore, D. S., & McCabe, G. (2005). *Introduction to the practice of statistics* (5th ed.). New York: W. H. Freeman & Co.

Moore, T. L. (Ed.) (2000). *Teaching statistics: Resources for undergraduate instructors.* (MAA Notes Number 52). Washington DC: Mathematics Association of America.

Moore-Russo, D. A., Cortes-Figueroa, J. E., & Schuman, M. J. (2006, Jan). Using a CBL unit, a temperature sensor, and a graphing calculator to model the kinetics of consecutive first-order reactions as safe in-class demonstrations. *Journal of Chemical Education, 83*(1), 64–68.

Moritz, J. B. (2004). Reasoning about covariation. In D. Ben-Zvi & J. Garfield (Eds.), *The challenge of developing statistical literacy, reasoning and thinking* (pp. 227–256). Dordrecht, The Netherlands: Kluwer Academic Publishers.

Morris, E. (1997). *An investigation of students' conceptions and procedural skills in the statistical topic correlation.* CITE Report No. 230. Centre for Information Technology in Education, Institute of Educational Technology, Walton Hall: The Open University.

Mosteller, F., & Tukey, J. W. (1977). *Data analysis and regression: A second course in statistics* (Addison-Wesley Series in Behavioral Science). Reading, MA: Addison-Wesley.

Mosteller, F., Kruskal, W. H., Pieters, R. S., & Rising, G. R. (Eds.) (1973a). *Statistics by example: Exploring data.* Reading, MA: Addison-Wesley.

Mosteller, F., Kruskal, W. H., Pieters, R. S., & Rising, G. R. (Eds.) (1973b). *Statistics by example: Weighing chances.* Reading, MA: Addison-Wesley.

Mosteller, F., Kruskal, W. H., Pieters, R. S., & Rising, G. R. (Eds.) (1973c). *Statistics by example: Detecting patterns.* Reading, MA: Addison-Wesley.

Mosteller, F., Kruskal, W. H., Pieters, R. S., & Rising, G. R. (Eds.) (1973d). *Statistics by example: Finding models.* Reading, MA: Addison-Wesley.

Mosteller, F., Rourke, R. E.K., & Thomas, G. B. (1973). *Probability with statistical applications.* Reading, MA: Addison-Wesley.

Nash, J. C., & Quon, T. K. (1996). Issues in teaching statistical thinking with spreadsheets. *Journal of Statistics Education, 4*(1). Retrieved December 4, 2006, from http://www.amstat.org/ publications/jse/v4n1/nash.html

National Council of Teachers of Mathematics (1989). *Curriculum and evaluation standards for school mathematics.* Reston, VA: National Council of Teachers of Mathematics (NCTM).

National Council of Teachers of Mathematics (1995). *Assessment standards for school mathematics.* Reston, VA: National Council of Teachers of Mathematics (NCTM).

National Council of Teachers of Mathematics (2000). *Principles and standards for school mathematics.* Reston, VA: National Council of Teachers of Mathematics (NCTM).

Nemirovsky, R. (1996). A functional approach to algebra: Two issues that emerge. In N. Dedrarg, C. Kieran & L. Lee (Eds.), *Approaches to algebra: Perspectives for research and teaching* (pp. 295–313). Boston: Kluwer Academic Publishers.

Nicholson, J., Ridgway, J., & McCusker, S. (2006). Reasoning with data – Time for a rethink? *Teaching Statistics*, 28(1), 2–9.

Nisbett, R. E. (1993). *Rules for reasoning.* Hillsdale, NJ: Erlbaum.

Noll, J. (2007). *Statistics teaching assistants' statistical content knowledge and pedagogical content knowledge.* Unpublished doctoral dissertation, Portland State University, Portland, OR. Retrieved April 20, 2008, from http://www.stat.auckland.ac.nz/~iase/publications/dissertations/07.Noll.Dissertation.pdf

O'Connell, A. A. (1999). Errors in probability problem-solving. *Educational Research and Evaluation, 5*(1), 1–21.

O'Connell, A. A., & Corter, J. E. (1993). Student misconceptions in probability problem-solving. *Proceedings of the AERA conference* (pp. 1–15). Atlanta.

O'Donnell, A. M. (2006). The role of peers and group learning. In P. A. Alexander & P. H. Winne (Eds.), *Handbook of educational psychology* (2nd ed.) (pp. 781–802). Mahwah, NJ: Lawrence Erlbaum Associates.

Oakes, M. (1986). *Statistical inference: A commentary for the social and behavioral sciences.* New York: Wiley.

Ooms, A. (2005). *Assessment in first courses of statistics: Today and tomorrow.* Paper presented at the Joint Statistical Meetings, Minneapolis, MN.

Orlandi, L. R. (1971). Evaluation of learning in secondary school social studies. In B. S. Bloom, J. T. Hastings & G. Madaus (Eds.*), Handbook on formative and summative evaluation of student learning*. New York: McGraw Hill.

Ormell, C. P. (1974). Bloom's taxonomy and the objectives of education. *Educational Research, 17*, 3–18.

Osvaldo Fabbroni, E., Chang Lee, J., Lievesley, D., Martin-Guzman, P., Morettin, P. A., & Rossman, A. J. (2007, July). *The challenges for cooperation in statistics education.* Panel discussion at the seventh International conference on the teaching of statistics (ICOTS-7), Salvador, Bahia, Brazil.

Palincsar, A. S., & Brown, A. L. (1984). Reciprocal teaching of comprehension-fostering and comprehension-monitoring activities. *Cognition and Instruction*, *2*(2), 117–175.

Pea, R. D. (1987). Cognitive technologies for mathematics education. In A. H. Schoenfeld (Ed.), *Cognitive science and mathematics education* (pp. 89–122). Hillsdale, NJ: Lawrence Erlbaum Associates.

Peck, R., & Chance, B. L. (2005). Assessing effectiveness at the program level: Undergraduate statistics program evaluation. Proceedings of the 2005 Joint Statistical Meetings, Minneapolis, MN.

Peck, R., Casella, G., Cobb, G., Hoerl, R., Nolan, D., Starbuck, R., et al. (Eds.). (2006). *Statistics: A guide to the unknown* (4th ed.). Belmont, CA: Duxbury Press.

Peck, R., Olsen, C., & Devore, J. L. (2007). *Introduction to statistics and data analysis* (3rd ed.). Belmont, CA: Duxbury Press.

Pellegrino, J. W., Chudowsky, N., Glaser, R. (Eds.) (2001). *Knowing what students know: The science and design of educational assessment.* Washington, DC: National Academy Press.

Pereira-Mendoza, L., Kea, L. S., Kee, T. W., & Wong, W. K. (Eds.) (1998). *Proceedings of the fifth international conference on the teaching of statistics* (ICOTS-5), Singapore. Voorburg, The Netherlands: International Statistical Institute. Retrieved December 4, 2006, from http://www.stat.auckland.ac.nz/ ~iase/publications.php?show=2

Perkins, D. V., & Saris, R. N. (2001). A "jigsaw classroom" technique for undergraduate statistics courses. *Teaching of Psychology, 28,* 111–113.

Peterson, C. (1980). Recognition of noncontingency. *Journal of Personality and Social Psychology, 38*, 727–734.

Petrosino, A. J., Lehrer, R., & Schauble, L. (2003). Structuring error and experimental variation as distribution in the fourth grade. *Mathematical Thinking and Learning, 5*, 131–156.

Pfannkuch, M. (2005a, July). Informal inferential reasoning: A case study. In K. Makar (Ed.), *Reasoning about distribution: a collection of research studies.* proceedings of the fourth international research forum on statistical reasoning, thinking, and literacy (SRTL-4), University of Auckland, New Zealand, 2–7 July. Brisbane: University of Queensland, [CD-ROM].

Pfannkuch, M. (2005b). Probability and statistical inference: How can teachers enable learners to make the connection? In G. A. Jones (Ed.), *Exploring probability in school: Challenges for teaching and learning* (pp. 267–294). New York: Springer.

Pfannkuch, M. (2006a). Informal inferential reasoning. In A. J. Rossman & B. L. Chance (Eds.), *Proceedings of the seventh international conference on the teaching of statistics* (ICOTS-7), Salvador, Bahia, Brazil, [CD-ROM]. Voorburg, The Netherlands: International Statistical Institute. Retrieved July 15, 2007, from http://www.stat.auckland.ac.nz/~iase/ publications/17/6A2_PFAN.pdf

Pfannkuch, M. (2006b, November). Comparing box plot distributions: A teacher's reasoning. *Statistical Education Research Journal, 5*(2), 27–45. Retrieved July 15, 2007, from http://www.stat.auckland.ac.nz/~iase/ serj/SERJ5(2)_Pfannkuch.pdf

Pfannkuch, M., & Reading, C. (2006, November). Reasoning about distribution: A complex process. *Statistical Education Research Journal, 5*(2), 4–9. Retrieved July 15, 2007, from http://www.stat.auckland.ac.nz/~iase/serj/SERJ5(2).pdf

Phillips, B. (Ed.). (2002). Developing a statistically literate society. *Proceedings of the sixth international conference on the teaching of statistics* (ICOTS-6), Cape Town, South Africa, [CD-ROM]. Voorburg, The Netherlands: International Statistical Institute. Retrieved December 4, 2006, from http://www.stat.auckland.ac.nz/~iase/publications.php?show=1

Phillips, D. C., & Kelly, M. E. (1975). Hierarchical theories of development in education and psychology. *Harvard Educational Review, 45*, 351–375.

Piaget, J. (1978). *Success and understanding*. Cambridge MA: Harvard University Press.

Piaget, J., & Inhelder, B. (1951). *La génese de l'idée de hasard chez l'enfant.* Paris: Presses Universitaires de France.

Piaget, J., & Inhelder, B. (1975). *The origin of the idea of chance in children*. London: Routledge & Kegan Paul.

Pollatsek, A., Konold, C., Well, A. D., & Lima, S. (1984). Beliefs underlying random sampling. *Memory and Cognition, 12*, 395–401.

Pollatsek, A., Lima, S., & Well, A. D. (1981). Concept or computation: Students' misconceptions of the mean. *Educational Studies in Mathematics, 12*, 191–204.

Pollatsek, A., Well, A. D., Konold, C., & Hardiman, P. (1987). Understanding conditional probabilities. *Organizational Behavior and Human Decision Processes, 40*, 255–269.

Poole, R. L. (1971). Characteristics of the taxonomy of educational objectives: Cognitive domain. *Psychology in the Schools, 8*, 379–385.

Poole, R. L. (1972). Characteristics of the taxonomy of educational objectives: Cognitive domain–A replication. *Psychology in the Schools, 9*, 83–88.

Posner, G. J., Strike, K. A., Hewson, P. W., & Gertzog, W. A. (1982). Accommodation of a scientific conception: Toward a theory of conceptual change. *Science Education, 66*(2), 211–227.

Potthast, M. J. (1999). Outcomes of using small-group cooperative learning experiences in introductory statistics courses. *College Student Journal, 33*, 34–42.

Pratt, D. (2007). How do teachers foster students' understanding of probability? In G. A. Jones (Ed.), *Exploring probability in school: Challenges for teaching and learning* (pp. 182–201). The Netherlands: Kluwer Academic Publishers.

Prodromou, T., & Pratt, D. (2006, November). The role of causality in the co-ordination of two perspectives on distribution within a virtual simulation. *Statistical Education Research Journal, 5*(2), 69–88. Retrieved July 15, 2007, from http://www.stat.auckland.ac.nz/~iase/serj/SERJ5(2).pdf

Quilici, J. L., & Mayer, R. E. (2002). Teaching students to recognize structural similarities between statistics word problems. *Applied Cognitive Psychology, 16*(3), 325–342.

Ramsey, F. L., & Schafer, D. W. (2002). *The statistical sleuth, a course in methods of data analysis* (2nd ed.). Belmont, CA: Duxbury Press.

Rau, W., & Heyl, B. (1990). Humanizing the college classroom: Collaborative learning and social organization among students. *Teaching Sociology, 18*, 141–155.

Reading, C. (2004, November). Student description of variation while working with weather data. *Statistics Education Research Journal, 3*(2), 84–105. Retrieved July 15, 2007, from http://www.stat.auckland.ac.nz/~iase/serj/SERJ3(2)_Reading.pdf

Reading, C., & Reid, J. (2006, November). An emerging hierarchy of reasoning about distribution: From a variation perspective. *Statistical Education Research Journal, 5*(2), 46–68. Retrieved July 15, 2007, from http://www.stat.auckland.ac.nz/~iase/serj/SERJ5(2)_reading_reid.pdf

Reading, C., & Shaughnessy, J. M. (2004). Reasoning about variation. In D. Ben-Zvi & J. Garfield (Eds.), *The challenge of developing statistical literacy, reasoning, and thinking* (pp. 201–226). Dordrecht, The Netherlands: Kluwer Academic Publishers.

Resnick, L. (1988). Treating mathematics as an ill-structured discipline. In R. Charles & E. Silver (Eds.), *The teaching and assessing of mathematical problem solving* (pp. 32–60). Reston, VA: National Council of Teachers of Mathematics.

Ritchie, D. (1996). The administrative role in integration of technology. *Bulletin of the National Association of Secondary School Principals, 80*(582), 42–52.

Roback, P., Chance, B. L., Legler, J., & Moore, T. L. (2006, July). Applying Japanese lesson study principles to an upper-level undergraduate statistics course. *Journal of Statistics Education, 14*(2). Retrieved December 28, 2006, from http://www.amstat.org/publications/jse/v14n2/roback.html

Rohrbeck, C. A., Ginsburg-Block, M. D., Fantuzzo, J. W., & Miller, T. R. (2003). Peer-assisted learning interventions with elementary school students: A meta-analytic review. *Journal of Educational Psychology, 95,* 240–257.

Rosenshine, B., Meister, C., & Chapman, S. (1996). Teaching students to generate questions: A review of the intervention studies. *Review of Educational Research 66*(2), 181–221.

Roseth, C. J., Johnson, D. W., & Johnson, R. T. (2008). Promoting early adolescents' achievement and peer relationships: The effects of cooperative, competitive and individualistic goal structures. *Psychological Bulletin, 134*, 223-246.

Ross, J. A., & Cousins, J. B. (1993). Patterns of student growth in reasoning about correlational problems. *Journal of Educational Psychology, 85*, 49–65.

Rossman, A. J., & Chance, B. L. (1999). Teaching the reasoning of statistical inference: A "top ten" list. *College Mathematics Journal, 30*(4), 297–305. Retrieved November 29, 2006, from http://rossmanchance.com/papers/topten.html

Rossman, A. J., & Chance, B. L. (2001). *Workshop statistics: Discovery with data, second edition.* Emeryville, CA: Key College Publishing.

Rossman, A. J., & Chance, B. L. (2002). A data-oriented, active-learning, post-calculus introduction to statistical concepts, applications, and theory. In B. Phillips (Ed.), *Proceedings of the sixth international conference on the teaching of statistics* (ICOTS-6), Cape Town, South Africa, [CD-Rom]. Voorburg, The Netherlands: International Statistical Institute. Retrieved September 28, 2007, from http://www.stat.auckland.ac.nz/~iase/publications/1/3i2_ross.pdf

Rossman, A. J., & Chance, B. L. (2004). *INsight into Statistical Practice, Instruction and REasoning (INSPIRE) 2004 workshop online materials.* Retrieved December 6, 2006, from http://statweb.calpoly.edu/chance/inspire/

Rossman, A. J., & Chance, B. L. (2005). *Workshop statistics student toolkit.* Emeryville, CA: Key College Publishing.

Rossman, A. J., & Chance, B. L. (Eds.) (2006). Working cooperatively in statistics education. *Proceedings of the seventh international conference on the teaching of statistics* (ICOTS-7), Salvador, Bahia, Brazil, [CD-ROM]. Voorburg, The Netherlands: International Statistical Institute. Retrieved December 4, 2006, from http://www.stat.auckland.ac.nz/~iase/publications.php?show=17

Rossman, A. J., Chance, B. L., & Lock, R. (2001). *Workshop statistics: Discovery with data and Fathom.* Emeryville, CA: Key College Publishing.

Rossman, A. J., Chance, B. L., & Medina, E. (2006). Some important comparisons between statistics and mathematics, and why teachers should care. In G. F. Burrill (Ed.), *Thinking and reasoning about data and chance: Sixty-eighth NCTM Yearbook* (pp. 323–334). Reston, VA: National Council of Teachers of Mathematics.

Rossman, A. J., & Garfield, J. (2001). *Teaching introductory statistics: Activities, technology, assessment.* Handbook for Eastern Kentucky University Workshop.

Rubin, A. (2001). *Visualizing statistical relationships.* Project funded by Research on Learning and Education, National Science Foundation.

Rubin, A., & Bruce, B. (1991). Using computers to support students' understanding of statistical inference. *New England Mathematics Journal, 23*, 9–20.

Rubin, A., Bruce, B., & Tenney, Y. (1991). Learning about sampling: Trouble at the core of statistics. In D. Vere-Jones (Ed.), *Proceedings of the third international conference on the teaching of statistics* (ICOTS-3), Dunedin, New Zealand (Vol. 1, pp. 314–319). Voorburg, The Netherlands: International Statistical Institute. Retrieved September 28, 2007, from http://www.stat.auckland.ac.nz/~iase/publications/18/BOOK1/A9–4.pdf

Rubin, A., Hammerman, J. K., Campbell, C., & Puttick, G. (2005a, July). The effect of distributional shape on group comparison strategies (SRTL-4). In K. Makar (Ed.), *Reasoning about distribution: A collection of current research studies.* Proceedings of the fourth international research forum on statistical reasoning, thinking, and literacy (SRTL-4), University of Auckland, New Zealand, 2–7 July. Brisbane: University of Queensland.

Rubin, A., Hammerman, J. K., & Konold, C. (2006). Exploring informal inference with interactive visualization software. In A. J. Rossman & B. L. Chance (Eds.), *Proceedings of the seventh international conference on the teaching of statistics* (ICOTS-7), Salvador, Bahia, Brazil, [CD-ROM]. Voorburg, The Netherlands: International Statistical Institute. Retrieved July 15, 2007, from http://www.stat.auckland.ac.nz/~iase/publications/17/2D3_RUBI.pdf

Rumsey, D. J. (1998). A cooperative teaching approach to introductory statistics. *Journal of Statistics Education, 6*(1). Retrieved December 28, 2006, from http://www.amstat.org/publications/jse/v6n1/rumsey.html

Rumsey, D. J. (2002, November). Statistical literacy as a goal for introductory statistics courses. *Journal of Statistics Education, 10*(3). Retrieved July 15, 2007, from http://www.amstat.org/publications/jse/v10n3/rumsey2.html

Russell, S. J. (1990). Issues in training teachers to teach statistics in the elementary school: A world of uncertainty. In A. Hawkins (Ed.), *Training teachers to teach statistics: Proceedings of the international statistical institute roundtable conference* (pp. 59–71). Voorburg, The Netherlands: International Statistical Institute.

Russell, S. J., & Mokros, J. (1996). What do children understand about average? *Teaching Children Mathematics, 2,* 360–364.

Saldanha, L. A., & Thompson, P. W. (2002). Conceptions of sample and their relationship to statistical inference. *Educational Studies in Mathematics, 51*(3), 257–270.

Saldanha, L. A., & Thompson, P. W. (2003). Conceptions of sample and their relationship to statistical inference. *Educational Studies in Mathematics, 51*, 257–270.

Sánchez, E. S. (2002). Teachers' beliefs about usefulness of simulations with the educational software Fathom for developing probability concepts in statistics classroom. In B. Phillips (Ed.), *Proceedings of the sixth international conference on the teaching of statistics* (ICOTS-6), Cape Town, South Africa, [CD-ROM]. Voorburg, The Netherlands: International Statistical Institute. Retrieved September 28, 2007, from http://www.stat.auckland.ac.nz/~iase/publications/1/6e2_sanc.pdf

Sánchez, F. T. (1999). Significado de la correlación y regression para los estudiantes universitarios [Meanings of correlation and regression for undergraduates]. Unpublished doctoral dissertation, University of Granada.

Sax, G., Eilenberg, E. G., & Klockars, A. J. (1972). Achievement as a function of test item complexity and difficulty. *Journal of Educational Psychology, J 6*, 89–103.

Scardamalia. M., & Bereiter, C. (1985). Fostering the development of self-regulation in children's knowledge processing. In S. F. Chipman, J. W. Segal & R. Glaser (Eds.), *Thinking and learning skills: Research and open questions.* Hillsdale, NJ: Lawrence Erlbaum Associates.

Schafer, D. W., & Ramsey, F. L. (2003, March). Teaching the craft of data analysis. *Journal of Statistics Education Volume, 11*(1). Retrieved December 4, 2006, from http://www.amstat.org/publications/jse/v11n1/schafer.html

Schau, C., & Mattern, N. (1997a). Assessing students' connected understanding of statistical relationships. In I. Gal & J. B. Garfield (Eds.), *The assessment challenge in statistics education* (pp. 91–104). Amsterdam, Netherlands: IOS Press. Retrieved April 27, 2008, from http://www.stat.auckland.ac.nz/ ~iase/publications/assessbk/chapter08.pdf

Schauble, L. (1996). The development of scientific reasoning in knowledge-rich contexts. *Developmental Psychology, 32*(1), 102–119.

Scheaffer, R. L. (1990). The ASA-NCTM quantitative literacy project: An overview. In D. Vere-Jones (Ed.), *Proceedings of the third international conference on the teaching of statistics* (ICOTS-3), Dunedin, New Zealand (Vol. 1, pp. 45–49). Dunedin, New Zealand. Retrieved December 12, 2006, from http://www.stat.auckland.ac.nz/~iase/publications/18/BOOK1/A1–2.pdf.

Scheaffer, R. L. (1992). Data, discernment and decisions: An empirical approach to introductory statistics. In F. Gordon & S. Gordon (Eds.), *Statistics for the twenty-first century* (pp. 69–82). Washington, DC: The Mathematical Association of America.

Scheaffer, R. L. (2000). NSF-funded workshop points the way to undergraduate symposium. *Amstat News, 276*, 18–19.

Scheaffer, R. L. (2001, Winter). Statistics education: Perusing the past, embracing the present, and charting the future. *Newsletter for the section on statistical education, 7*(1). Retrieved November 6, 2006, from http://www.amstat. org/sections/educ/newsletter/v7n1/Perusing.html

Scheaffer, R. L. (2007). *Using statistics effectively in mathematics education research: A report from a series of workshops organized by the American statistical association.* Available: http://www.amstat.org/research_grants/pdfs/ SMERReport.pdf

Scheaffer, R. L., & Stasney, E. A. (2004). The state of undergraduate education in statistics: A report from the CBMS 2000. *The American Statistician, 56*(4), 165–271.

Scheaffer, R. L., Watkins, A. E., Witmer, J., & Gnanadesikan, M., (2004a). *Activity-based statistics: Instructor resources* (2nd ed., revised by Tim Erickson). Emeryville, CA: Key College Publishing.

Scheaffer, R. L., Watkins, A. E., Witmer, J., & Gnanadesikan, M., (2004b). *Activity-based statistics: Student guide* (2nd ed., revised by Tim Erickson). Emeryville, CA: Key College Publishing.

Scheaffer, R. L., Watkins, A. E., & Landwehr, J. M. (1998). What every high-school graduate should know about statistics. In S. P. Lajoie (Ed.), *Reflections on statistics: Learning, teaching, and assessment in grades K-12* (pp. 3–31). Mahwah, NJ: Lawrence Erlbaum Associates.

Schoenfeld, A. H. (1983). Episodes and executive decisions in mathematical problem solving skills. In R. Lesh & M. Landau (Eds.), *Acquisition of mathematical concepts and processes* (pp. 345–395). New York: Academic Press.

Schoenfeld, A. H. (1987). What's all the fuss about metacognition? In A. H. Schoenfeld (Ed.), *Cognitive science and mathematics education* (pp. 189–215). Hillsdale, NJ: Lawrence Erlbaum Associates.

Schoenfeld, A. H. (1992). Learning to think mathematically: Problem solving, metacognition, and sensemaking in mathematics. In D. Grouws. (Ed). *Handbook of research on mathematics teaching and learning*. New York: MacMillan.

Schwartz, D. L., Goldman, S. R., Vye, N. J., & Barron, B. J. (1998). Aligning everyday and mathematical reasoning: The case of sampling assumptions. In S. P. Lajoie (Ed.), *Reflections on statistics: Learning, teaching, and assessment in grades K-12* (pp. 233–273). Mahwah, NJ: Lawrence Erlbaum.

Schwartz, D. L., Sears, D., & Chang, J. (2007). Reconsidering prior knowledge. In M. C. Lovett & P. Shah (Eds.), *Thinking with data* (pp. 319–344). Mahwah, NJ: Lawrence Erlbaum Associates. Retrieved April 26, 2008, from http://aaalab.stanford.edu/papers/ Schwartz_Reconsidering_Prior_K.pdf

Seddon, G. M. (1978). The properties of Bloom's taxonomy of educational objectives for the cognitive domain. *Review of Educational Research, 48*(2), 303–323.

Sedlmeier, P. (1999). *Improving statistical reasoning: Theoretical models and practical implications*. Hillsdale, NJ: Erlbaum.

Sedlmeier, P., & Gigerenzer, G. (1997). Intuitions about sample size: The empirical law of large numbers. *Journal of Behavioral Decision Making, 10*, 33–51.

Seier, E., & Robe, C. (2002). Ducks and green – an introduction to the ideas of hypothesis testing. *Teaching Statistics, 24*(3), 82–86.

Seggie, J., & Endersby, H. (1972). The empirical implications of Piaget's concept of correlation. *Australian Journal of Psychology, 24*, 3–8.

Sfard, A. (1991). On the dual nature of mathematical conceptions: Reflections on processes and objects as different sides of the same coin. *Education Studies in Mathematics, 22*(1), 1–36.

Shaughnessy, J. M. (1977). Misconceptions of probability: An experiment with a small-group, activity-based, model building approach to introductory probability at the college level. *Educational Studies in Mathematics, 8*, 285–316.

Shaughnessy, J. M. (1992). Research in probability and statistics: Reflections and directions. In D. Grouws (Ed.), *Handbook of research on mathematics teaching and learning* (pp. 465–494). New York: Macmillan.

Shaughnessy, J. M. (1997). Missed opportunities in research on the teaching and learning of data and chance. In F. Biddulph & K. Carr (Eds.), *People in mathematics education* (Proceedings of the 20th annual meetings of the mathematics education research group of Australasia) (pp. 6–22). Rotorua, New Zealand: MERGA.

Shaughnessy, J. M. (2003). Research on students' understanding of probability. In J. Kilpatrick, W. G. Martin & D. Schifter (Eds.), *A research companion to principles and standards for school mathematics* (pp. 216–226). Reston, VA: National Council of Teachers of Mathematics.

Shaughnessy, J. M. (2007). Research on statistics learning and reasoning. In F. K. Lester (Ed.), *The second handbook of research on mathematics* (pp. 957–1010). Information Age Pub Inc.

Shaughnessy J. M., & Chance, B. L. (2005). *Statistical questions from the classroom.* Reston, VA: National Council of Teachers of Mathematics.

Shaughnessy, J. M., Ciancetta, M., Best, K., & Canada, D. (2004, April). *Students' attention to variability when comparing distributions.* Paper presented at the research pre-session of the 82nd annual meeting of the National Council of Teachers of Mathematics (NCTM), Philadelphia, PA.

Shaughnessy, J. M., & Dick, T. P. (1991). Monty's dilemma: Should you stick or switch? *The Mathematics Teacher, 84*, 252–256.

Shaughnessy, J. M., Garfield, J., & Greer, B. (1996). Data handling. In A. J. Bishop, K. Clements, C. Keitel, J. Kilpatrick & C. Laborde (Eds.), *International handbook of mathematics education* (Vol. 1, pp. 205–237). Dordrecht, Netherlands: Kluwer.

Shaughnessy, J. M., Watson, J. M., Moritz, J. B., & Reading C. (1999). *School mathematics students' acknowledgment of statistical variation: There's more to life than centers.* Paper presented at the research pre-session of the 77th annual meeting of the National Council of Teachers of Mathematics (NCTM), San Francisco, CA.

Short, T. H., & Pigeon, J. G. (1998, March). Protocols and pilot studies: Taking data collection projects seriously. *Journal of Statistics Education, 6*(1). Retrieved December 6, 2006, from http://www.amstat.org/publications/jse/v6n1/short.html

Simon, J. L. (1994, November). What some puzzling problems teach about the theory of simulation and the use of resampling. *The American Statistician, 48*(4), 290–293.

Simon, J. L., Atkinson, D., & Shevokas, C. (1976). Probability and statistics: Experimental results of a radically different teaching method. *American Mathematical Monthly, 83*(9), 733–739.

Simon, M. A. (1995). Reconstructing mathematics pedagogy from a constructivist perspective. *Journal for Research in Mathematics Education, 26*(2), 114–145.

Slavin, R. E. (1980). Cooperative learning. *Review of Educational Research, 50,* 315–342.

Slavin, R. E. (1983). When does cooperative learning increase student achievement? *Psychological Bulletin, 94,* 429–445.

Slavin, R. E. (1995). *Cooperative learning* (2nd ed.). Boston: Allyn & Bacon.

Smith, G. (1998, November). Learning statistics by doing statistics. *Journal of Statistics Education, 6*(3). Retrieved November 6, 2006, from http://www.amstat.org/publications/jse/v6n3/smith.html

Smith, K. A. (2000). Going deeper: Formal small-group learning in large classes. *New directions for teaching and learning, 81,* 25–46.

Smith, M. H. (2004, July). A sample/population size activity: Is it the sample size of the sample as a fraction of the population that matters?. *Journal of Statistics Education, 12*(2). Retrieved November 12, 2006, from http://www.amstat. org/publications/jse/v12n2/smith.html

Smith, R., Clark, T., & Blomeyer, R. L. (2005). A synthesis of new research on K–12 online learning. Retrieved March 20, 2007, from http://www.ncrel. org/tech/synthesis/

Snedecor, G. W. (1937). *Statistical methods.* Ames, Iowa: Collegiate Press.

Snell, J. L. (1999). *Using chance media to promote statistical literacy.* Paper presented at the 1999 Joint Statistical Meetings, Dallas, TX.

Snell, J. L., Peterson, B., Moore, T. L., & Garfield, J. (1998). *Teaching chance: An instructor guide for teaching a chance course*. Hanover, NH: Dartmouth College. Retrieved October 21, 2007, from http://www.dartmouth.edu/ ~chance/teaching_aids/teaching_aids.html

Spurrier, J. D., Edwards, D. G., & Thombs, L. A. (1995). Elementary statistics laboratory manual, DOS version. Belmont, CA: Wadsworth.

Stanley, J. C., & Bolton, D. T. (1957). A review of Bloom's "Taxonomy of educational objectives" and J. R. Gerberich's "Specimen objective test items, a guide to achievement test construction." *Educational and Psychological Measurement, 17*, 631–634.

Starkings, S. (1997). Assessing student projects. In I. Gal & J. B. Garfield (Eds.), *The assessment challenge in statistics education* (pp. 139–151). Netherlands: IOS Press. Retrieved April 27, 2008, from http://www.stat.auckland.ac.nz/ ~iase/publications/assessbk/chapter11.pdf

Steffe, L. P., & Thompson, P. W. (2000). Teaching experiment methodology: Underlying principles and essential elements. In A. E. Kelly & R. A. Lesh (Eds.), *Handbook of research design in mathematics and science education* (pp. 267–307). Hillsdale, NJ: Lawrence Erlbaum Associates, Publishers.

Steinberg, L., Brown, B., & Dornbusch, S. (1996). *Beyond the classroom: Why school reform has failed and what parents need to do*. New York: Simon and Schuster.

Stigler, J. W., & Hiebert, J. (1999). *The teaching gap: Best ideas from the world's teachers for improving education in the classroom*. New York, NY: Summit Books.

Stigler, S. M. (1986). *The history of statistics: the measurement of uncertainty before 1900*. Cambridge, MA: Harvard University Press.

Stockburger, D. W. (1982). Evaluation of three simulation exercises in an introductory statistics course. *Contemporary Educational Psychology, 7*(4), 365–370.

Stohl, H., & Tarr, J. E. (2002). Developing notions of inference with probability simulation tools. *Journal of Mathematical Behavior, 21*(3), 319–337.

Stohl, H. (2005). Probability in teacher education and development. In G. A. Jones (Ed.), *Exploring probability in school: Challenges for teaching and learning* (pp. 345–366). NY: Springer.

Strauss, S. (1987). Educational-developmental psychology: Some case studies. In L. Liben (Ed.), *Development and learning: Conflict or convergence?* (pp. 133–157). Hillsdale, NJ: Erlbaum.

Strauss, S., & Bichler, E. (1988). The development of children's concepts of the arithmetic average. *Journal for Research in Mathematics Education, 19*(1), 64–80.

Symanzik, J., & Vukasinovic, N. (2002). Teaching statistics with electronic textbooks. In W. Härdle and B. Rönz (Eds.), *COMPSTAT 2002: Proceedings in computational statistics* (pp. 79–90). Heidelberg: Physica-Verlag.

Symanzik, J., & Vukasinovic, N. (2003). Comparative review of ActivStats, CyberStats, and MM*Stat. *MSOR (Maths, Stats & Operational Research) Connections, 3*(1), 37–42. Retrieved December 18, 2006, from http://ltsn.mathstore.ac.uk/newsletter/feb2003/pdf/activstatscyberstatsmmstat.pdf

Symanzik, J., & Vukasinovic, N. (2006, March). Teaching an introductory statistics course with CyberStats, an electronic textbook. *Journal of Statistics Education, 14*(1). Retrieved December 18, 2006, from http://www.amstat. org/publications/jse/v14n1/symanzik.html

Tanur, J. M., Mosteller, F., Kruskal, W. H., Link, R. F., Pieters, R. S., Rising, G. R., & Lehmann, E. L., (Eds.) (1972). *Statistics: A guide to the unknown* (3rd ed.). Duxbury Press.

Tarr, J. E., & Lannin, J. K. (2005). How can teachers build notions of conditional probability and independence? In G. A. Jones (Ed.), *Exploring probability in school: Challenges for teaching and learning* (pp. 215–238). NY: Springer.

Tempelaar, D. T., Gijselaers, W. H., & van der Loeff, S. S. (2006, March). Puzzles in statistical reasoning. *Journal of Statistics Education, 14*(1). Retrieved July 15, 2007, from http://www.amstat.org/publications/jse/v14n1/tempelaar.html

The Design-Based Research Collective (2003). Design-based research: An emerging paradigm for educational inquiry. *Educational Researcher, 32*(1), 5–8.

Thistead, R. A., & Velleman, P. F. (1992). Computers and modern statistics. In D. Hoaglin & D. S. Moore (Eds.), *Perspectives on contemporary statistics* (pp. 41–53). Washington, DC: Mathematics Association of America.

Thompson, C. L., & Zeuli, J. S. (1999). The frame and the tapestry: Standards-based reform and professional development. In L. Darling-Hammond & G. Sykes (Eds.), *Teaching as the learning profession: handbook of policy and practice* (pp. 341–375). San Francisco: Jossey-Bass.

Thompson, P. W. (1994). Images of rate and operational understanding of the fundamental theorem of calculus. *Educational Studies in Mathematics, 26*, 229–274.

Thompson, P. W., Saldanha, L. A., Liu, Y. (2004). *Why statistical inference is hard to understand.* Paper presented at the Annual Meeting of the American Educational Research Association, San Diego.

Truran, J. M. (1997). Understanding of association and regression by first year economics students from two different countries as revealed in responses to the same examination questions. In J. Garfield & J. M. Truran (Eds.), *Research papers on stochastics educations from 1997* (pp. 205–212). Minneapolis, MN: University of Minnesota.

Tudor G. E. (2006). Teaching introductory statistics online – satisfying the students. *Journal of Statistics Education, 14*(3). Retrieved December 18, 2006, from www.amstat.org/publications/jse/v14n3/tudor.html

Tukey, J. W. (1977). *Exploratory data analysis.* Reading, Mass: Addison-Wesley.

Tversky, A., & Kahneman, D. (1971). Belief in the law of small numbers. *Psychological Bulletin, 76*, 105–110. (Reprinted in D. Kahneman, P. Slovic & A. Tversky (1982) *Judgment under uncertainty: Heuristics and biases*. Cambridge University Press.)

U. S. Congress, Office of Technology Assessment (1995, April). *Teachers and technology: Making the connection*. OTA-EHR-616. Washington DC: US Government Printing Office. Retrieved April 26, 2008, from http://www.coedu.usf.edu/ itphdsem/eme7938/ota9541.pdf

Utts, J. (2003). What educated citizens should know about statistics and probability. *The American Statistician, 57*(2), 74–79.

Utts, J., Sommer, B., Acredolo, C., Maher, M. W., & Matthews, H. R. (2003). A study comparing traditional and hybrid internet-based instruction in introductory statistics classes. *Journal of Statistics Education, 11*(3). Retrieved December 18, 2006, from http://www.amstat.org/publications/jse/v11n3/utts.html

Utts, J. (2004). *Seeing through statistics* (3rd ed.). Duxbury Press.

Vallecillos, J. A., & Holmes, P. (1994). Students' understanding of the logic of hypothesis testing. In the *Proceedings of the fourth international conference on the teaching of statistics* (ICOTS-4). Marrakech, Morocco: International Statistical Institute.

Velleman, P. F. (1998). *Learning data analysis with Data Desk.* Reading, MA: Addison-Wesley.

Velleman, P. F., & Hoaglin, D. C. (1981). *Applications, basics, and computing of exploratory data analysis*. Boston, MA: Duxbury Press.

Verzani, J. (2005). *Using R for introductory statistics*. Chapman & Hall/CRC.

Vygotsky, L. (1978). *Mind in society*. Cambridge, MA: Harvard University Press.

Vygotsky, L. (1986). *Thought and language*. Cambridge, MA: The MIT Press.

Walvoord, B. F. (1986). *Helping students write well: A guide for teachers in all disciplines.* (2nd ed.) New York: Modern Language Association.

Ward, B. (2004, November). The best of both worlds: a hybrid statistics course. *Journal of Statistics Education, 12*(3), 74–79. Retrieved July 15, 2007, from http://www.amstat.org/publications/jse/v12n3/ward.html

Ward, W., & Jenkins, H. (1965). The display of information and the judgment of contingency. *Canadian Journal of Experimental Psychology, 19*, 231–241.

Watkins, A. E., Scheaffer, R. L., & Cobb, G. W. (2004). *Statistics in action: Understanding a world of data*. Emeryville, CA: Key Curriculum Press.

Watson, J. M. (2004). Developing reasoning about samples. In D. Ben-Zvi & J. Garfield (Eds.), *The challenge of developing statistical literacy, reasoning and thinking* (pp. 277–294). Dordrecht, The Netherlands: Kluwer Academic Publishers.

Watson, J. M. (2005, July). Developing an awareness of distribution. In K. Makar (Ed.), *Reasoning about distribution: A collection of research studies*. Proceedings of the fourth international research forum on statistical reasoning, thinking, and literacy (SRTL-4), University of Auckland, New Zealand, 2–7 July, 2005. Brisbane: University of Queensland, [CD-ROM].

Watson, J. M. (2006). *Statistical literacy at school: Growth and goals* (Studies in mathematical thinking and learning series). Mahwah, New Jersey: Lawrence Erlbaum Associates.

Watson, J. M., & Callingham, R. (2003). Statistical literacy: A complex hierarchical construct. *Statistics Education Research Journal, 2*, 3–46. Retrieved April 26, 2008, from http://www.stat.auckland.ac.nz/ ~iase/serj/SERJ2(2)_Watson_Callingham.pdf

Watson, J. M., Kelly, B. A., Callingham, R. A., & Shaughnessy, J. M. (2003). The measurement of school students' understanding of statistical variation. *International Journal of Mathematical Education in Science and Technology, 34*, 1–29.

Watson, J. M., & Moritz, J. B. (1999). The beginning of statistical inference: Comparing two data sets. *Educational Studies in Mathematics, 37*(2), 145–168.

Watson, J. M., & Moritz, J. B. (2000a). Developing concepts of sampling. *Journal for Research in Mathematics Education, 31*, 44–70.

Watson, J. M., & Moritz, J. B. (2000b). Development of understanding of sampling for statistical literacy. *Journal of Mathematical Behavior, 19*(1), 109–136.

Watson, J. M., & Moritz, J. B. (2000c). The longitudinal development of understanding of average. *Mathematical Thinking and Learning, 2*, 11–50.

Watson, J. M., & Moritz, J. B. (2000d). The development of concepts of average. *Focus on Learning Problems in Mathematics, 21*, 15–39

Watson, J. M., & Shaughnessy, J. M. (2004). Proportional reasoning: Lessons from research in data and chance. *Mathematics Teaching in the Middle School, 10*, 104–109.

Well, A. D., Pollatsek, A., & Boyce, S. J. (1990). Understanding the effects of sample size on the variability of the mean. *Journal of Organizational Behavior and Human Decision Processes, 47*, 289–312.

Wertsch, J. V. (1991). *Voices of the mind: A sociocultural approach to mediated action*. Cambridge, MA: Harvard University Press.

Wertsch, J. V. (1998). *Mind as action*. New York: Oxford University Press.

West, W. R., Wu, Y., & Heydt, D. (2004). An introduction to StatCrunch 3.0. *Journal of Statistical Software, 9*(5).

Wetzel, N. (2004, August). *Class data collection and analysis – assessment*. Paper presented at the ARTIST Roundtable Conference on Assessment in Statistics. Available from http://www.rossmanchance.com/artist/proceedings/wetzel.pdf

White, B. C., & Frederiksen, J. (1998). Inquiry, modeling, and metacognition: Making science accessible to all students. *Cognition and Instruction, 16*(1), 3–118.

Wiggins, G., & McTighe, J. (1998). *Understanding by design*. Alexandria, VA: Association for Supervision and Curriculum Development.

Wild, C. J. (2006, November). The concept of distribution. *Statistics Education Research Journal, 5*(2), 10–26. Retrieved July 15, 2007, from http://www.stat.auckland.ac.nz/~iase/serj/SERJ5(2)_Wild.pdf

Wild, C. J., & Pfannkuch, M. (1999). Statistical thinking in empirical enquiry. *International Statistical Review, 67*(3), 223–265.

Wild, C. J., Triggs, C., & Pfannkuch, M. (1997). Assessment on a budget: Using traditional methods imaginatively. In I. Gal & J. B. Garfield (Eds.), *The assessment challenge in statistics education*. Netherlands: IOS Press. Retrieved April 26, 2008, from https://app.gen.umn.edu/artist/articles/Wild.pdf

Wilensky, U. (1995). Learning probability through building computational models. In D. Carraher & L. Meira (Eds.), *Proceedings of the 19th PME conference* (Vol. 3, pp. 152–159). Recife, Brazil: Organizing Committee. Retrieved April 26, 2008, from http://www.ccl.sesp.northwestern.edu/papers/pme19/

Wilensky, U. (1997). What is normal anyway? Therapy for epistemological anxiety. *Educational Studies in Mathematics, 33*, 171–202. Retrieved April 26, 2008, from http://www.ccl.sesp.northwestern.edu/papers/normal/

Wilkerson, M., & Olson, J. R. (1997). Misconceptions about sample size, statistical significance, and treatment effect. *The Journal of Psychology, 131*(6), 627–631.

Williams, A. M. (1997). Students' understanding of hypothesis testing: The case of the significance concept. In F. Biddulph & K. Carr (Eds.), *People in mathematics education: Proceedings of the twentieth annual conference of the mathematics education research group of Australasia* (pp. 585–591). Rotorua, New Zealand: MERGA.

Williams, A. M. (1999). Novice students' conceptual knowledge of statistical hypothesis testing. In J. M. Truran & K. M. Truran (Eds.), *Making the difference: Proceedings of the twenty-second annual conference of the mathematics education research group of Australasia* (pp. 554–560). Adelaide, South Australia: MERGA.

Wise, S. (1985). The development and validation of a scale measuring attitudes toward statistics. *Educational and Psychological Measurement, 45*(2), 401–405.

Wisenbaker, J., & Scott, J. (1997). *Modeling aspects of students' attitudes and achievement in introductory statistics courses*. Paper presented at annual meeting of the American Educational Research Association (AERA), Chicago, IL.

Yu, C. H., & Behrens, J. T. (1995). Applications of multivariate visualization to behavioral sciences. *Behavior Research Methods, Instruments, and Computers, 27*(2), 264–271.

Zawojewski, J. S., & Shaughnessy, J. M. (2000). Mean and median: Are they really so easy? *Mathematics Teaching in the Middle School, 5*(7), 436–440.

Zieffler, S. A. (2006). *A longitudinal investigation of the development of college students' reasoning about bivariate data during an introductory statistics course*. Unpublished Ph.D. Thesis. University of Minnesota. Retrieved July 15, 2007, from http://www.stat.auckland.ac.nz/~iase/publications/dissertations/06.Zieffler.Dissertation.

Zieffler, S. A., Garfield, J., delMas, R., & Gould, R. (2007, August). *Studying the development of college students' reasoning about statistical inference*. Paper presented at the fifth international research forum on statistical reasoning, thinking, and literacy (SRTL-5). University of Warwick, UK.

Zimmerman, C. (2005). *The development of scientific reasoning: What psychologists contribute to an understanding of elementary science learning*. Paper commissioned by the National Academies of Science (National Research Council's Board of Science Education, Consensus Study on Learning Science, Kindergarten through Eighth Grade). Retrieved April 26, 2008, from http://www7.nationalacademies.org/bose/Corinne_Zimmerman_Final_Paper.pdf

Resources

This section provides useful information on major books, conferences, and publications in statistics education.

Chapter 1: The Discipline of Statistics Education

A. Professional Organizations that Support Statistics Education

- The Consortium for the Advancement of Undergraduate Statistics Education (CAUSE)

http://www.causeweb.org/

Arising from a strategic initiative of the American Statistical Association, CAUSE is a national organization whose mission is to support and advance undergraduate statistics education in four target areas: resources, professional development, outreach, and research.

- The American Statistical Association, Section on Statistical Education

http://www.amstat.org/sections/educ/

This is an active group of statisticians and statistics educators who are involved in and dedicated to the teaching of statistics. They organize sessions on statistics education at the annual *Joint Statistics Meetings* (see below) and produce a newsletter on current activities and news related to teaching and learning statistics. They oversee nominations and awards for excellence in teaching statistics.

- The International Association for Statistical Education (IASE)

http://www.stat.auckland.ac.nz/~iase/

IASE (a section of The International Statistical Institute) is the international umbrella organization for statistics education. IASE seeks to promote, support, and improve statistical education at all levels everywhere around the world. It fosters international cooperation and stimulates discussion and research. It disseminates ideas, strategies, research findings, materials and information using publications, international conferences (such as ICOTS, the International Conference On Teaching Statistics, every four years), and has a rich Website.

- The Mathematics Association of America, Special Interest Group in Statistics Education (SIGMAA)

http://www.pasles.com/sigmaastat/

This is a special interest group for mathematicians who teach statistics. The purpose of this group is to facilitate the exchange of ideas through meetings, sessions, publications, and electronic media about teaching statistics and the undergraduate curricula; to foster increased understanding of statistics among members of the Mathematics Association of America (MAA) and among broader constituencies; to promote the discipline of statistics among students; and to work cooperatively with other organizations to encourage effective teaching and learning of statistics.

- The National Council of Teachers of Mathematics (NCTM)

http://www.nctm.org/

Data and Chance are among the *five* main content areas in the *Principles and Standards for School Mathematics* document (National Council of Teachers of Mathematics, 2000). The NCTM has published many books about teaching statistics at the school level (some of them are mentioned below).

B. Publications

I. Books on Teaching Statistics

- Ben-Zvi, D., & Garfield, J. (Eds.) (2004). *The challenge of developing statistical literacy, reasoning, and thinking*. Dordrecht, The Netherlands: Kluwer Academic Publishers.

This book collects, presents, and synthesizes cutting edge research on different aspects of statistical reasoning, and applies this research to the teaching of statistics to students at all educational levels. It presents the research foundation on which teaching should be based. The chapters in this volume are written by leading researchers in statistics education.

- Burrill, G. F. (Ed.) (2006). *Thinking and reasoning with data and chance*: Sixty-eighth NCTM yearbook. Reston, VA: National Council of Teachers of Mathematics.

The sixty-eighth NCTM yearbook (2006) focuses on students' and teachers' learning and reasoning about data and chance. Topics include the relation between mathematics and statistics, the development and enrichment of mathematical concepts through the use of statistics, and a discussion of the research related to teaching and learning statistics. The accompanying CD offers support material for many of the articles, including lessons, software demonstrations, and video clips of classrooms.

- Garfield, J. (Ed.) (2005). *Innovations in teaching statistics* (MAA Notes Volume 65). Washington, DC: Mathematics Association of America.

A book of stories about teaching statistics. These stories are told by fourteen different instructors of innovative statistics courses, who demonstrate that learning statistics can be a positive, meaningful, and even exciting experience. In the classes of the instructors whose stories fill this book, students are engaged in learning, are empowered to do statistics, and appreciate the instructional methods of their teachers. Each chapter begins by describing how the author became a teacher of statistics, then provides details about the courses they currently teach, describing their teaching methods, textbook, types of student assessments, and uses of technology. One typical class is described in detail, to provide a snapshot of what each person's teaching looks like. The writers then tell the story of the process they went through in developing an innovative course, and conclude their chapters with a discussion of their future plans for course revision or development.

- Gelman, A., & Nolan, D. (2002). *Teaching statistics: A bag of tricks.* New York: Oxford University Press.

This book provides a wealth of demonstrations, examples, and projects that involve active student participation. Part I of the book presents a large selection of activities for introductory statistics and Part II gives tips on what does and what does not work in class. Part III presents material for more advanced courses on topics such as decision theory, Bayesian statistics, and sampling.

- Gordon, F., & Gordon, S. (Eds.) (1992). *Statistics for the twenty-first century* (MAA Notes Volume 26). Washington DC: Mathematical Association of America.

This book suggests innovative ways of bringing an introductory statistics course to life. The articles focus on current developments in the field, and how to make the subject attractive and relevant to students. All articles provide suggestions, ideas, and a list of resources to faculty teaching a wide variety of introductory statistics courses. Some of the ideas presented include exploratory data analysis, computer simulations of probabilistic and statistical principles, "real world" experiments with probability models, and individual statistical research projects to reinforce statistical methods and concepts.

- Moore, T. L. (Ed.) (2000). *Teaching statistics: Resources for undergraduate instructors* (MAA Notes Volume 52). Washington DC: Mathematics Association of America.

This book is a collection of articles on various aspects of statistics education along with a collection of descriptions of several effective and innovative projects. The book opens with a classic article produced by the MAA Focus Group on Statistics Education during the infancy of the statistics education reform movement. Following sections include motivation for and advice on how to use real data in teaching, how to choose a textbook at the introductory or mathematical statistics level, how to make effective use of technology, and how to more effectively assess students by going beyond the reliance on in-class examinations.

- Shaughnessy J. M., & Chance, B. L. (2005). *Statistical questions from the classroom*. Reston, VA: National Council of Teachers of Mathematics.

This book deals with the teaching of some of the more difficult conceptual conundrums in teaching introductory statistics.

- *The Best of Teaching Statistics*: Three collections of articles from *Teaching Statistics.*

Some of the best articles from Teaching Statistics have been put together in the following three publications.

1. *The Best of Teaching Statistics* – published in 1986 and available online: http://www.rsscse.org.uk/ts/bts/contents.html.
2. *Teaching Statistics at its Best* – 50 of the best articles from Volumes 6–14 (Edited by D. Green, 1994). Some of the articles are available online: http://www.rsscse.org.uk/ts/best.html.
3. *Getting the Best from Teaching Statistics* – The latest anthology with articles from Volumes 15–21. Available online: http://www.rsscse.org.uk/ts/gtb/contents.html.

- The *Navigations* Series on Data Analysis and Probability, published by the *National Council of Teachers of Mathematics*, Reston, VA.

Grade-band books with activities and materials to implement ideas from the NCTM *Principles and Standards for School Mathematics* (2000). There are four books – *Navigating through Data Analysis and Probability* (prekindergarten–grade 2, grades 3–5, 6–8, and 9–12), and two books – *Navigating through Probability* (grades 6–8, and 9–12).

II. Journals and Newsletters

- *Statistics Education Research Journal* (SERJ)

http://www.stat.auckland.ac.nz/~iase/serj

SERJ is a peer-reviewed electronic journal of the International Association for Statistical Education (IASE) and the International Statistical Institute (ISI). SERJ is published electronically twice a year and is free. *SERJ* aims to advance research-based knowledge that can help to improve the teaching, learning, and understanding of statistics or probability at all educational levels and in both formal (classroom-based) and informal (out-of-classroom) contexts. Such research may examine, for example, cognitive, motivational, attitudinal, curricular, teaching-related, technology-related, organizational, or societal factors and processes that are related to the development and understanding of statistical knowledge. In addition, research may focus on how people use or apply statistical and probabilistic information and ideas, broadly viewed.

- *Journal of Statistical Education* (JSE)

http://www.amstat.org/publications/jse/

The *JSE* disseminates knowledge for the improvement of statistics education at all levels, including elementary, secondary, post-secondary, post-graduate, continuing and workplace education. It is distributed electronically and, in accord with its broad focus, publishes articles that enhance the exchange of a diversity of interesting and useful information among educators, practitioners, and researchers around the world. The intended audience includes anyone who teaches statistics, as well as those interested in research on statistical and probabilistic reasoning. All submissions are rigorously refereed using a double-blind peer review process.

- *Teaching Statistics*

http://www.rsscse.org.uk/ts/

Teaching Statistics seeks to help those teaching any type of statistics to pupils aged 9–19 by showing how statistical ideas can illuminate their work and how to make proper use of statistics in their teaching. It is also directed toward those who teach statistics as a separate subject and to those who teach statistics in conjunction with mathematics courses. In the United States, teachers will find it useful in teaching the data-handling aspects of the *Principles and Standards for School Mathematics* (NCTM, 2000).

- *STATS* Magazine

http://www.amstat.org/advertising/index.cfm?fuseaction=stats

Stats is a lively magazine, directed toward student members of the American Statistical Association (ASA) and the ASA school membership. *Stats* features career information, student experiences, current problems, case studies, first person stories from leaders in the field, and humor.

- *Statistics Teacher Network*

http://www.amstat.org/education/stn/

The *Statistics Teacher Network* (STN) is a newsletter published three times a year by the American Statistical Association and the National Council of Teachers of Mathematics Joint Committee on Curriculum in Statistics and Probability for Grades K-12. STN is a free publication whose purpose is to keep grades K-12 teachers informed of statistical workshops, programs, and reviews of books, software, and calculators. In addition, articles are included describing statistical activities that have been successful in the classroom.

- *Newsletter of the Section on Statistical Education of the American Statistical Association*

http://www.amstat.org/sections/educ/newsletter/index.html

The newsletter provides section members with: (1) short descriptions and references to resources where they can learn about new ideas in how to teach or how people learn statistics; (2) news items about current happenings in the teaching of statistics that are of interest to teachers of statistics but are not directly applicable to classroom practice; and (3) actual descriptions of teaching ideas.

- *International Statistical Review*

http://isi.cbs.nl/isr.htm

The International Statistical Review provides a comprehensive review of work in statistics, over the whole spectrum of the statistical profession, including the most relevant aspects of probability. It publishes original research papers of wide interest; integrated critical surveys of particular fields of statistics and probability; and reports on recent developments in statistics, computer facilities, survey programs, teaching methods and experience.

III. Articles on the Introductory Statistics Course

These are important articles on undergraduate statistics education that often provide historical insights into developments and issues regarding the introductory statistics course.

Cobb, G. W. (1993, July). Reconsidering statistics education: a National Science Foundation conference. *Journal of Statistics Education, 1*(1). Retrieved November 6, 2006, from http://www.amstat.org/publications/jse/v1n1/cobb.html

Cobb, G. W., & Moore, D. S. (1997). Mathematics, statistics, and teaching. *American Mathematical Monthly, 104*, 801–823.

Garfield, J., Hogg, B., Schau, C., & Whittinghill, D. (2002, July). First courses in statistical science: the status of educational reform efforts. *Journal of Statistics Education, 10*(2). Retrieved November 6, 2006, from http://www.amstat.org/publications/jse/v10n2/garfield.html

Moore, D. S. (1992). Teaching statistics as a respectable subject. In F. Gordon & S. Gordon (Eds.), *Statistics for the Twenty-First century* (pp. 14–25). Washington DC: The Mathematical Association of America.

Moore, D. S. (1995, January). The craft of teaching. *MAA FOCUS, 15*(2), 5–8. Retrieved November 6, 2006, from http://www.stat.purdue.edu/ ~dsmoore/articles/Craft.pdf

Moore, D. S. (1997). New pedagogy and new content: The case of statistics. *International Statistical Review, 65*, 123–137. Retrieved November 6, 2006, from http://www.stat.purdue.edu/~dsmoore/articles/PedagogyContent.pdf

Moore, D. S. (1998, December). Statistics among the liberal arts. *Journal of the American Statistical Association, 93*(144), 1253–1259. Retrieved November 6, 2006, from http://www.stat.purdue.edu/~dsmoore/articles/LibArts.pdf

Scheaffer, R. L. (2001, Winter). Statistics education: perusing the past, embracing the present, and charting the future. *Newsletter for the Section on Statistical Education*, *7*(1). Retrieved November 6, 2006, from http://www.amstat.org/ sections/educ/newsletter/v7n1/Perusing.html

C. Conferences

- ICOTS – International Conference on the Teaching of Statistics (2010, every four years)

http://www.stat.auckland.ac.nz/~iase/conferences

The ICOTS conferences, held by IASE, are the most important events on the international statistics education calendar. ICOTS-7, Brazil, 2006, see http://www.maths.otago.ac.nz/icots7/icots7.php; ICOTS-8, Slovenia, 2010, see http://icots8.org/.

- SRTL – Statistical Reasoning Thinking and Literacy International Research Forums (2009, every two years)

http://srtl.stat.auckland.ac.nz/

The SRTL series of biennial research forums brings together researchers working in the fields of statistical reasoning, thinking, and literacy. SRTL5 on "reasoning about statistical inference – innovative ways of connecting chance and data", The University of Warwick, United Kingdom, 2007, see http://srtl.stat.auckland.ac.nz/srtl5/research_forums. SRTL6 on "the role of context and evidence in informal inferential reasoning", The University of Queensland, Brisbane, Australia, 2009, see http://srtl.stat.auckland.ac.nz/srtl6/research_forums.

- IASE Roundtable (2008, every 4 years)

http://www.stat.auckland.ac.nz/~iase/conferences

These are small workshop conferences that bring together a select international group of experts to address a particular theme and to make recommendations from which institutions and individuals engaged in statistical education and training (in developed and developing countries) may benefit. The 2004 Roundtable addressed Curricular Development in Statistics Education. The 2008 will address Statistics Education in School Mathematics: Challenges for Teaching and Teacher Education. IASE Roundtables are held near the site of ICME conferences (listed below).

- IASE Satellite Conferences (2007, every two years)

http://www.stat.auckland.ac.nz/~iase/conferences

These are themed conferences held in close proximity to the ISI congresses. The 2005 conference focused on Statistics Education and the Communication of Statistics, and the 2007 conference on Assessing Student Learning in Statistics.

- American Statistical Association with Mathematical Association of America Joint Statistical Meetings (JSM) – (every year)

http://www.amstat.org/meetings/jsm/2007/index.cfm

JSM is the largest gathering of statisticians held in North America. The Section on Statistical Education of the American Statistical Association organizes a wide variety of invited and contributed paper sessions as well as roundtable discussions and posters.

- USCOTS – United States Conference on Teaching Statistics (2009, every two years)

http://www.causeweb.org/uscots/uscots05/

USCOTS is a U.S. conference that focuses on undergraduate level statistics education (including Advanced Placement Statistics), targeting statistics teachers. It aims at sharing ideas, methods, and research results regarding what teachers want to know about teaching statistics; facilitating incorporating new ideas, methods, and resources into existing courses and programs; and promoting connections between all teachers of undergraduate level statistics throughout the United States

- The International Statistical Institute (ISI) Session (2009, every two years)

http://isi.cbs.nl/

The biannual scientific conference of the International Statistical Institute (ISI) has been held since 1853, recent sessions attracting in excess of 2,000 delegates. Participants include academics, government and private sector statisticians and related experts from various institutes. ISI Sessions provide an opportunity for statisticians to attend scientific meetings focusing on their own specialty and at the same time absorb new research in other statistical fields that may have unanticipated applications to one's own specialty.

- Joint Mathematics Meetings (every year)

http://www.ams.org/amsmtgs/national.html

The Joint Mathematics Meetings are held for the purpose of advancing mathematical achievement, encouraging research, and providing the communication necessary to progress in the field. The SIGMAA for Statistics Education (listed above) organizes several sessions on teaching statistics and has its annual business meeting at this conference.

- ICME – International Congress on Mathematical Education (2008, held every four years)

http://www.mathunion.org/o/Organization/ICMI/ICME_congress.html

A major event in the life of the international mathematics education community is formed by the quadrennial International Congress on Mathematical Education, ICME, held under the auspices of the International Commission on Mathematical Instruction (ICMI, http://www.mathunion.org/ICMI/). This major scientific gathering includes several sessions on statistics education.

D. Websites

- CAUSE (Consortium for the Advancement of Undergraduate Statistical Education)

http://www.causeweb.org/

Arising from a strategic initiative of the American Statistical Association, CAUSE is a national organization whose mission is to support and advance undergraduate statistics education, in four target areas: resources, professional development, outreach, and research.

- International Statistical Literacy Project

http://www.stat.auckland.ac.nz/~iase/islp

The mission of the International Statistical Literacy Project (ISLP) is to provide those interested in statistical literacy with information and resources, and to aid them in the development of statistical literacy around the world. It replaces the World Numeracy Project of the International Statistical Institute (ISI).

- The International Association for Statistical Education (IASE) Website

http://www.stat.auckland.ac.nz/~iase/

The IASE Website disseminates ideas, strategies, research findings, publications, materials, and information related to statistics education.

- Adapting and Implementing Innovative Material in Statistics Project (AIMS)

http://www.tc.umn.edu/~aims/

This project is about adapting and implementing innovative materials for introductory statistics courses. These materials include textbooks, software, Web resources, and special simulation tools, lesson plans, and student activity guides. The suggested lessons are designed to involve students in small and large group discussion, computer explorations, and hands-on activities.

- The Guidelines for Assessment and Instruction in Statistics Education (GAISE)

GAISE College Report:
http://www.amstat.org/Education/gaise/GAISECollege.htm
GAISE PreK-12 Report:
http://www.amstat.org/education/gaise/GAISEPreK-12.htm

Chapter 2: Research on Teaching and Learning Statistics

A. Books

- Ben-Zvi, D., & Garfield, J. (2004). *The challenge of developing statistical literacy, reasoning, and thinking*. Dordrecht, the Netherlands: Kluwer Academic Publishers.

This book collects, presents, and synthesizes cutting edge research on different aspects of statistical reasoning and applies this research to the teaching of statistics to students at all educational levels. It presents the research foundation on which teaching should be based. The chapters in this volume are written by leading researchers in statistics education.

- Lajoie, S. P. (1998). *Reflections on statistics: Learning, teaching, and assessment in grades K-12* (studies in mathematical thinking and learning series). Mahwah, NJ: Lawrence Erlbaum Associates.

This volume represents the emerging findings of an interdisciplinary collaboration among a group of mathematics educators, cognitive scientists, teachers, and statisticians to construct an understanding of how to introduce statistics education and assessment for students in elementary and secondary schools. A premise of this volume is that when students are introduced to statistics at the K-12 level and provided with opportunities to do statistics that are related to actual life situations, they will be better prepared for decision making in the real world. The book is organized around four interdependent themes: content, teaching, learning, and assessment.

- Lovett, M. C., & Shah, P. (Eds.) (2007). *Thinking with data* (Carnegie Mellon Symposia on Cognition Series). Mahwah, NJ: Lawrence Erlbaum Associates.

A collection of papers presented at the 33rd Carnegie Symposium on Cognition: Thinking with Data. This volume is organized around three themes: (a) reasoning about uncertainty and variation (b) statistical reasoning and data analysis, and (c) learning from and making decisions with data.

B. Articles

These are important articles on the nature, scope, and main themes of research in statistics education.

Batanero, C., Garfield, J., Ottaviani, M. G., & Truran, J. (2000, May). Research in statistical education: Some priority questions. *Statistical Education Research Newsletter, 1*(2), 2–6. Retrieved December 3, 2006, from http://www.stat.auckland.ac.nz/~iase/serj/newsmay00.pdf

Joliffe, F. (1998). What is research in statistics education? In L. Pereira-Mendoza (Ed.), *Proceedings of the fifth international conference on teaching statistics* (pp. 801–806). Singapore: International Statistical Institute. Retrieved April 20, 2008, from http://www.stat.auckland.ac.nz/~iase/publications/2/Topic6x.pdf

Jones, G. A., & Thornton, C. A. (2005). An overview of research into the teaching and learning of probability. In G. A. Jones (Ed.), *Exploring probability in school: Challenges for teaching and learning*, (pp. 65–92). New York: Springer.

Shaughnessy, J. M. (2007). Research on statistics learning and reasoning. In F. K. Lester (Ed.), *Second handbook of research on mathematics teaching and learning* (pp. 957–1010). Information Age Pub Inc.

C. Websites

- CAUSE (Consortium for the Advancement of Undergraduate Statistical Education): Research Section

http://www.causeweb.org/research/

These Web pages include lists of important reading in or related to research in statistics education, a searchable data base of abstracts of over 2000 articles related to the research in statistics education, and practical advice about conducting statistics education research. Links are provided to journals that publish research in this area, to conferences where statistics education research is presented, and to Websites of active statistics education research groups and their projects.

Chapter 3: Creating a Statistical Reasoning Learning Environment

A. Websites

- Webinars of the Consortium for the Advancement of Undergraduate Statistics Education (CAUSE)

http://www.causeweb.org/webinar/

Free monthly Web-based seminars for up to 25 statistics educators. Sessions are recorded for others to view later.

B. Books and Articles

- Burrill, G. F. (Ed.) (2006) *Thinking and reasoning with data and chance: The sixty-eighth NCTM yearbook*. Reston, VA: National Council of Teachers of Mathematics.

This book focuses on students' and teachers' learning and reasoning about data and chance. Topics include the relationship between mathematics and statistics, the development and enrichment of mathematical concepts through the use of statistics, and a discussion of the research related to teaching and learning statistics. The accompanying CD offers support material for many of the articles, including lessons, software demonstrations, and video clips of classrooms.

- Davis, B. G. (1993). *Tools for teaching* (Jossey-Bass Higher and Adult Education Series). San Francisco: Jossey-Bass.

A compendium of classroom-tested strategies and suggestions designed to improve the teaching practices of all college instructors, including beginning, mid-career, and senior faculty members. The book describes 49 teaching tools that cover both traditional practical tasks – writing a course syllabus, delivering an effective lecture – as well as newer, broader concerns, such as responding to diversity on campus and coping with budget constraints.

- Moore, D. S. (1995). The craft of teaching. *Mathematical Association of America FOCUS, 15*(2), 5–8. Retrieved April 20, 2008, from http://www.stat.purdue.edu/~dsmoore/articles/Craft.pdf

This paper presents David Moore's philosophy of teaching statistics, composed on the occasion of winning a major teaching award. David Moore is a well-known statistician (and former President of the American Statistical Association) who has had a major impact on the reform of the introductory statistics course.

- Moore, T. L. (Ed.) (2000). *Teaching statistics: Resources for undergraduate instructors*. (MAA Notes Number 52). Washington DC: Mathematics Association of America.

This book is a collection of articles on various aspects of statistics education along with a collection of descriptions of several effective and innovative projects. The book opens with a classic article produced by the MAA Focus Group on Statistics Education during the infancy of the statistics education reform movement. Subsequent sections include motivation for and advice on how to use real data in teaching, how to choose a textbook at the introductory or mathematical statistics level, how to make effective use of technology, and how to more effectively assess students by going beyond the reliance on in-class examinations.

- Shaughnessy J. M., & Chance, B. L. (2005). *Statistical questions from the classroom*. Reston, VA: National Council of Teachers of Mathematics.

This small book presents eleven short discussions of some of the most frequently asked questions about statistics. Some of the questions, such as "What is the difference between a sample and a sampling distribution?" involve major concepts in statistics. Other questions such as "Why are deviations squared?" deal briefly with some of the more technical aspects of the mathematics in statistical theory. The authors offer teachers of statistics some quick insight and support in understanding these issues and explaining these ideas to their own students.

Chapter 4: Assessment in Statistics Education

A. Books

- Angelo, T., & Cross, K. P. (1993). *A handbook of classroom assessment techniques for college teachers* (2nd ed.). San Francisco: Jossey-Bass.

A collection of practical ways to assess student learning and for giving the instructor student feedback on various course components; includes ideas like the minute paper and the punctuated lecture.

- Gal, I., & Garfield, J. (Eds.) (1997). *The assessment challenge in statistics education*. Amsterdam: IOS Press. Retrieved April 26, 2008, from http://www.stat.auckland.ac.nz/ ~iase/publications/assessbk/

A collection of articles, both conceptual and practical, on issues of assessment in statistics education.

- Gold, B., Keith, S., & Marion, W. (Eds.) (1999). *Assessment practices in undergraduate mathematics*, MAA Notes #49. Washington, D.C.: Mathematical Association of America.

A collection of articles discussing assessment practices and techniques ranging from program assessment to classroom assessment to assessment of teaching.

- Pellegrino, J. W., Chudowsky, N., & Glaser, R. (Eds.) (2001). *Knowing what students know: the science and design of educational assessment.* National Research Council. Available from http://www.nap.edu/catalog.php? record_id=10019

This report reviews and synthesizes advances in the cognitive sciences and measurement and explores their implications for improving educational assessment. It addresses assessments used in both classroom and large-scale contexts for three broad purposes: to assist learning, to measure individual achievement, and to evaluate programs.

- Wiggins, G., & McTighe, J. (2006). *Understanding by design* (2nd ed.). Englewood NJ: Prentice Hall.

This book brings to bear the author's understanding of teaching for understanding and its implications for performance-based assessment including pragmatic advice, background, and extensive references. An accompanying workbook is the *Understanding by design: professional development workbook* (2004, workbook ed.) by the same authors, published by the Association for Supervision and Curriculum Development (ASCD).

B. Articles

Below are two important articles on alternative methods of assessment in the introductory statistics course. In addition, there are many readings at the ARTIST Website (https://app.gen.umn.edu/artist/).

Chance, B. L. (2000). Experiences with authentic assessment techniques in an introductory statistics course. In T. L. Moore (Ed.) *Teaching statistics: resources for undergraduate instructors* (pp. 209–218). Washington D.C: Mathematical Association of America.

Garfield, J. (2000). Beyond testing and grading: New ways to use assessment to improve student learning. In T. L. Moore (Ed.) *Teaching statistics: resources for undergraduate instructors* (pp. 201–208). Washington DC: Mathematical Association of America.

C. Assessment Items

- ARTIST

https://app.gen.umn.edu/artist/

One of the broadest endeavors is the ARTIST project, which provides a variety of assessment resources including sample assignments, research instruments, discussion of implementation issues, and the *Assessment Builder*, which provides free downloadable access to over 1,000 high-quality assessment items that focus on statistical literacy, reasoning, and thinking.

- Survey of Attitudes Toward Statistics (SATS)

http://www.unm.edu/~cschau/satshomepage.htm

This is a tool for assessing student's attitudes towards statistics.

D. Research Instruments

A collection of research instruments can be found at the ARTIST Website (https://app.gen.umn.edu/artist/).

Chapter 5: Using Technology to Improve Student Learning of Statistics

A. Websites

- Consortium for the Advancement of Undergraduate Statistics Education (CAUSE)

http://www.causeweb.org

A collection of resources and services aimed at supporting and advancing undergraduate statistics education in the areas of resources: peer-reviewed collections of examples, datasets, activities, tools, professional development, outreach, and research.

- CHANCE Database

http://www.dartmouth.edu/~chance

A collection of materials related to a quantitative literacy course developed around current news stories involving probability and statistics. Resources include an ongoing archive of articles, videos, data, activities, and other teaching aids.

- The Data and Story Library (DASL)

http://lib.stat.cmu.edu/DASL

A library of datafiles and stories illustrating basic statistical methods. Searches can be conducted by topic, statistical method, or data subjects.

- Journal of Statistics Education(JSE) Data Archive

http://www.amstat.org/publications/jse/jse_data_archive.html

A collection of datasets submitted by instructors with background details and instructions for classroom use.

- Rice Virtual Lab in Statistics

http://www.ruf.rice.edu/~lane/rvls.html

Includes java applets that can be downloaded for off line demonstrations, an online statistics text, case studies, and an analysis lab of basic statistical analysis tools.

- Rossmanchance Applet Collection

http://www.rossmanchance.com/applets/index.html

An informal collection of applets useful for data analysis, sampling distribution simulations, probability, and inference.

- Tools for Teaching and Assessing Statistical Inference

http://www.tc.umn.edu/~delma001/stat_tools

A collection of materials and software geared to helping students understand core concepts underlying statistical inference, including sampling distributions, confidence intervals, and P-values.

- Illuminations: The National Council of Teachers of Mathematics (NCTM)

http://illuminations.nctm.org/

A selected Web Links that are useful school mathematics education resources on the Internet. A section on data analysis & probability is provided. Each resource has been approved by an NCTM editorial board.

- The National Library of Virtual Manipulatives (NLVM)

http://nlvm.usu.edu/en/nav/vlibrary.html

A library of interactive, Web-based virtual manipulatives or concept tutorials, mostly in the form of Java applets, for mathematics instruction (K-12 emphasis). The project in Utah State University includes data analysis & probability learning tools related to the *Principles and Standards for School Mathematics* (NCTM, 2000).

- Statistics Online Computational Resource (SOCR)

http://www.socr.ucla.edu/

This resource provides portable online aids for probability and statistics education, technology-based instruction, and statistical computing. SOCR tools and resources include a repository of interactive applets, computational and graphing tools, instructional and course materials.

- Interactive: Shodor Education Foundation

http://www.shodor.org/interactivate/activities/tools.html

A collection of interactive Java-based courseware for exploration in science and mathematics that includes sections on statistics and probability.

B. Books and Journals

- Technology Innovations in Statistics Education (TISE)

http://repositories.cdlib.org/uclastat/cts/tise

An online journal that reports on studies of the use of technology to improve statistics learning at all levels, from kindergarten to graduate school and professional development.

- The 1996 IASE Roundtable on the Role of Technology

http://www.stat.auckland.ac.nz/~iase/publications.php?show=8

The proceedings of the 1996 International Association for Statistical Education (IASE) Roundtable that discussed the current state of research on the role of technology in statistics education (Garfield & Burrill, 1997).

- The 2003 IASE conference on Statistics Education and the Internet

http://www.stat.auckland.ac.nz/~iase/publications.php?show=6

The proceedings of a conference that was dedicated to the recent increase in the use of the Internet as a resource for helping teach statistics.

C. Articles

Garfield, J., Chance, B. L., & Snell, J. L. (2000). Technology in college statistics courses. In D. Holton et al. (Eds.), *The teaching and learning of mathematics at university level: An ICMI study* (pp. 357–370). Dordrecht, The Netherlands: Kluwer Academic Publishers. Retrieved April 20, 2008, from http://www.dartmouth.edu/~chance/teaching_aids/books_articles/technology.htm

Malone, C. J., & Bilder, C. R. (2001, July). Statistics course Web sites: beyond syllabus.html. *Journal of Statistics Education, 9*(2). Retrieved December 18, 2006, from http://www.amstat.org/publications/jse/v9n2/malone.html

Mulekar, M. (2000, July). Internet resources for AP statistics teachers. *Journal of Statistics Education 8*(2). Retrieved December 18, 2006, from http://www.amstat.org/publications/jse/secure/v8n2/mulekar.cfm

Rubin, A. (2007). Much has changed; little has changed: Revisiting the role of technology in statistics education 1992-2007. *Technology Innovations in Statistics Education, 1*(1), Article 6. Retrieved April 20, 2008, from http://repositories.cdlib.org/uclastat/cts/tise/vol1/iss1/art6.

Velleman, P. F., & Moore, D. S. (1996). Multimedia for teaching statistics: Promises and pitfalls. *The American Statistician*, 50, 217–225.

D. Statistical Software for Teaching Statistics

The following are well-known statistical software packages but by no means is an exhaustive list.

- DataDesk (http://www.datadesk.com)
- Fathom (http://www.keypress.com/fathom)
- JMP (http://www.jmp.com)
- Minitab (http://www.minitab.com)
- Model Chance (http://www.umass.edu/srri/serg/projects/ModelChance)
- ProbSim (http://www.umass.edu/srri/serg/software/download-chanceplus/ProbSimdl.html)
- R (http://www.r-project.org)
- StatCrunch (http://www.statcrunch.com)
- TinkerPlots (http://www.keypress.com/tinkerplots)

E. Other Statistical Software

- ActivStats (http://www.activstats.com)
- CyberStats (http://www.cyberk.com)
- Excel (http://office.microsoft.com)
- SAS (http://www.sas.com)
- S-plus (http://www.insightful.com)
- SPSS (http://www.spss.com)
- STATA (http://www.stata.com)

F. Class Management Software

- Blackboard (http://www.webct.com)
- Moodle (http://moodle.org)

Chapter 15: Collaboration in the Statistics Classroom

A. Websites

- Collaborative Learning: Group Work and Study Teams

http://teaching.berkeley.edu/bgd/collaborative.html

Barbara Gross Davis (University of California Berkeley) presents a helpful set of guidelines and advice on using collaborative learning in the college classroom.

B. Articles

Dunn, D. S. (1996). Collaborative writing in a statistics and research methods course. *Teaching of Psychology, 23*(1), 38–40.

Verkoeijen, P. P. J. L., Imbos, Tj., van de Wiel, M. W. J., Berger, M. P. F., & Schmidt, H. G. (2002, July). Assessing knowledge structures in a constructive statistical learning environment. *Journal of Statistics Education 10*(2). Retrieved December 30, 2006, from http://www.amstat.org/publications/jse/v10n2/verkoeijen.html

Chapter 16: Collaboration in Teaching and Research

Books

- Lewis, C. (2002). *Lesson study: A handbook of teacher-led instructional improvement.* Philadelphia: Research for Better Schools.

This handbook illuminates both the key ideas underlying lesson study and the practical support needed to make it succeed in any subject area. It addresses topics including the basic steps of lesson study, supports, misconceptions, system impact, how to pioneer lesson study in your setting, schedules, data collection examples, protocols for lesson discussion and observation, and instructional plans.

Appendix: Tables of Activities

Table 1 Table of activities by chapter[1]

Topic and Chapter	Lesson Name and Number	Activity Title	Credits and References
Chapter 6 Data	1. Data and Variability	Meet and Greet	
		Developing a Class Survey	
		Variables on Backs	Rossman and Chance (2004)
	2. Avoiding Bias	How you Ask a Question	Rossman and Garfield(2001)
		Critiquing the Student Survey	
	3. Random Sampling	Gettysburg Address	Chance and Rossman (2005)
		Student Survey Sampling	
	4. Randomized Experiments	Taste Test	Snell, Peterson, Moore, and Garfield (1998)
Chapter 7 Statistical Models and Modeling	1. Using Models to Simulate Data	One-Son Modeling Activity	Konold (1994a) Also mentioned in simon (1994)
		Let's Make a Deal Simulation	Snell et al. (1998) Shaughnessy and Dick (1991)
	2. Modeling Random Variables	Coins, Cards, and Dice	
	3. The Normal Distribution as a Model	What is Normal?	
		Normal Distribution Applications	

[1] Note: all activities without specific references in this table were developed by members of the EPSY 3264 Team: Joan Garfield, Andy Zieffler and Sharon Lane-Getaz.

Table 1 (continued)

Topic and Chapter	Lesson Name and Number	Activity Title	Credits and References
Chapter 8 Distribution	1. Distinguishing Distributions	Distinguishing Distributions	Rossman and Chance (2002)
		Growing a Distribution	Bakker (2004b)
	2. Exploring and Sorting Distributions	What is a Histogram?	
		Stretching Histograms	
		Exploring Different Representations of Data	
		Sorting Histograms	Garfield (2002a)
		Matching Histograms to Variable Descriptions	Rossman and Chance(2002)
Chapter 9 Center	1. Reasoning about Measures of Center	What does the Mean Mean?	
		What does the Median Mean?	
		Means and Medians	Erickson (2002)
	2. Choosing Appropriate Measures	What is Typical?	
		Choosing an Appropriate Measure of Center	
Chapter 10 Variability	1. Variation	How Big is Your Head?	
	2. Reasoning about the Standard Deviation	Comparing Hand Spans	Watkins, Scheaffer and Cobb (2004)
		What Makes the Standard Deviation Larger or Smaller?	delMas (2001b)
Chapter 11 Comparing Groups	1. Understanding Boxplots	How Many Raisins in a Box?	
	2. Comparing Groups with Boxplots	Gummy Bears	Scheaffer et al. (2004a, b)
		Comparing Boxplots	
	3. Reasoning about Boxplots	Interpreting Boxplots	
		Matching Histograms to Boxplots	Scheaffer et al. (2004a, b)
	4. Comparing Groups with Histograms, Boxplots, and Statistics	How do Students Spend Their Time?	Garfield et al. 2007

Table 1 (continued)

Topic and Chapter	Lesson Name and Number	Activity Title	Credits and References
Chapter 12 Samples and Sampling Distributions	1. Sampling from a Population	Reese's Pieces	Rossman and Chance(2002)
	2. Generating Sampling Distributions	Body Temperatures Sampling Words Sampling Pennies	
	3. Describing the Predictable Pattern: The Central Limit Theorem	Central Limit Theorem	Garfield, delMas and Chance (2000)
Chapter 13 Statistical Inference	1. Testing Statistical Hypotheses	Modeling Coin Tosses	
		Balancing Coins	Scheaffer et al. (2004a, b)
	2. P-values and Estimation	P-values	Seier and Robe (2002)
		Types of Errors	Seier and Robe (2002)
		Introduction to Confidence Intervals	
	3. Reasoning about Confidence Intervals	Estimating with Confidence Estimating Word Lengths	
		What Does the 95% Mean?	Garfield et al. (2000)
	4. Using Inference in an Experiment	Gummy Bears Revisited	Scheaffer et al. (2004a, b)
	5. Solving Statistical Problems Involving Statistical Inference	Research Questions Involving Statistical Methods	
Chapter 14 Covariation	1. Reasoning about Scatterplots and Correlation	Credit Questions	Cook and Weisberg(1999)
		Interpreting Scatterplots	Cook and Weisberg(1999)
		Reasoning about the Correlation Coefficient Guessing Correlations	
	2. Fitting a Line to Data	Diamond Rings	Chu (2001)
		da Vinci and Body Measurements	Watkins et al. (2004)
	3. Inferences Involving Bivariate Data	Testing Relationships: Admissions Variables Testing Relationships: Baseball Variables	

Table 2 The sequence of activities table

This table provides a list of all activities in Table 1 in the actual order in which they can be taught

Topic	Lesson Name and Number	Activity Title
Data	1. Data and Variability	• Meet and Greet • Developing a Class Survey • Variables on Backs
Statistical Models and Modeling	1. Using Models to Simulate Data	• One-Son Modeling Activity • Let's Make a Deal Simulation
Data	2. Avoiding Bias	• How you Ask a Question • Critiquing the Student Survey
	3. Random Sampling	• Gettysburg Address • Student Survey Sampling
	4. Randomized Experiments	• Taste Test
Distribution	1. Distributions	• Distinguishing Distributions • Growing a Distribution
	2. Exploring and Sorting Distributions	• What is a Histogram? • Sorting Histograms • Matching Histograms to Variable Descriptions • Creating graphs for variables without data • Exploring Different Representations of the Same Data
Center	1. Reasoning about Measures of Center	• What does a Mean Mean? • What does a Median Mean? • Means and Medians
	2. Choosing Appropriate Measures	• What is Typical? • Choosing an Appropriate Measure of Center
Variability	1. Variation	• How Big is Your Head?
	2. Reasoning about the Standard Deviation	• Comparing Hand Spans • What Makes the Standard Deviation Larger or Smaller?
Comparing Groups	1. Understanding Boxplots	• How Many Raisins in a Box?
	2. Comparing Groups with Boxplots	• Gummy Bears • Comparing Boxplots
	3. Reasoning about Boxplots	• Interpreting Boxplots • Matching Histograms to Boxplots
	4. Comparing Groups with Histograms, Boxplots, and Statistics	• How do Students Spend Their Time?
Statistical Models and Modeling	2. Modeling Random Variables	• Coins, Cards, and Dice
	3. The Normal Distribution as a Model	• What is Normal? • Normal Distribution Applications

Table 2 (continued)

Topic	Lesson Name and Number	Activity Title
Samples and Sampling Distributions	1. Sampling from a Population	• Reece's Pieces
	2. Generating Sampling Distributions	• Body Temperature • Sampling words • Sampling Pennies
	3. Describing the Predictable Pattern: The Central Limit Theorem	• Central Limit Theorem
Statistical Inference	1. Testing Statistical Hypotheses	• Modeling Coin Tosses • Balancing Coins
	2. *P*-values and Estimation	• *P*-values • Types of Errors • Introduction to Confidence Intervals
	3. Reasoning about Confidence Intervals	• Estimating with Confidence • Estimating Word Lengths • What Does the 95% Mean?
	4. Using Inference in an Experiment	• Gummy Bears Revisited
Covariation	1. Reasoning about Scatterplots and Correlation	• Credit Questions • Interpreting Scatterplots • Reasoning about the Correlation Coefficient • Guessing Correlations
	2. Fitting a Line to Data	• Diamond Rings • da Vinci and Body Measurements
	3. Inferences involving Bivariate Data	• Testing Relationships: Admissions Variables • Testing Relationships: Baseball Variables
Statistical Inference	5. Solving Statistical Problems Involving Statistical Inference	• Research Questions Involving Statistical Methods

Author Index

Subject Index

Printed in the United States
135588LV00006B/26/P

9 781402 083822